The Farthest Shore

A 21st Century Guide to Space

The Farthest Shore - A 21st Century Guide to Space
ISBN 9781-926592-07-7 - ISSN 1496-6921
 We acknowledge the financial support of the Government of Canada through the Book Publishing Industry Development Program for our publishing activities.

Published by Apogee Books, Box 62034, Burlington, Ontario, Canada, L7R 4K2
http://www.apogeebooks.com
Cover: Robert Godwin
Printed and bound in Canada

The Farthest Shore

A 21st Century Guide to Space

Executive Editors
Joseph N. Pelton and Angelia P. Bukley

Manuscript Editor
Michael Rycroft

An Apogee Books Publication

Dedication

This book is dedicated to the International Space University. In particular it is dedicated to the ISU students and alumni, as well as this university's exceptional faculty, leadership and staff around the world and in Strasbourg, France. In short, we wish to recognize everyone in the ISU community for all they do to advance the cause of space exploration, applications and science.

Contents

Acknowledgments

The executive editors wish to acknowledge the efforts of the editorial board members and all of those individuals who contributed content to this book. What began as an idea of a few International Space University faculty members talking over coffee in the faculty office during the 2007 Space Studies Program in Beijing, China has, over the last couple of years, blossomed into what is probably one of the most comprehensive and "user friendly" books addressing the many facets of space exploration. We are grateful to ISU President Michael Simpson and Dean Walter Peeters, who both enthusiastically endorsed the idea of generating this book and promoted its development to the ISU Board of Trustees, who readily approved the effort. They have been supportive throughout our quest and have graciously provided us with the communications tools we needed to coordinate with each other to realize this work. As our team was scattered across 12 time zones, this support was invaluable. We sincerely and greatly appreciate the efforts of Mr. Jean-Marc Seiler, Secretary General of the ISU, who was of great assistance in negotiating the contract with the publisher. We also appreciate the efforts of Ms. Madge Gibson and Ms. Charlotte Henley who provided help and advice towards formatting the final version of the manuscript. Finally, we wish to express our profound gratitude to Dr. Michael Rycroft whose meticulous editing has led to the quality product you now hold in your hands. Dr. Rycroft has made many contributions to ISU, including editing as well as writing portions of *Keys to Space*, the first ISU textbook. His tireless and consistent efforts are most appreciated.

Joe Pelton and Angie Bukley

11 March 2009

Arlington, VA, USA

Foreword

Michael K. Simpson, President, International Space University

Not too long ago, in October 2007, we celebrated the 50th anniversary of the space age. Commonly dated from the launch of Sputnik on October 4, 1957, the space age saw an extraordinary evolution from one simple, human-made sphere orbiting the Earth for four months to an age where hundreds of satellites orbit permanently, providing a range of services without which modern life would be very different. Along the way, humankind perfected the art of lofting representatives of our species into near Earth orbit, sending a few early voyagers to the surface of the Moon, and maintaining a permanent human presence on board the International Space Station.

Some of the authors contributing to this book were very early participants in the creation of the space age while others joined it along the way. To this extent at least, this volume is multi-generational. As such it may well point to one of the interesting questions of our time: is the first space age giving way to a new one? I believe that the answer is yes and when asked to explain what the difference between the two "ages" might be, I respond as follows:

> *In the first space age, human kind managed to send representatives into space and to apply novel space technologies in service to human society on Earth. In the next space age we will manage to send human society itself into space.*

As our technologies improve to the point where we can already talk of the possibility of space tourism as an industry, we necessarily approach a time when groups of people might be willing to accept the rigors of permanent migration beyond our home planet. I do not expect that vision to be realized in my lifetime, but I fully expect that it will be a choice available to my grandchildren and their families. In this way, too, the prospects of space are intergenerational.

Whether your interest in space is the result of a casual curiosity or of a deep desire to explore unknown places and ideas, I invite you to use this book as a tool to become more aware of the vast realm lying beyond Earth's shallow atmosphere. I also invite you to reflect on the extraordinary confluence of professional disciplines, generational contributions, nationalities, and cultures that have combined and continue to combine to make this strange and wonderful realm a little more understandable each day.

In 1938, writing in his book, *The Evolution of Physics*, Albert Einstein confessed his resignation with humankind's confinement to the Earth:

> *Unfortunately we cannot place ourselves between the Sun and the Earth... This can only be done in imagination. All our experiments must be performed on the Earth on which we are compelled to live.*

Fifty-one years later, human beings were walking on the Moon.

As you use this book to extend your knowledge beyond what even Einstein could know in 1938 understand that for you, too, there will be a constant need to renew your understanding and rethink the limits of our species.

When members of the ISU community describe the values that have bonded them to this special school, they invariably refer to the "3i's." Their experience whether as teacher, student, staff member, trustee, or sponsor has been forged in an environment that was intentionally international, intercultural and interdisciplinary. Increasingly with a commitment to continuous renewal of the vision, many of us add a fourth "i" as well: intergenerational.

The International Space University could hardly live up to its international mission were it merely an assembly of people from a few scattered countries. In fact its Masters and Space Studies Programs (SSP) so regularly enroll students from many countries that ISU is increasingly sought after as a host for special symposia and workshops whose themes demand the broadest global perspectives. With a global faculty representing 15 nationalities and students from more than 25 each year in the Masters and the SSP, ISU has managed in only 21 years to develop a network of graduates representing 100 different countries.

Equally as important as these numbers is that ISU has maximized the opportunity for people of different national origins to work closely with each other, solve problems collaboratively, and discover the rich benefits of diverse perspective as a problem-solving tool. Through intense team projects and hands-on experiences, ISU provides its students with the chance to work with one another on complex problems. With common objectives but often widely differing methods of work, ISU students come to terms with their differences and grow beyond the boundaries of their national backgrounds.

Similarly, they learn to work within their diversity of culture by bringing to bear habits of mind and problem solving approaches that often differ widely even within the same nationality. Learning to speak openly about their assumptions and thoughtfully about their perspectives, ISU students discover how to form bonds of synergy that often escape professionals who may never have had the experience of seeing difference as a fount of opportunity rather than a cause of error.

In many ways the habits of mind we acquire in our professions and in the specialized training that prepares us for them represent a particular kind of culture, but one so special as to merit separate attention. Generations of ISU students have come away from their experience in this school with the impression that of all the challenges inherent in the 3i approach, the challenge of interdisciplinary cooperation is often the most difficult. Lawyers think differently than do engineers. Natural scientists often look at a problem very differently than do economists. Social scientists have habits of communication that are very different than those of health professionals. Yet, with the right opportunity to see the value that multiple perspectives and professional paradigms can bring to the process of seeking solutions to complex problems, a well-rooted faith in the power of synergy is possible.

It is from this vision that this volume is born. From the multiple beauties and myriad challenges of Earth to the farthest shore of humankind's ability to dream and to venture, this book seeks to present space as both a distant destination for our species and an intriguing neighborhood for our fragile planet. It emerges from the collaboration of many people, trained in many cultures and professional traditions. It is rooted in the belief that the challenge of tomorrow is in using and exploring the opportunities of space. But we will only succeed in this leap toward the Farthest Shore by harnessing the talent of men and women from around the world.

Enjoy the journey.

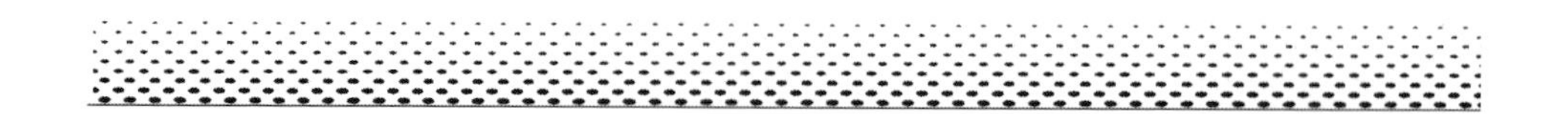

Exploration in the tradition of the past (Courtesy of ESA).

Preface

Dr. Ernst Messerschmid

Professor, University of Stuttgart, and former German Astronaut

When I applied in the early 1980s to become an astronaut on behalf of Germany for one of the first Spacelab missions on the U.S. Space Shuttle, I certainly knew that it was a "long shot". Yet the pull of scientific interest and the desire for a bit of adventure were too strong an incentive not to give it a try. Even the remotest possibility that I might be picked to see the Earth from above, without the filtering and blurring effects of the atmosphere, spurred me into action. To my own surprise I was chosen and flew on the first German Spacelab D1 mission in 1985. Worldwide, thousands of people apply to become astronauts. I suspect that millions would actually try if the odds were somewhat better. And, a billion people around the globe follow with interest human space flight events on television. Young people flock to public appearances by astronauts in order to hear of their experiences in outer space. Indeed in *The Farthest Shore* you will read of very real plans to bring thousands of people into space as "citizen astronauts" within a few decades.

Are there intangible reasons that drive such a tremendous interest in space? Virtually all of us, we *Homo sapiens*, look out from the Earth, into space, and wonder from where we come? To explore seems to be the human condition - our signature - almost a part of our DNA. We humans seem to have the inevitable urge restlessly to move on, to explore and to exploit. Natural selection seems to provide humans with a gut imperative that plays out physically and culturally. Our constant goal remains the expansion of human horizons.

Human history has favored both spatial and cultural expansion. Fresh prospects yield new perspectives. Life springing from the sea to the land was similarly favored. We now stand on a beach, our small world, timidly dipping a toe into the sea, which is the Universe. We stare into this ocean of night and imagine that we are the new Columbus generation.

Two thousand years ago Greek philosophers wrote science fiction stories on air and space travel. The French writer Jules Verne foresaw in remarkable detail what happened a century later in the Apollo program. More than a hundred years ago, Otto Lilienthal and the Wright brothers changed the world with their daring ingenuity that gave us the first aircraft.

No one then could have predicted the explosive growth of aeronautics and how it would come to dominate global transportation in the twentieth century. Today, no one can predict where space travel, and the uses of outer space, will take us tomorrow.

With air travel commonplace and with technology progressing at stunning speed, space travel by the middle of the twentieth century seemed just around the corner - and indeed it was. Worldwide interest in space was ignited when the Soviet Union opened the Space Age with the launch of Sputnik in 1957. The U.S. responded to the challenge with what became the Apollo program to the Moon. This clearly defined program, mandated by President Kennedy, was a logical sequence of missions focused on a clear destination, and carried out with a national willpower and political commitment never seen before or since.

The Russian lunar program, although hidden behind a curtain of secrecy, was no less ambitious. More than anything else, the race to the Moon was a demonstration to the world of the almost unlimited potential of scientists and engineers when given a challenge and supported by their country's leaders. And, having set foot on another planetary body, humanity's potential for exploring, utilizing, and even colonizing the solar system seemed unlimited. But the wave of progress broke after Apollo, and the world's human spaceflight programs have since stayed frustratingly close to the shore. Little of the promise of those early decades has been realized. In the view of many, human spaceflight has been mired in Earth orbit, and the public has grown apathetic.

Strong sentiments have recently emerged that there must be a clear destination and purpose for human spaceflight. We propose a global guide to space built on human needs, scientific knowledge, technological challenge, and the sense of discovery and progress that only space exploration can provide. Others recognize that space applications can provide vital knowledge to deal with life and death issues such as global warming, worldwide drought, and holes in the ozone layer that could lead to genetic mutations that may ultimately endanger life on Earth. A well-conceived international program of human space exploration, space science and space applications can advance discovery, understanding, and cooperation. It can lift our sights, and fuel our dreams.

Thus it is time to develop a logical, systematic, and evolutionary architecture for human expansion into the solar system, with an approach leading ultimately to the human exploration of Mars and a permanent human presence in the solar system. Likewise it is time for international cooperation to use space to unlock new scientific knowledge and to use space technology to improve the human condition. Within this framework, we need to identify clear scientific objectives, solve complex problems, enhance our technological prowess, and set sail to new destinations, as human explorers to the farthest shore. If we have learnt one thing from human history it is that each successive destination, each new frontier, and each new set of capabilities, is a stepping-stone to the future.

The International Academy Astronautics[1] has identified several imperatives that have drawn humankind into space, and that now provide the impetus for human exploration of the solar system. In contrast to the era of the Space Race and the Cold War, no single imperative is sufficient today to motivate the investments and national willpower required for human exploration beyond low Earth orbit. Rather, these factors, in some combination, must inspire us to take action.

To Explore: This human imperative embodies the characteristic drive to expand beyond set boundaries and to explore the unknown:

- To expand the frontiers of human experience
- To fulfill the human need to advance and learn
- To inspire, educate, and engage our youth and the public.

To Understand: This scientific imperative drives the desire to understand our natural world and the Universe around us. The public continues to view space as an exciting new frontier for exploration, and has demonstrated that it will support space exploration in terms that are deeply rooted in our consciousness:

- To seek knowledge and understanding of what surrounds us in space
- To find answers to fundamental questions of our origins and destiny
- To advance and sustain human experience and technological progress.

To Unify and To Prosper: This political imperative reflects the desire of nations to compete for technological superiority; but now, hopefully, it can represent a unifying context within which interested nations can work together:

- To strive for worldwide cooperation and to enhance the "global commons" we call Earth
- To achieve mutual security through challenging enterprise

- To seek wise utilization of the resources of the Earth and, in time, the Cosmos.

As we move forward in space, and as we seek "The Farthest Shore", fundamental questions will also guide our way.

Where do we come from?

- Can we find out how the Universe of stars and planets began and evolved?

- Can we determine the origin and evolution of Earth and its biosphere?

What will happen to us in the future?

- Can we understand the nature of the space environment and cosmic hazards to Earth?

- Can we pursue the potential for permanent human presence in space?

Are we alone?

- Can we determine if there is, or ever has been, other intelligent life in the universe?

- Can we search for life-bearing planets around other stars?

Frontiers breed liberty. They make possible a freedom of movement and ideas. Room to breathe and think anew is not sufficient to ensure progress, but it is essential. The pages that follow in *The Farthest Shore* will not reveal all the answers to these fundamental questions, nor show us the magic combination of "imperatives" that will lead us to new heights in space exploration in the twenty first century. But the rich interdisciplinary blend of knowledge about each and every aspect of space can, and will, offer us new insights about new ways forward. Enjoy!

1"The next steps in exploring deep space" as proposed by International Academy of Astronautics in September 2004

Chapter One

The Farthest Shore: A 21st Century Guide to Space

"That's one small step for man—one giant leap for mankind"

— Neil Armstrong, US astronaut, the first human being to step onto the Moon

Why is there an Earth, and why is our planet unique and perhaps alone in the Universe? As thinking human beings we explore space in our quest for understanding everything, both large and small, everything on our planet, within our solar system, throughout our galaxy, and indeed in the Universe beyond. Since the time of Rene Descartes, who famously expressed the thought "cogito ergo sum" (I think, therefore I am), humans have striven for a better understanding of the Universe. In the age of the "enlightenment" the role of "science" was expanded to find out if the "rules of life and human existence" were the same everywhere throughout the Universe. We continue today to strive to understand why universal rules about time, space, energy and mass obey common laws of behavior throughout the cosmos. Just as Albert Einstein improved on the knowledge provided to us by Isaac Newton, space explorers may some day help us to achieve new insights even beyond those of Einstein. Exploring space is about comprehending the role, purpose and opportunity of our species on our beautiful yet endangered planet — in the past, present and, especially, in the future. It is also about understanding what the human condition might become once our species moves beyond the bonds of Earth's gravity to other parts of our solar system and, ultimately, into the great Milky Way.

Since the beginning of time, humans have wondered about the stars and the "wanderers" in the sky called planets. Ancient cultures built temples, monuments and cities based on their understanding of the skies, the seasons, and of the Sun and the Moon. For many millennia those who searched the skies made what seemed to be a logical assumption that the stars, the Sun and the wandering planets somehow moved above the fixed Earth. The Greek astronomer living in Egypt, Claudius Ptolemy, developed a complex model that showed how this might be possible, and this concept of the heavens persisted for many, many centuries. It became a part of religious dogma that made consideration of other theories that much more difficult. Finally, after a number of people, including Galileo Galilei, who was severely punished for the offense, had questioned whether such a complicated explanation made sense, the Polish astronomer, Nikolai Copernicus, developed a heliocentric (Sun-centered) model of our solar system. Once improved by the calculations of Johannes Kepler, which showed that the best explanation was that the planets traveled in elliptical orbits, we began to understand that the Sun was indeed the center of our solar system. Then most people still believed that our Sun must also be the center of the entire Universe.

In time, with the development of increasingly sensitive telescopes and other new concepts, we began to understand that we lived towards the outskirts of a huge and vast galaxy of stars. When the astronomer Edwin Hubble developed the breakthrough concept of a "red shift" of the light from receding stars, we began to understand that the Universe was much, much larger and much, much older than we had ever imagined. NASA's Cosmic Background Explorer (COBE) satellite launched in 1989 precisely documented the existence of the cosmic microwave background noise, radiation left over from the "Big Bang" origin of the Universe that occurred about 13.7 billion years ago. The reason why all the stars are moving away from us was seemingly explained. But the stars of the Universe are not all moving away from our planet or solar system - they are all flying away from the "singularity" of the "Big Bang". All parts of the Universe have been moving away from each other ever since this mega-explosion.

However, as we, decade by decade, century by century, through obser-

Figure 1.1. Artist's conception of the planet Fomalhaut b (Courtesy of NASA).

vation, experimentation, study and thought began to understand that the Earth was a six million million million million ton planet (i.e. six sextillion tons) orbiting as the third rock from the Sun in a solar system, in a vast galaxy, part of a still vaster Universe, we developed new tools to explore our world and the worlds about us. We developed areas of knowledge that included physics, chemistry, medical science, agriculture, cartography, mechanical and other types of engineering, mining and transportation, and a host of new skills and tools of analysis, and we devised economic and legal systems. These areas of knowledge enabled us to evolve more rapidly than before. Agriculture allowed us to create permanent settlements and build towns, cities, and ports; we built bridges, and walls for fortification. After five million years of human biological evolution, farming and agriculture that started some ten thousand years ago allowed human society and knowledge to move ahead with remarkable speed. The observation and study of the skies that began while humans were exclusively hunter-gatherer nomads led to a human populace that became able to design tools and to write down our language, all knowledge that could be applied to building a permanent human culture.

Imagine a gigantic building that is 10,000 stories (some 32 kilometers) high, representing five million years of human evolution since Australopithecus Man. When humans began farming we find ourselves on floor 9,980. At the time represented by our recent technologies of modern rockets, satellites, computers, lasers and transistors we are only a few tens of centimeters from the ceiling of the top floor. Yet if we consider this same 10,000-story skyscraper in terms of information and our human capabilities, only the bottom 1,500 floors preceded the launch of Sputnik 1 in 1957. The top 6,000 floors of acquired global information have come since the start of the Internet in the mid 1980s. If the current doubling of information continues apace this "Skyscraper of Human Knowledge" would extend more than 500 kilometers into space by 2020.

The point is that our pursuits in space, the development of science and human discoveries, have moved in step for many years. Sometimes the driving force has been military (such as the V-2 rocket, or the development of nuclear missiles or surveillance satellites) and sometimes civilian, for scientific objectives. Today the exploration of space, plus space science and applications, has spread to a dizzying array of activities. Whilst we now know that our space systems are imperiled by magnetic storms, we use space as a vantage point for making key observations to understand more about environmental issues such as global warming or the holes in the protective ozone layer. We now use space for astronomy, telecommunications and broadcasting, navigation and surveying, education and medical training, new developments in materials, and a growing array of business services. Technology transfer from space generates new applications in sports, medical care, ground transportation, urban planning and construction.

Today's tennis rackets and fishing rods are stronger because of materials developed for the space program. Air conditioners and heating systems are more efficient and insulation systems for homes and offices work more effectively. Laser surgery and health monitoring equipment developed for space programs are now employed in hospitals around the world. The next generation of transportation systems will be safer, stronger, more fuel efficient or faster as a result of space research spin-offs. And many of the improvements here on planet Earth are directly dependent on space applications—not just industrial spin-offs. Airplanes take off and land more safely because they can use satellite navigation to know exactly where they are. Many rural and isolated parts of the world have access to broadband communications, and the Internet, because of satellite communications. Many, many thousands of lives have been saved because meteorological satellites have provided timely warnings of deadly storms. Other satellites have helped locate and save downed pilots, stranded hikers, and ship's crews and passengers that have sunk or been swamped. One can go to the web sites of NASA or the European Space Agency and find literally hundreds of space spin-offs and applications that are critical to the world's economy today.

Nor have all of the benefits of space programs yet been realized. Thanks to the space programs of today and tomorrow, buildings of the future will be constructed of lighter and stronger materials that will age better and demonstrate greater energy efficiency. Space scientists are trying to develop new techniques to use "smart satellites" to help us predict coming natural disasters and to assist with recovery when such emergencies occur. It is possible that solar power satellites might be able to provide clean energy from space and new Earth observation satellites and scientific satellites can help us to restore the ozone hole that increases the risks of genetic mutation. Economic studies have shown that, in several areas, money invested in space applications has yielded a twenty-fold return on investment in terms of new goods, products, services and improved economic output. Today as we face significant peril from coming climate change, space technology in its many dimensions will be critical in saving our planet from the destructive path followed by Venus when greenhouse gases trapped in its atmosphere destroyed all possibility of life on our sister planet. When someone asks: Why do we need to spend money on space?" There is a really good and short answer. We need space systems, space science and space applications if we humans—and indeed all flora and fauna on the planet—are going to survive another century or two.

Figure 1.2. Space City: Applications from satellite communications to new materials for building started with space research (Courtesy of NASA).

The "Space Age" which began in October 1957 with the launch of Sputnik 1 was clearly seen in the context of the Cold War between the Soviet Union and the United States. The competition spurred a number of space firsts and led to the Apollo Program, with astronauts landing on the Moon from 1969 to 1972. The early days of the Space Age focused on nuclear missiles, spy satellites, monitoring systems for atomic blasts, and military uses of space. The last thirty years, however, have seen civil space programs grow and spread to an ever-expanding number of nations. Today it is incredibly exciting to examine, explore and exploit space. We are now planning to return to the Moon, to establish a continuing presence there and even to go to Mars.

Today only some 500 people have gone into space. Almost all of these have been official explorers of the national space agencies. The USA, Russia (and the former Soviet Union) and China have provided the launches of astronauts from many nations. But to date six private or "citizen" astronauts have gone into space to visit the International Space Station (ISS). Now there are even plans for a private astronaut to orbit the Moon. Private space enterprises are developing space planes so that private space tourists experience a sub-orbital flight into the dark sky of space, to see the Big Blue Marble from a height of over 100 kilometers, high enough for them to be designated "astronauts". Over 40 companies in more than a dozen countries have already tried or are now seeking to make commercial space tourism a viable industry, and to do so within the next five years. Private space stations, hotels in space, reusable space planes and private commercial orbital transportation systems (COTS) to the ISS are all under development, and have the necessary commercial capital. Space activities already involve over US $100 billion, and in the coming decades this might grow to a trillion dollar enterprise.

Space is not a fringe activity – it is an activity for all. And our quest in space today is international, intercultural and interdisciplinary, to mention the mantra of the International Space University (ISU), in Strasbourg, France. As one of the premier space education institutions in the world, the ISU is dedicated to bringing space not only to specialized students in field, but also to the world at large. Betty in Peoria, Indu in Agra, Habib in Dubai, Francine in Lyon, Zhengzhu in Shanghai, Renaldo in Santiago, and Takashi in Kobe, this book is for you. Whilst we have avoided complicated formulas and too much technical jargon, we have tried to define acronyms and technical phrases as we have gone along.

The Farthest Shore: A 21st Century Guide to Space is a different kind of book about space. First of all, it is not just a book, but a web-based interactive teaching and learning system that will be continuously updated over time. Secondly, exactly like the International Space University, it is aggressively interdisciplinary in its scope. We hope that you will find that it is written in a lucid, expository style that favors clarity rather than complexity. *The Farthest Shore* addresses not only the past, present and future of space activities as well as related space research and exploration but also seeks to explore the myriad scientific, technological, legal, commercial and even social and literary elements of the field today.

Being linked with the ISU, we know that space is exciting, challenging, forward looking and essential for our survival. If we are to combat global warning, avoid societal destruction from a wayward asteroid or comet, use satellite technology to share international health and educational systems, locate key resources, farm and fish more efficiently, discover the secrets of the Universe and evade the most destructive forces of Mother Nature, we must use space and space systems. The understanding of space of the future requires a broad understanding, one that is international, intercultural and interdisciplinary in approach. The breadth of our understanding of space today is thus presented in the following nine fields of study.

Space and Society: Since the dawn of human awareness, the cosmos has fascinated humankind, stirred the imagination of artists and scientists, and helped to lead us forward to a new understanding of the human condition. Futurist, storyteller and interpreter of the arts *par excellence,* Dr

James Dator examines the rich history of how the stars and the arts and humanities have been interlinked, from tales of science fiction to the fine arts.

Stories from Space: Here we hear from two astronauts and follow their stories. We begin with the earliest days of space travel and the transcendent story of Apollo astronaut Rusty Schweickart. Next we hear from an astronaut who had the very tough assignment of repairing the Hubble Space Telescope (HST) the first time around. Astronaut Dr Jeff Hoffman, who is one of the most traveled of all human space farers, has made a career out of capturing and repairing spacecraft and telescopes in orbit. We then hear from the scientists and engineers who are preparing exotic space probes to gather precious new information so that one day humans will explore the surface of Mars. Finally, we hear the stories of the billionaires who are financing commercial space developments, and the progress that is now being made to establish the incipient new space tourism business. All these people who have made amazing advances in space are, in fact, social and warm individual human beings. They feel that sharing their experiences with others was, and still is, an essential part of their job. It is a way to repay society for the special honor of leading the way forward into a deeper knowledge and understanding of the cosmos.

The Future of Space: The future of humankind and the future of space are inherently intertwined. We need to use space systems to understand weather, escape the full fury of hurricanes and monsoons, cope with global warming, and to understand how the Earth, the solar system, and indeed the entire cosmos came into being. We need satellites to aid us when communicating, broadcasting, navigating, and remote sensing; we rely on space telescopes and observatories to unlock the mysteries of the Universe. What does the future hold? Could space elevators put satellites into orbit? Could mirrors in space cancel global warming? Could space colonies on the Moon or Mars be followed by the terraforming of planetary atmospheres? Could von Neumann machines help us to explore the cosmos or space probes be used to divert an asteroid from wreaking havoc on planet Earth? Or could Solar Power Satellites bring us clean energy from space? Only if we manage to overcome the many dangerous challenges of the 21st century will we know the potential of humans and our roles in outer space. The potential is enormous, however.

Understanding the Cosmos: The centuries since the early astronomical findings of Nikolai Copernicus, Tycho Brahe, and Galileo Galilei have produced a great body of science. We now know much about the Big Bang, the huge number of galaxies in the Universe, and how they are arranged, the existence of black holes and of dark matter. We know of the four basic force fields that exist everywhere and which we believe obey the same laws of physics throughout the Universe. Yet for all that we know there is much that we still do not know. We do not know if the Universe is open or closed, or how many dimensions there may be beyond the three familiar dimensions of space and the one time dimension. The great wonder and interest of science in and from space is that the more we know the more questions we have to solve.

Figure 1.3. The Hubble Space Telescope image of the starburst galaxy NGC 1569

(Courtesy of NASA).

Space Applications: The term satellus from the ancient Greek language means servant. In its first usage the Moon was seen to be at the command of Earth's pull of gravity as it moved through the heavens. Today over a thousand satellites are in orbit around the Earth providing valuable services to people everywhere. Over 10,000 television channels are provided around the globe via satellite as well as countless telephone and data links. Satellites today are vital for global business, banking and financial transactions, as well as for weather forecasting, aircraft landing and taking off, and for monitoring weapons systems to enforce strategic arms limitation treaties. Satellites are necessary for almost every aspect of our lives nowadays.

Space Engineering and Space Systems: The many people concerned with space exploration and space science instruments rely upon the scientists and engineers who design, manufacture, launch and operate increasingly elaborate space systems. Over the past fifty years space engineers have designed better, more efficient satellites and safer launch vehicles that have made all of our advances in space possible. Only an overview of the complexities and intricacies of orbital mechanics, new propulsion and steering systems, advances in various materials, and space operations will be given here. However, we shall discuss why and how we build space systems and space vehicles, and how these various demanding programs are executed.

Space Medicine and Life Sciences: Some may think that the only difference between living on the ground on Earth and living in space is to experience living in a weightless condition - something like learning how to swim. The truth is that if you put a human, or indeed any plant or animal, into space everything behaves in a dramatically different way. The way that bones and hair grow, the way that ears keep our balance, the way that the heart and the blood circulatory system function, and thousands of other things about "normal human biological operations" are different. Sometimes the differences are dramatic and at other times only minor adjustments have to be made. Understanding these differences and why they occur are the key to understanding new possibilities for the future. In time these might help us to cure diseases, extend human lives, or develop new drugs for use here on Earth.

Space Business and Commerce: Various forms of communications satellite services and related industries now represent a US $100 billion enterprise worldwide. Navigation, broadcasting, and remote sensing and Earth observation systems all support growing industries. Remote sensing now supports fishing, "smart" farming, detecting crop diseases, developing urban landscapes, managing water resources, controlling pollution in the atmosphere and the seas; all these impact the global economy in important ways. New systems being developed to support tele-education and telehealth services, to bring broadband Internet services to remote areas, to take people into space as "space tourists" and even to deploy space habitats. Entrepreneurial people, as well as companies like Google, Virgin Galactic or O3b (for the Other 3 billion people on Earth), popularize commercial space travel or bring broadband services to the developing world using the latest space technologies.

Space Law and Security: Sometimes overlooked elements of space relate to the legal, regulatory and policy aspects; space programs do not run on science and technology alone. Budgetary allocations, the regulation of radio transmissions to and from space, conventions on the status of the Moon and celestial bodies, international cooperation on space programs and international treaties on the peaceful uses of outer space need to be worked out by politicians, lawyers and policy experts. Institutions such as the United Nations Security Council and General Assembly, plus the UN Committee on the Peaceful Uses of Outer Space, the International Telecommunication Union, UNESCO, standards bodies, and other regional bodies collaborate to promote global cooperation in space. These entities, along with national governments and professional organizations, devise the international "rules of the road" for peaceful cooperation in space, create rules to limit the increase of orbital debris in space, and develop ways to regulate commercial space flights. One of the most critical elements of space policy is the use of space for strategic purposes. Ever since the development of ballistic missiles, space policy has had to consider the possible use of space and space systems for military purposes. Capabilities such as anti-satellite weapons, multiple re-entry vehicles, space targeting systems and space-based weapon systems are just some of the many strategic concerns that have to be addressed.

Diversity and Rewards of Space

To explain the multi-disciplinary nature of space becomes more and more of a challenge every year. This book is designed boldly to explore the true richness of every aspect of space. The International Space University, headquartered in Strasbourg, France, is uniquely organized to include all the disciplines of space. It prides itself in its pursuit of the so-called "3i's" of space - interdisciplinary, intercultural and international. Space is a field that is rewarding, with the need to develop knowledge in a rich diversity of disciplines from astrophysics to zoology, as well as to engage in complex international ventures. The International Space Station, for instance, has participants from eighteen nations, so everyone has to be sensitive to intercultural differences.

The chapters that follow span the many dimensions of space and all of the wonders that they entail. There is a wide range of the arts and sciences, engineering, business opportunities, cultural riches, plus a treasure trove of applications, and a smattering of policy and law, not to mention plenty of excitement and wonder in the pages that follow. A diverse group of space experts from around the world has joined in creating the panoply of space knowledge and information that follows. Enjoy your trip through the many interesting and challenging facets of space!

Chapter Two

Humans and Space: Stories, Images, Music and Dance

When told that the launching of Sputnik in 1957 meant that humans had finally gone into space, Buckminster Fuller is said to have replied: *"We are all already in space! We are on spaceship Earth."*

"I think a future flight should include a poet, a priest and a philosopher . . . we might get a much better idea of what we saw." — Michael Collins, Astronaut

Dreams Lead to Technology Leads to Dreams ...

Dreams of space — of not-Earth — have inspired humanity over the ages. Dream-inspired humans made space travel a reality. Without our dreams, there would be no space programs anywhere. Artistic, religious, philosophical, and ethical perspectives are not frills or mere add-ons to space activities. They are absolutely essential parts of all aspects of all space endeavors.

At the same time, without the science and technology that enables humans to loose the bonds of Earth, humans would still only be dreaming of space while never going to the Moon and beyond. But wait! Science and technology are themselves the products of human dreams and desires. Dreams, beliefs, science, and technology — along with natural and human resources harnessed by human will and labor — are all required to attain and maintain space activities

Stories About How the Earth and Humans Were Created

All cultures have stories about the origin of Earth and humans. Almost all creation stories say that Earth, humans, and all living things on Earth were created by nonhumans or superhumans, many of whom either live in the sky, or descended from the sky to create Earth and life on Earth. Sometimes the stories tell of conflict between competing nonhumans with one kind being victorious over the others. I am not aware of any creation stories that describe something even vaguely similar to Darwinian evolution.[1]

All creation stories leave unanswered (and unasked) the most important creation question — who or what created the nonhumans who then created humans and the Earth? The stories of all cultures say that nonhumans existed before and created humans and that they came from and returned to places that are not-Earth — typically places somewhere distant "up in the sky." Most cultural stories and many religions also describe nonhumans who exist on Earth and with whom humans interact in various ways. These may be spirits, or ghosts, or angels, or various forms of microvita, many of whom also have not-Earthly origins and homes.

Most cultures and religions also have stories that tell of voyages between Earth and not-Earth, sometimes taken by humans but often involving nonhumans as well or instead. Moreover, some stories say that humans exist somewhere else before they are born and live on Earth, and almost all religions and cultures state that humans go somewhere not-Earth when they die, either to a different place, such as Heaven and Hell, or a different plane of existence, such as Nirvana.

So the idea that there are worlds and beings beyond Earth, humans, and the other creatures of Earth is apparently universal, and clearly precedes science fiction, space fiction, and other forms of space art.

Some Stories About Voyages To and From Space

One of the oldest and best-known Japanese stories is about *Kaguya Hime* (often translated as the Moon Princess) as told in the *Taketori Monogatari.* A bamboo cutter (a *taketori*) discovered a baby girl inside a bamboo shoot. He took her home, and he and his delighted wife reared the baby as their own. She grew into an extremely beautiful — and very large — woman. Her father wanted her to marry, and five princes came to woo her. But she did not want to marry them and gave them impossible tasks to perform. Even the Emperor could not persuade her to marry him. Then she revealed she was not of this world, and was transported back to the Moon from which she came by an array of otherworldly attendants.[2]

In the summer of 2007, Japan launched a lunar orbiter from Tanegashima Space Center. The name originally given to the orbiter was "Selene" for (SELenological and ENgineering Explorer). *Selene* is the name of the Greek goddess of the Moon. However, the Japanese national space agency, JAXA, asked the Japanese public what nickname they thought should be given to the orbiter. *Kaguya*, and *Kaguya Hime*, were overwhelmingly the number one and two choices. JAXA thereafter called the explorer *Kaguya* in Japanese language reports, though English-language reports still often referred to the orbiter as *Selene*.[3]

Another well-known story that came to Japan from China and Korea is recalled during the annual summer *Tanabata* festival (called *Qi Xi*, in Chinese, *Chil-wol chil-seok* in Korean). It is held on the seventh night of the seventh month in celebration of the annual meeting of the stars Vega (or *Orihime*, the seamstress) and Altair (or *Hikoboshi*, the ox-herder) that are separated by the Milky Way (or *Amanogawa*) and only allowed to be together once a year for a brief time. The reasons vary as to why the two lovers are separated and allowed to meet each other only briefly once a year, but they are all based on plots that have humans being punished for doing what they *want* to do rather than doing what they know they *ought* to do.[4]

The Japanese space agency, then known as NASDA, named a pair of satellites *Orihime* and *Hikoboshi*. The two were launched together as ETS7 (*Kiku* 7) and separated before coming together again in 1999. It was the first time such automated docking had ever been carried out in space. So JAXA was following good precedence in choosing to name its lunar orbiter *Kaguya*.[5]

But consider the names NASA has given to some of its space vehicles — Mercury, Apollo, Gemini, Saturn, Jupiter, Orion, Titan, Ares, Altair — all named after Greek or Roman gods.

Who and what is this (see Figure 2.1)?

Figure 2.1. Icarus (Artwork Courtesy of NASA).

Yes, Icarus. Icarus the Greek who, with his father, Daedalus, fashioned themselves wings so that they could fly like the birds, something they were totally unable to do with their wingless, flightless natural bodies. So they first imagined flight, and then developed the technologies that would enable it. Icarus also received a warning from his father before he took off. According to Ovid's *Metamorphoses* (10 AD), Daedalus

> "Then thus instructs his child: My boy, take care
> To wing your course along the middle air;
> If low, the surges wet your flagging plumes;
> If high, the sun the melting wax consumes:
> Steer between both: Nor to the northern skies,
> Nor south Orion turn your giddy eyes;
> But follow me: let me before you lay
> Rules for the flight, and mark the pathless way."[6]

But Icarus, being a typical son — and a typical human — disregarded his father's warning. He did fly. But then he sailed too close to the Sun, which melted the wax on his wings and cast Icarus to his death in the ocean below.

The daring and hubris of Icarus has been an extremely popular theme in Western art and literature, warning us of the eternal tension between what we *want* to do and then *can* do because we develop technological capabilities, on the one hand, in contrast with what we *ought* to do, given our ethical limitations and frailties, on the other.

Precursors to Science and Space Fiction

Space fiction, almost by definition, involves boldly going. Well before the modern era, most cultures had their own stories about voyages of discovery. Heroes leave home, travel through strange times and places, overcome many adversities and have many exceptional experiences before returning home again, enlightened by the process. The basic archetypical stories for Western cultures are *Gilgamesh,* and the *Iliad* and the *Odyssey. Gilgamesh* is about the adventures of a king of Uruk, named Gilgamesh, and his half-wild (and perhaps gay) friend, Enkidu, somewhere between 2750 and 2500 BCE.[7] The *Iliad* and the *Odyssey* are thought to have been first composed between 800 and 600 BCE. The *Odyssey,* recounting twenty years of travel by Odysseus (Ulysses, in Latin), is a prime example of a voyage of discovery.[8]

In the *Bible*, the first book, *Genesis*, is immediately followed by *Exodus*: departure happens soon after creation. The story of Moses leading the Jewish people to the Promised Land — and the belief in the existence of a

Promised Land that is rightly theirs — is an unfinished narrative of travel and travail. In Christian belief, Wise Men traveled far to find the Messiah. Muslim faithful must travel to Mecca. There appears to be an almost irresistible urge for humans Boldly To Go — or at least for some people, usually men, to go. In each of the stories above there are those who warn against the journey, and/or who patiently stay at home waiting for the hero's eventual return, such as Ulysses' Penelope.[9]

Around 150 A.D. the Greek philosopher Lucian of Samosata wrote what might be the first two western works of space fiction, *Icaromenippus* and the *True History*. In the first, written as a play, Menippus (specifically wanting to avoid the failure of Icarus) took one wing from an eagle and another from a vulture, and fashioned them so he could fly from Mount Olympus to the Moon. There he visited with many famous dead Greeks and discovered that they were all pretty much losers. Having also confirmed that the Earth is round, Menippus then tried to fly to the Sun. He got as far as heaven where he met the great god Zeus and some other lesser gods. Zeus questioned him about what was happening on Earth and why mortals didn't worship and fear him any more. The gods eventually realized that their weakening powers would be completely lost if humans were able to fly up to see how they really lived, and so they confiscated the wings of Menippus and sent him back to Earth. They ordered him never to fly again, thereby re-emphasizing one of the continuing themes of space fiction — hubris: the fact that humans try to do things beyond their natural physical and ethical ability, and are punished for it.[10]

In Lucian's second story, *The True History*, a ship exploring the Atlantic was carried by a waterspout to the Moon. Its crew found the Moon-King at war with the Sun-King over which would rule the planet Jupiter. While on the Moon, they saw various marvels such as a combination telescope/telephone which Lucian described as "a large mirror suspended over a well of no great depth; anyone going down the well can hear every word spoken on our Earth; and if he looks at the mirror, he sees every city and nation as plainly as though he were standing close above each." After various adventures, the voyagers eventually returned to Earth.[11]

Lucian's stories are more than mere fanciful tales. In them, Lucian ridiculed humanity, its tired old philosophies and pagan beliefs, and the emptiness of the intellectual life of the time — all themes that we find repeatedly in space fiction.

Though written as theological works, St. Augustine's De *Civitate Dei* in the 5th Century,[12] and Joachim of Flora (or Fiore)'s *Liber Concordiae Novi ac Veteris Testamenti* and *Expositio in Apocalipsim* in the early 13th

Century, each exhibited futures-oriented utopian thinking.[13]

The great Muslim scholar, Ibn Kahldun, wrote in the 14th Century. He is considered one of the fathers of sociology, presenting a sophisticated philosophy of history and society in his *Muqaddimah* — again, not a work of fantasy or fiction, but rather exhibiting, perhaps for the first time, a way of thinking about humans and their past and future that greatly influenced the emergence of modern attitudes towards social change, and hence, modern science fiction.[14]

Another contribution to modern ways of thinking was indeed the emergence of utopian literature that actually goes back at least to Plato's *Republic*[15] and was also exhibited in Joachim of Fiore's writing and flourished in Europe from the 17th through 19th centuries. *Utopia* is a Greek word that literally means "no place". A *utopia* is a perfect society by some understanding of perfection. A contrasting term is *dystopia* — a largely negative and undesirable place of some kind or another. There is a third term, *eutopia*, which means a good (but not perfect) place — what I call "the best possible real world given the challenges and opportunities of the future" — in other words, "a preferred future".

Unfortunately, as we will see, a great deal of current science and space fiction is dystopian. Little is either utopian or eutopian. This is of concern since images of the future are major components in shaping the eventual future itself. Many people see the future as grim because of the way it is portrayed in modern science fiction. They therefore may avoid thinking more usefully about the future and how they can shape it — and so the future may be grimmer than it needs to be. On the other hand, utopian thinking is equally dangerous, historically often leading to totalitarianism if taken and acted on too seriously. We need eutopias!

Early modern European society saw an outburst of utopian literature. Thomas More's *Utopia* (1516)[16] described his view of a perfect society: a democratic yet paternalistic agrarian community set on an island, impregnable to foreign attack. In this communal state, people alternated between living in cities and farms. Food was abundant. Diversity of religious belief was permitted, gold and silver were despised and used for chamber pots, and all people wore the same coarse clothing, symbolizing their equality.

Francis Bacon's *New Atlantis* (1627)[17] is more nearly a true work of utopian science fiction since he imagined a perfect society based specifically on the application of the scientific method to everything, including government and family life — an idea replicated by many people subsequently, most powerfully by the behavioral psychologist, B. F. Skinner, in *Walden Two* (1948).[18]

Though not a utopia, Johannes Kepler's *Somnium* ("A Dream" 1634)[19] is clearly a work of space fiction, describing a person who was transported to the Moon by demons where he discovered that each half of the Moon is inhabited by two different kinds of beings. Interestingly, Kepler's book, though written in Latin, was published the same year that Lucian's ancient *True History* first appeared in English translation. Once new ideas began to appear in languages understood by relatively unlearned people; they could spread and provoke social change more widely and rapidly.

Francis Godwin's *The Man in the Moone* (1638)[20] combines both genres. Birds carried a shipwrecked Spaniard, Domingo Gonsales, to the Moon where he found a simple utopian world with no waste of any kind. Food grew without labor, and clothing and housing were provided virtually without work. All women were extraordinarily beautiful and yet no man desired any woman but his wife — truly a utopia. There was no crime. Occasionally, imperfect children were born, but they were shipped off to America.

Shortly thereafter, Cyrano de Bergerac wrote two classic space travel stories *Histoires comique des etat et empires de la lune* (1648) and *Histoires comique des etats et empires du soleil* (1662)...." (Voyages to the Moon and Sun) After failing to be lifted to the Moon by evaporating dew (though he did make it to Canada) Cyrano used firecrackers and eventually landed on the Moon on the Tree of Knowledge where he met and talked with early biblical characters.[21]

Bernard le Bovier de Fontenelle's Entretiens *sur la Pluralité des Mondes* (Conversations on the Plurality of Worlds, 1686) was written — in French, not Latin — as a fictional conversation between two people walking in a garden late at night and looking at the stars.[22] It was a daring conscious defense of the Copernican worldview and speculated about the possibilities of extra-terrestrial life and intelligence. It did so as a work of fiction in order to avoid censure by the Church. It is worth noting also that Kepler had written *Somnium* as fiction for the same reason. These are two of many examples of science fiction used to express unpopular or forbidden views with relative safety.

Somewhat later, classic satires and critiques of contemporary society, safely told in the form of miraculous voyages, appeared, such as Jonathan Swift's *Gulliver's Travels* (1726)[23], Ludvig Holberg's *Nicolai Klimii Iter Subterraneum (*Nicolai Limmi's Underground Journey, 1741)[24], and Voltaire's *Micromegas* (1752).[25]

Micromegas was influenced by Swift and featured huge intelligent creatures from Sirius who were on a tour of the solar system. They walked around Earth in 36 hours, at first thinking whales were the dominant intelligence, and only later finally spotting the tiny humans. They encountered philosophers arguing about unimportant things and left in disgust, giving them a book of philosophy they thought humans might be able to understand — since it had only blank pages: yet another common theme in space literature — humans focus on unimportant and petty issues while great world-changing events swirl around them.

In 1781 a story describing a newly-discovered part of the Earth where free love is allowed was published: *La decouverte australe par un homme volant, ou le Dedale francais*, (Southern Discovery by a Flying Man, or the French Daedalus) by Nicolas Restif de la Bretonne. This is sometimes said to be the first space pornography, a genre, though extensive, we will not otherwise consider.[26]

Balancing "Science" and "Fiction"

There have been many definitions of "science fiction." Science fiction is seldom fiction about science. Rather, it is stories that arise when some people become aware of the fact and possibility of social change. Most societies are, or at least seem to their members to be, fixed and unchanging. In the past, a few societies went through sufficiently extended periods of social change that some people living in them realized that the present was different from the past, and that the future could be different from the present in important and uncertain ways. They would then speculate what a different future might be like. These stories are important precursors to science fiction and tell important stories in their own right. But science fiction, *per se*, is a product of the scientific, technological, and industrial revolution that was made possible in Europe with the Black Plague, the Reformation, and the Renaissance between the 14th and 17th Centuries, and then bloomed during the late 18th, 19th and 20th Centuries. Science fiction proper emerged in the mid 19th Century first in Europe, then in the UK, then in the US, Japan, China, India (and perhaps elsewhere) simultaneously later as the social and environmental consequences of the scientific, technological and industrial revolution spread.[27]

Fiction is stories about imaginary worlds and people. A work of fiction does not intend to be "true" (though it may deal in truths). Humans have invented, lived in, and believed in imaginary worlds at least since they learned to speak, perhaps 40,000 years ago. Indeed, the world constructed by each language is an artificial world — though an extremely powerful one!

In contrast, *science* intends to tell the "truth" about the world and people; truths that others, using the same evidence and methods, can confirm or deny. However, few final and absolute truths are ever declared scientifically. Science is constantly revising old ideas and establishing new bases for thought and useful action. Moreover, science is not new. Science is also as old as speech and focused thought, but it was the invention of writing that made science more powerful by enabling ideas to be decontextualized, analyzed, and categorized more carefully than is possible only with speech.

Among many others, the Sri Lankan scientist and futurist, Susantha Goonatilake, has made it clear that what is called "science" now is not something invented in Europe in recent centuries.[28] Rather, European scholars adopted and refined theories and methods that had been honed over long periods of time previously, primarily but not exclusively in Asia and northern Africa. European scholars also used or invented instruments that enabled them to discover things otherwise unknowable to humans. Gonnatilake argues convincingly that there have always been scientists, and scientific ways of knowing, who have contested against commonsense, folk knowledge and religion, and that present-day science is simply the latest evolutionary manifestation in a continuing contest between different ways of knowing and acting.

Most science fiction is more about technology than it is about science. Most science fiction is about how humans might behave and how society might change as new technologies come along. Often the "science" in science fiction is quite unscientific while the behavior said to result from new technologies is sometimes more plausible. But much science fiction is bad social science as well as bad natural science — and not very good fiction either. Science fiction and space fiction thus exhibit a tension between two modes of knowing — one scientific, the other fictional. Some science fiction is closer to science than to fiction, and thus often boring though factual. Most science fiction is closer to fiction than to science, thus exciting, but misleading. The best science fiction finds a balance between both.

The literature of science fiction and space fiction dealing with new technologies and technological change has generally been of one of two kinds. Jules Verne and many others were basically optimists, believing in inevitable progress through technological change. This optimistic view of the future permeated much early science fiction and space fiction. Some science fiction writers and futurists still share this view today. But, from the beginning, other science fiction writers had more of a love-hate relation with technology and often wrote of that relationship critically and ironically. One of the leading examples of this ironic — if not pessimistic — school was Alfred Jarry, a French author who invented the term *'pata-*

physics, "the science of imaginary solutions" (*la 'pataphysique ou science des solutions imaginaries).* It has been said that "'pataphysics is to metaphysics what metaphysics is to physics." Jarry is also considered to be a forerunner of the "Theater of the Absurd."[29] The 'pataphysic tradition remains strong today in old as well as in various new forms, as we will see.

Space and Science Fiction in Europe

Without a doubt, the most important single figure in the origins of science fiction and space fiction is the French author, Jules Verne. His books, *Voyage au centre de la Terre* (Journey to the Center of the Earth, 1864) and *De la terre à la lune* (From the Earth to Moon, 1865) and many more, were translated into every major language of the world.[30] During his lifetime Verne was perhaps the most widely read author in the world, and his books are still popular. Almost all early pioneers in space reality and space fiction said that they were inspired by Verne.

The importance of space fiction in creating space reality — and vice versa — cannot be overstated.

The extraordinary pioneer of Russian space flight and modern science fiction, Konstantin Tsiolkovsky, said his enthusiasm for space came from reading Jules Verne. In addition to his vital role in envisioning and enabling actual space flight, Tsiolkovsky himself wrote classics of Russian science fiction including one that the world's first cosmonaut, Yuri Gagarin, said was his favorite: *Vne zemli* (*Beyond the Planet Earth,* 1896).[31] Alexei Tolstoy was inspired by Verne and Tsiolkovsky to write what became a Russian space fiction classic, *Aelita* (1922) about a trip to Mars.[32]

In Germany, Kurd Lasswitz said Verne's example encouraged him to write what became the very popular *Auf Zwei Planeten* (*Two Planets,* 1897) describing an advanced civilization on Mars. Having a background in mathematics and engineering, much of what Lasswitz described as techniques of space travel were accurate or at least reasonable.[33]

The Czech, Josef Capek, brother of Karel Capek, is credited with suggesting the term, "robot" (which means "forced labor" in Czech), to describe autonomous or semi-autonomous artificial life in Karel's book titled *R.U.R.* (1921). The abbreviations stand for *Rossumovi Univerzální Roboti* ("Rossum's Universal Robots").[34]

There was considerable science fiction produced in the Soviet Union — and much anti-Soviet science fiction as well, such as *We* (1920) by the Russian émigré Yevgeny Zamyatin.[35] Ivan Efremov's popular novel,

Tumannost Andromedy ("The Andromeda Nebula," 1957), was Efremov's pro-Soviet answer to anti-utopian books by Zamyatin and others.[36] Efremov described in great detail an ideal Communist society engaged in the exploration of outer space where a network of civilizations exists. It was translated into many languages and was highly influential. In 1967 the first part of the novel was made into a film, *Tumannost Andromedy. Plenniki Zheleznoi Zvezdy* ("The Andromeda Nebula. Prisoners of the Iron Star"). The film was a popular success.[37]

Stanislaw Lem of Poland, who some critics consider to be one of the best science fiction writers of all time, published his masterpiece *Solaris* in 1961.[38] It features a planet covered by a mysterious ocean that scientists from Earth are exploring. However, it turns out that the ocean is also examining the scientists and is able to become whatever deep anxiety or feeling of guilt each one of the scientists has. *Solaris* was also made into an excellent, if slow-moving, film in 1972 by the Soviet director, Andrei Tarkovsky. It was also made into a very bad US movie in 2002.

Although the earliest science and space fiction originated in Europe, there were clear precursors in other countries. Science and space fiction *per se* emerged rapidly wherever industrialization and technologically induced social change spread in the world.

Space and Science Fiction in India[39]

The roots of science fiction in India are from 1500 BCE in the ancient Vedic literature. In these texts there are many descriptions of unidentified flying objects referred to as *vimanas*.[40]They were of two types: "manmade crafts that fly with the aid of birdlike wings, or odd shapes that fly in a mysterious manner and are not made by human beings". Yet, despite — or maybe because of — these early science fiction-like images, the influence of this genre on Indian literature and culture is very recent and slight. However, science fiction did emerge in India where, as one author says, "the effects of the industrial revolution were being felt in urban India in the 19th century just as keenly as they were in Europe and the U.S." The earliest notable Bengali space fiction was Jagadananda Roy's *Shukra Bhraman* ("Travels to Jupiter"), written in 1857 and published in 1879. This story is of particular interest as it described a journey to another planet, while the existence of the creatures seen there was explained using evolutionary concepts. It should be noted that this story was published well before H. G. Wells' *The War of the Worlds* (1898) in which Wells described an invasion from Mars. The father of Indian literary fiction in Hindi is often considered to be Acharya Caturasena (1891-1960). He wrote more than 400 books during his lifetime, of which most were novels based on historical events, mythology, or social issues. Three were science fiction novels: *Khagras* (The Eclipsed Moon), *Neelmani* (The Sapphire), and *Adbhut Manav* (The Amazing Man).

An extraordinary early work was *Sultana's Dream* (1905) by Rokeya Hosain. It is an intriguing example of a feminist utopia — a world where women are socially and politically dominant over men, and where that dominance is seen as natural. It may have been the first utopian novel of any kind written in India.[41]

Even in Indian cinema, which churns out hundreds of movies a year, the influence of science fiction is slight. In 1987, Shekhar Kapur's *Mr. India* is a story of a young man who discovers his father's invisibility device and battles a madman's attempt to rule the world. Rakesh Roshan's *Koi Mil Gaya* ("I found someone") deals with the rise and ultimate demise of a mentally challenged man who befriends an alien being.[42]

Space and Science Fiction in China[43]

Wu Dingbo states that the "era of science nurtures scientific literature and art which, in turn, reflect the era of science. Science fiction as a modern genre that emerged after the industrial revolution has developed along with the development of science." "Science fiction, however, is also a cultural phenomenon which has to develop in accordance with the specific conditions in a given nation."[44]

Chinese creation stories typically have themes involving space, the cosmos, and chaos, such as Pan Gu who separated the heavens from Earth and Nu Wa who patched up the falling heavens. Similar themes are found in the earliest Chinese literature, such as *Tian wen* (Questioning Heaven) by Qu Yuan (347-278 BCE) and *Hou Yi sheri* (Hou Yi Shooting the Suns) and *Chang E benyue* (Chang E Goes to the moon) by Lu An (197-122BCE). "Questioning Heaven'" is a series of questions about the creation of the world and the nature of the Sun, Moon and stars. "Hou Yi Shooting the Suns" tells how Hou Yi shot down nine of the ten suns in the sky, thus saving the world from scorching, whereas the story of Chang E is about Hou Yi's wife who stole and drank an immortality elixir that enabled her to fly to and live on the Moon. Wu says that this is probably the world's earliest story about space travel. The first Chinese lunar orbiter, launched in October 2007, was named Chang'e-1.[45]

Zhang Zhan's story *Tanwen*, written in the 4th century, details the antics of a robot that sings and dances as well as any human. Wu says this is the world's first story about robots. They also appeared in Zhang Zhou's *Chao ye qian zai* (The Complete Records of the Court and the Commoners) in the seventh century, as well as in Shen Kou (1031-1095) *Meng zi bi tan* (Sketches and Notes by Dream Creek) that described a robot that killed rats.[46]

Nonetheless, Zhao states that, "prior to the concept of modernity being imported into China there had been no fiction about the future. In traditional China, history did not have directionality."[47] The introduction, first of ideas about "progress" and "development", and then of Marxism, changed that. The young intellectuals of the late 19th century in China sought to change their "backward" country into a modern nation-state. To do that — they learned from Japan and the West — science and technology was necessary. So "in order to stimulate people's interest in science and technology, some enlightened intellectuals discovered science fiction and began introducing this new literature to the Chinese reading public. Lu Xun (1881-1936) is one of them".[48] Lu is considered to be "the father of Chinese science fiction". In his preface to his 1903 translation of Jules Verne's novel *From the Earth to the Moon,* Lu lamented that, while China had every other kind of literature in the world, "science fiction is as rare as unicorn horns, which shows in a way the intellectual poverty of our time. In order to fill in the gap in the translation circles and encourage the Chinese people to make concerted efforts, it is imperative to start with science fiction."[49] Lu's translation of Verne's novel is posted on the official website devoted to the Chang'e-1 lunar orbiter.

It is usually said that modern Chinese science fiction began in 1904 with the serialization of *Yueqiu zhimindi xiaoshou* (Tales of Moon Colonization) in *Xiuxiang xiaoshuo* (Portrait Fiction).[50] It was written under the pseudonym Xu Nianci and the real author is unknown.

The mission of Chinese science fiction has always been to help people understand what science and technology is and to encourage them to help their country develop into a modern world power. Thus, as Wu points out "Chinese science fiction seldom tackles the subjects of space colonization, galactic empires, alternative histories, cataclysms, apocalyptic visions, telepathy, cybernetics, religion, sex, and taboos. On the whole, Chinese science fiction is optimistic. People always get the upper hand over nature, science, evil, or whatever enemy or obstacle they may face. The hero is supposed to succeed, emerging triumphant and unscathed from difficulties. Visions of the future are always bright and promising, although a spectrum of possibilities for that bright future is projected. As a result, Chinese science fiction stories mostly depict the near, foreseeable future. In China, science fiction's main function is utilitarian rather than aesthetic. It aims to create interesting stories in a simple and effective prose and to teach moral lessons…."[51]

Currently, Chinese science fiction is attracting adult audiences as it brings a basic understanding of space to the Chinese people, developing in them once again a desire to explore the Universe. Traditionally, space (or the sky, in ancient terms) was an arena controlled by the emperor. With the emperors gone, space is becoming one more locale for exploitation and development by anyone.[52]

During a discussion of the role of ethics in science fiction, Han Song stated, "the unconscious descriptions of ethical problems in Chinese science fiction probably hint at a future reality in the country's space program. The coming conflicts between China and other space giants might evolve from our ethical differences rather than our technological gaps." Understanding one another's science fiction might lead to better relationships among all space programs, he concluded.[53]

Space and Science Fiction in Japan[54]

We have seen that some Japanese folk tales were explicitly about creatures that came to the Earth from not-Earth, and eventually returned. In addition, at least one very well known story can be interpreted as depicting the hero, *Momotaro* (The Peach Boy), as coming to Earth in a spaceship that seemed to be a giant peach. While there appears to have been no early utopian literature in Japan, Pure Land (*Jodo*) Buddhism urges humans to strive for life in a Pure Land that is something between a distant heaven in another dimension and a perfect society at a later time and place on Earth. The appeal of a Pure Land was very powerful during the medieval period of "The Warring States".

The first western utopian work was translated into Japanese in 1868, the year marking the "opening" of Japan following the end of feudalism and the 300-year long isolation of the Tokugawa era. It was *Anno 2056* (1865) by the Dutch author, Alexander Bikkers. More's *Utopia* was translated in 1882. Translations of Verne and other early utopian and science fiction writers were extremely popular in Japan from the 1880s onward, with about forty translations of various works appearing during the Meiji era (1868-1912). These translations inspired a number of local writers, including Tetcho Suehiro whose *Setchubai* (Pure Blossoms in the Snow, 1885) described a very prosperous, modern and internationally powerful Japan in 2040, while his *23-nen Miraiki* (The Future of the Year 23 [1890], 1886) elaborates the same theme.[55] Nakae Chomin's *Sansuijin Keiron Mondo* (A Discourse by Three Drunkards on Government, 1888) argued the merits of democracy over traditional *samurai* values, illustrating once again that unpopular ideas are often expressed in the form of satire or a utopia.

There was in Japan an explosion of science and space fiction after the Second World War. Many of the stories exhibit profound uncertainty as to the morality of humans venturing in space. Concern is often expressed about the destruction of nonhuman life by humans during exploration rather of the loss of human life itself. Indeed a major point of many of the stories is that humans should not place themselves above other forms of life, on Earth or elsewhere, but should learn to live in respectful harmony with everything. In these, and in almost all other Japanese space fiction stories of this era, the emphasis is on harmony, peace, environmental protection, concern about science gone wild, and about maintaining good interpersonal relations.

Robert Matthew, in a review of Japanese science fiction, reminds us of a very important difference between Asian and western philosophy, religion, and fiction. Asian cultures generally maintain that humans are basically good, and can be made to act better by proper education and environmental reinforcements. By the same token, humans can be led to do bad things by circumstance or personal choice, but they are not fundamentally evil, *per se*. In contrast, people living in cultures steeped in the Jewish/Christian/Islamic heritage often believe that humans are basically sinful and greedy, and can never be made good. Repressive social institutions might control or even channel humans' evil behavior into social good, but in their hearts, humans only care about themselves. Matthew thus contrasts "the Christian concept of original sin versus the Confucian concept of original virtue" as distinctive features underlying the fiction — and policies — of these two traditions.[56]

More will be said about contemporary science fiction in Japan when we discuss the emergence of cyberpunk, *anime* and computer games later.[57]

Space and Science Fiction in the UK, North America, and Oceania

Science and space fiction written in English in the UK, North America, New Zealand and Australia is literally too numerous to summarize or even categorize fairly in this brief review, but H. G. Wells is to English science and space fiction what Jules Verne was to European — and the world's — science fiction. Among Wells' best-known stories are *The Time Machine* (1895), *The War of the Worlds* (1898), and *The First Men in the Moon* (1901).[58]

In the United States, science and space fiction's heyday is found in the "pulp" magazine, *Amazing Stories*, that began publication in 1926, and *Astounding Stories* that went through many name changes, ending up as *Analog* today. Many other pulp magazines devoted to science and space fiction existed from the 1930s to 50s, and almost all the great and not so great names of science fiction history published in them, defining the genre from that point onward.[59] Some of the most important contributors include Arthur C. Clarke, Robert Heinlein, Isaac Asimov, Frederik Pohl, A. E. Van Vogt, Poul Anderson, Larry Niven, Kim Stanley Robinson, Pamela Sargent and others whose work typically (but not always) celebrates scientific and technological achievements and progress though still often asking probing questions about humans and human purpose.

These authors contrast strongly with the powerful dystopian novels of the same period that may have had an even greater impact on thinking about the futures and space.[60] This was also the period of the Cold War rivalry between "Communism" and "Capitalism". As we noted above concerning space fiction in the Soviet Union, much space fiction in the United States was scarcely veiled propaganda.[61]

Space Fiction in Films and Television

Movies, and later, television, have almost always featured some kind of "space opera", from *Le Voyage dans la Lune*, ("A Trip to the Moon" 1902) through *Buck Rogers* and *Flash Gordon* in the 1930s and 40s, *Tetsuwan Atomu* after the Second World War, and many others onward. Even though most of these films were "B" grade movies at best, they and the pulp science fiction books of the era created themes that defined and have persisted in almost all space fiction everywhere in the world.

Stanley Kubrick's co-production with Arthur C. Clarke of *2001: A Space Odyssey* (1968), brought space fiction in films to a high level, while Roger Vadim's adaptation of the French comic strip character, *Barbarella* (1968), was a magnificent high camp film in the 'pataphysics mode. There is no more striking contrast in space films than between *2001*'s cold, barren, functional space ship (with the onboard intelligent computer, HAL, and the human Dave locked in a gripping though slowly-realized battle of wits to the death) and the nubile and naked Jane Fonda slithering across her fur-lined space ship. *Silent Running* (1971) was a superbly poignant movie featuring heart-tugging robots loyally tending the last remaining "world heritage" stock of plants and animals sent to space from an Earth rapidly approaching terminal over-pollution from human arrogance and neglect.

Dystopian movies of this period include *Metropolis* (1927), *A Clock-*

work Orange (1971), *Soylent Green* (1973), *Mad Max* (1979), *Mad Max II: The Road Warrior* (1982), and *Brazil* (1985).

On television, the quirky British space fiction show, *Dr. Who* (1963 onward), which was itself inspired by the earlier very popular *Quatermass* series of the 1950s, influenced many viewers' ideas about space. But for Americans, good space fiction on television began with the extremely popular TV series *Star Trek* in 1966, which dealt with current social, political, and ethical issues in the guise of exploring the Universe. After the blockbuster movie, *Star Wars* in 1977 (and its successors), space fiction on television for some time was dominated by triumphal voyages of ISU-like international, intercultural, and interdisciplinary members of a united Earth federation boldly exploring the cosmos, "doing good (or at least not doing evil while trying to obey 'the prime directive' of not interfering in the lives of other cultures)." These were magnificently written and produced shows that made you feel good after watching each episode, and sad and empty when the series ended. But new and better productions seemed to spew forth annually: *Star Trek: The Next Generation, Babylon 5,* and a new *Battlestar Galactica* series.

Cyberpunk, Anime, and Electronic Games

One of the most important developments in recent science fiction has been the emergence of cyberpunk literature spearheaded by William Gibson's *Neuromancer* (1984).[62] It was inspired by contemporary and emerging developments in electronic communication technologies, biotechnology, and nanotechnology, combined with deep anxiety about the environmental and social consequences of these and other developments. Cyberpunk treats science and technology critically and ironically, and is more in keeping with the European 'pataphysics tradition than with much American science fiction which often is either triumphant, scientistic and optimistic, or gloomily pessimistic and dystopian. Two excellent, though lamentably short-running, television cyberpunk series were the British-inspired *Max Headroom* (1987-88) and *Dark Angel* (2000-2002).

Recent films that exhibit cyberpunk perspectives include *BladeRunner* (1982; derived from Philip Dick's *Do Androids Dream of Electronic Sheep?), Johnny Mnemonic* (1995), and *The Matrix* (1999).

A rather Gothic style of cyberpunk pervades most interactive electronic games, a form of science and space fiction that may rapidly be replacing not only written literature but also cinema and television.[63]

Contemporary Japanese science fiction, especially in its *anime* and *manga* forms, is heavily cyberpunkish. Its origins lie in Japanese *ukiyo-e* woodblock prints of the Edo era (17th to 18th Centuries) that eventually blossomed into *Tetsuwan Atomu* (mistranslated as *Astro Boy* in English), originally a *manga* in 1951, that soon became one of the earliest examples of what would become a flood of Japanese science fiction *anime,* while the world famous *Godzilla*, a 1954 film of a monster dinosaur mistakenly brought back to life by radioactive waste, has resulted in more than thirty Godzilla-based films.

Space Fiction and the Alien Other

We have seen that space fiction is often used as a safe way to criticize contemporary society. That is especially apparent in recent work done by feminist, gay, and ethnic studies scholars who often also adopt a postmodern academic approach. Some of the best space fiction, and certainly best scholarship on space fiction, is of this kind. It also deserves more consideration than we are able to give it here.[64]

Space Poetry

So far, we have concentrated entirely on space stories in print and in the contemporary media of film, television and electronic games. But space has also inspired many works of poetry, drawing, painting, sculpture, music, and dance.

Poetry About Space[65]

There are countless poems and stories that humans have told upon looking up at the dark sky at night, such as this poem, *Drinking Alone Under the Moon*, by the Chinese poet, Li Po (701-762), one of scores he wrote on this theme:

> I take my wine jug out among the flowers/ to drink alone, without friends. / I raise my cup to entice the moon./ That, and my shadow, makes us three. / But the moon doesn't drink,/ and my shadow silently follows. / I will travel with moon and shadow,/ happy to the end of spring. / When I sing, the moon dances./ When I dance, my shadow dances, too. / We share life's joys when sober. / Drunk, each goes a separate way. / Constant friends, although we wander,/ we'll meet again in the Milky Way.[66]

I earlier stated that space fiction, as a branch of science fiction, emerged along with early modern thought, technology, and industry in Europe from the 16th Century onward, and spread globally with modernity. All other space art follows a similar trajectory for the same reasons. As scientific ways of thinking were beginning to challenge earlier cultural modes, art in all its forms was also influenced by the new ideas, technologies, and discoveries. The following illustrate this in western poetry.

William Drummond wrote *The Shadow of the Judgment* at a time (1630) when the discovery of new stars was viewed as an omen of the end of the world, since (according to the philosophical and religious thought of the time) the heavens should be fixed and unchanging:

> … They which dream / An everlastingness in world's vast frame, / Think well some region where they dwell may wrack, / But that the whole nor time nor force can shake; / Yet, frantic, muse to see heaven's stately lights, / Like drunkards, wayless reel amidst their heights.[67]

The telescope also led to much speculation about life elsewhere in the Universe — and the insignificance of the petty squabbles of humans on Earth in comparison:

> But if that infinite Suns we shall admit, / Then infinite worlds follow in reason right, / For every Sun with Planets must be fit, / And have some mark for his farre-shining shafts to hit...[From *Democritus Platonissans, or an Essay Upon the Infinity of Worlds* by Henry More (1647)].[68]

> All these illustrious worlds, and many more, / Which by the tube astronomers explore: / And millions which the glass can ne'er descry, / Lost in the wilds of vast immensity; / Are suns, are centres, whose superior sway / Planets of various magnitudes obey…. / We may pronounce each orb sustains a race / Of living things adapted to the place. /. / How many roll in ether, which the eye / Could n'er, till aided by the glass, descry; / And which no commerce with the Earth maintain! / Are all these glorious empires made in vain?[69]

One of America's greatest poets was Walt Whitman. Many passages from his then-revolutionary book of poetry, *Leaves of Grass* (1900), are about voyaging on Earth and space:

> On the beach at night alone, / As the old mother sways her to and fro singing her husky song, / As I watch the bright stars shining, I think a thought of the clef of the universes and of the future. / A vast similitude interlocks all / All spheres, grown, ungrown, small, large, suns, moons, planets, / All distances of place however wide, / All distances of time, all inanimate forms, / All souls, all living bodies though they be ever so different, or in different worlds, / All gaseous, watery, vegetable, mineral processes, the fishes and the brutes, / All nations, colors, barbarisms, civilizations, languages, / All identities that have existed or may exist on this globe, or any globe. / All lives and deaths, all of the past, present, future, / This vast similitude spans them, and always has spann'd / And shall forever span them and compactly hold and enclose them.[70]

By the 20th Century also, Einstein, Heisenberg, Schrodinger, Planck and others in physics and Eddington, Wheeler, Bell, Penrose and others in astronomy were putting the earlier physics of Newton and the astronomy of Brahe, Galileo *et al.* into a different light. We now understand them — and so many before them — to be pioneers who led humanity to its present understanding which they themselves did not have or anticipate. Alfred Noyes (1922) expresses this very well in one of the most elegant space/astronomy poems ever written:

> ...Then Tycho showed his tables of the stars, / Seven hundred stars, each noted in its place / With exquisite precision, the result / Of watching heaven for five-and-twenty years... / "In the time to come," / Said Tycho Brahe, "perhaps a hundred years, / Perhaps a thousand, when our own poor names / Are quite forgotten, and our kingdoms dust, / On one sure certain day, the torch-bearers / Will, at some point of contact, see a light / Moving upon this chaos. Though our eyes / Be shut forever in an iron sleep, / Their eyes shall see the kingdom of the law, / Our undiscovered cosmos. They shall see it, — / A new creation rising from the deep, / Beautiful, whole.
>
> We are like men that hear / Disjointed notes of some supernal choir. / Year after year, we patiently record / All we can gather. In that far-off time, / A people that we have not known shall hear them, / Moving like music to a single end."[71]

But in the mid 20th century, the British-American poet, T. S. Eliot (1942) (also radical in style for the time but conservative in sentiment) put all such boastful human pretensions into humbling perspective:

We shall not cease from exploration / And the end of all our exploring / Will be to arrive where we started / And to know the place for the first time.[72]

Space Poetry After Sputnik

All of the poems so far were written before Sputnik, Yuri Gagarin, and Neil Armstrong. The authors had neither themselves gone into space, save on spaceship Earth, nor had seen any direct evidence of what not-Earth is like. While there has been no great space poetry subsequently, there has been some pretty-good poetry inspired by the early days of the space age — and precious little since.

But as with science and space fiction, so also with space poetry, the contemporary mood is problematic and questioning, and not grand and heroic. Robinson Jeffers, a major 20th century American poet, is not alone in writing majestic poetry glorifying nature while roundly condemning humanity. In *Orca* he calls humans "a botched experiment that has run wild and ought to be stopped." In his poem *Love-Children* he observed "I'm never sorry to think that here's a planet/Will go on like this glen, perfectly whole and content, after mankind is scummed from the kettle. /No ghost will walk under the latter starlight."[73]

The poets and scientists mentioned so far were men. Few of the poets were scientists, and few scientists were poets (though there were some spectacular exceptions). One of the most outstanding contemporary astronomers who is also highly sensitive to the poetic aspect of her work is Dame Jocelyn Bell Burnell. In her essay on "Astronomy and Poetry" in *Contemporary poetry and contemporary science*, she describes the tremendous reaction she received when she first began introducing poetry into her lectures — especially the positive reaction by women in the audience.[74]

Gwyneth Lewis is from Wales. Her cousin, Joe Tanner, was an astronaut on the Hubble Repair Mission, STS-82. She wrote several poems to and about him, heavy with irony:

Last suppers, I fancy, are always wide-screen. / I see this one in snapshot, your brothers are rhymes / with you and each other. John has a shiner / from surfing. Already we've started counting time / backwards to zero. The Shuttle processed / out like an idol to its pagan pad. / It stands by its scaffold, being tended and blessed / by priestly technicians. You refuse to feel sad, / can't wait for your coming wedding with speed / out into weightlessness. We watch you dress / in your orange space suit, a Hindu bride, / with wires like henna for your loveliness. /

You carry your helmet like a severed head. / We think of you as already dead.[75]

But Is Any of This Really Space Poetry?

None of the poets cited so far has actually been in space (except on Earth!) and so they are all writing from imagination and from the images of others, not from their own direct experience. But there have been poems written by astronauts and cosmonauts.

During the nearly 67 hours his fellow astronauts Scott and Irwin were on the Moon during the Apollo 15 mission, Al Worden was in complete solitude, floating in space. He said the overwhelming experience of being alone in the Universe gave him a profound feeling of rejuvenation. In 1974 Worden wrote a book of poems entitled, *Hello Earth: Greetings from Endeavor.* As for so many others, Worden's experience in space changed his view of reality on Earth:

> Quietly, like a night bird, floating, soaring, wingless / We glide from shore to shore, curving and falling / but not quite touching; / Earth: a distant memory seen in an instant of repose, / crescent shaped, ethereal, beautiful, / I wonder which part is home, but I know it doesn't matter . . . / the bond is there in my mind and memory; / Earth: a small, bubbly balloon hanging delicately / in the nothingness of space.[76]

Another American Astronaut, Story Musgrave, has written a great deal of poetry based on his own experiences. Musgrave first flew on STS-6, 1983. He was also on STS-51F/Spacelab-2 in July 1985, and the first Hubble Space Telescope repair mission, STS-61. His last mission was on STS-80 in 1996. He is exceptionally well educated with degrees in mathematics, business administration, chemistry and biophysics, and a doctor of medicine. Finally coming to his senses, his last degree was a master of arts in literature:

Floating in a spaceship,
Falling through my heaven,
Through epic altitudes,
And higher latitudes
Falling into sleep,
Drifting into dreams,
Cosmic crashes in my eye,

Cosmic flashes in my brain
Cosmic rays and Wilson clouds,
Clear my consciousness.
Memories of infinity,
Particles of eternity
Starlettes pierce my eyes,
In my brain fire flies.
Periods of light,
Punctuate my night.
Cosmic Fireflies[77]

Space Tanka and Renshi

Japan was the first country to bring poetry officially into its space program. Japanese astronaut Chiaki Mukai (STS-95) began a *tanka* while she was in space on the shuttle. Thousands of Japanese on Earth completed it.[78] A *tanka* is a poem that is made up of five lines with a specific number of syllables in each line. It is customary for one person to start the first three lines of a *tanka*, and then for someone else to complete it.

The first three lines that Mukai said in Japanese, with the last two lines left for anyone to complete, are (in roman letters):
(5 syllables) *Chuugaeri,*
(7 syllables) *Nandemo dekiru,*
(5 syllables) *Mujuuryoku*

In English translation of the same number of syllables per line, this becomes:
(5 syllables) *Turn space somersaults*
(7 syllables) *Do as many as you want*
(5 syllables) *That is weightlessness*
(7 syllables) — ......
(7 syllables) — ......

Here are two *tanka* among those completed by participants of the summer program of the International Space University, in Thailand, 1999, from the sublime to the realistic:

> Turn space somersaults / Do as many as you want / That is weightlessness / Moon is where we can make love / Long cold night but love is warm (by Jim Burke)

> Turn space somersaults / Do as many as you want / That is weightless-

ness / Hard to know which way is up / I vomit copiously (by Christopher Connor)

Chain poetry (*renshi*) is another form developed from traditional Japanese linked verse (*renga* and *renku*) and popularized by the poet Makoto Ooka. Many people contribute to a *renshi*. Someone starts by writing a short verse. The next person repeats a word or phrase from the first poem and writes her own verse, while the next person picks up another phrase from the first or second poem, and so on, endlessly. *Renshi* are not of any fixed number of syllables.

JAXA sponsored a *renshi* project from 2006, and the resulting poem was recorded on DVD and sent to the International Space Station in the Japanese Experimental Module *Kibo* ("Hope"). Here are the first and some following verses of that *renshi*:

We children of the stars / children of space / born in the oceans and matured on land / have the history of the universe / its hundreds of billions of years / etched on our bodies / Look! Today too somewhere a tiny light (by Naoko Yamazaki)

The address is not a village nor a city nor a county / not even a country / The address is this planet
/ and the Milky Way Galaxy / Led on by light / sped along by an energy that lurks in the dark (by Shuntarou Tanikawa)[79]

JAXA has expanded its original *renshi* program by involving people from around the world and expects to continue the program indefinitely.[80]

Space Illustrations and Art

We went to the Moon as technicians; we returned as humanitarians. — Edgar Mitchell, Astronaut

When man conquers space it will not be with rockets, sputniks or spaceships, for in that case, we will remain tourists in space. Instead it will be by inhabiting it with sensitivity, that is, not merely by being in it but by imbuing ourselves with it through our solidarity with life itself, as represented by that space where the tranquil and formidable force of pure imagination reigns. — Yves Klein, in Ottinger, p. 284

Let's distinguish between "illustrations" and "art". "Illustrations" are meant either to depict some understanding of something, or to accompany and make visual a textual, mathematical, or verbal statement. In contrast, "art" is a visual expression, perhaps inspired by or drawn from some aspect or experience, that is evocative of the artist's own subjective perspective. "Art" is not intended to be a literal, "objective" rendering of something.

Of course, all illustrations in fact are subjectively rendered by some artist, and (when illustrating fictional texts, for example) may be purposely imaginative and not "factual". Moreover, some illustrations become "works of art" on their own, usually when they are separated from their original place and meaning, and given value in and of themselves. But when visuals are meant to be as "true" as technically and epistemologically possible to the thing being depicted, then I believe they should be called "illustrations", whereas when they are meant to be expressions of the artist's feelings in interaction with something, I prefer to call it "art". I will use those distinctions here.

Thus, most of what is called "space art" is in fact illustrations, while "art" — on any subject — is a relatively new thing. Art emerged with modernity, which values both individual subjective experience and the commodification of "art objects".

In the Beginning

Humans have made pictures of what we see in the night sky for tens of thousands of years. We have also made pictures of what we imagine to be above us but cannot actually see. Fundamental beliefs about humans and not-humans, and about Earth and not-Earth are illustrated in both such pictures.

For most of *homosapiens, sapiens* time on Earth, we have been nomadic hunters and gatherers. Fewer in number and physically weaker than many other creatures around us, we seem typically to have lived in conditions of environmental abundance called "subsistence affluence".[81] With the evolution of the ability to speak and the invention of language, possibly 40,000 years ago, humans were able to think about and organize ourselves and our environment in ways never before possible. As a consequence, we rapidly grew in numbers and in the impact we made on the Earth.[82]

Eventually it became necessary for some humans to settle down and farm. Land and property became important. Later, urban settlements, built on an agrarian base, arose. Especially with the invention of writing, probably less than 10,000 years ago, some people in the urban centers began to record events around them, noting the cyclical regularities of nature, and eventually how these cycles corresponded to the apparent movements of the Sun, Moon and stars above them. Ideas about how and why these coincidences occurred were recorded and transmitted from generation to generation in ways which only writing made possible. Writing enabled the colonization of both space and time — a process often called "civilization".[83]

There is little doubt that first-people worldwide made images of the sky and derived lessons for humans from them. But they left few examples of those images or, of those that do remain, what the images meant to them. Many early people also created huge structures out of stone or dirt that still remain. They may have had astronomical or astrological meaning at the time — it is a plausible explanation for some structures — but no one can be sure. Our interpretations of them may tell us more about our time and beliefs than about those of the people who constructed them.

Different cultures have seen and depicted different patterns in the stars and darkness of a night sky. What may look like a "Big Dipper" to a society that uses such an instrument (do *you* know what a "dipper" is?) might seem to be a plough, or a wagon, or a bear, or a salmon net, or a saucepan, or the Seven Great Sages, or merely the northern seven stars to another.

I cited earlier the East Asian story about the separated lover-stars (in Japanese, *Orihime* and *Hikoboshi*, or, Vega and Altair). Vega is also the central star in the constellation known as Lyra from the Greek interpretation of the form as being that of a lyre (harp). But others see a vulture in the same group of stars, whereas the Boorong of Australia see the *Loan* bird — a creation-being whose celestial behavior is used to instruct humans about proper parenting and gender roles, among other lessons. All cultures have their own way of "seeing" patterns in the sky and of interpreting them.[84]

The objects above us in the sky seem to move constantly, but often in some kind of a repeating pattern. Knowing what formation the stars take just before the seasons change is important information for any successful agricultural community. This might have been one of the earliest practical uses of astronomical information.

Clearly the apparent rising and setting of the Sun and Moon affected humans and the Earth. So also might the stars generally, it was often reasoned. Hence, in some cultures — perhaps first in Mesopotamia — there developed a method of predicting the future — of entire societies as well as of individuals in them — based on the movement of the stars, now known as astrology, but an important prelude to modern day astronomy as well.[85]

Eclipses of the Sun and Moon were especially perplexing since they seemed so out of the ordinary. Eventually careful observers and recorders of heavenly movements noted their regularity and began predicting them with great accuracy.

Comets, meteors and other "falling stars" played a special role in human history since they exhibit extraordinary astral behavior. In medieval and early modern Europe, and perhaps elsewhere, comets were typically seen as omens of bad things. Comets have been depicted in several surviving visualizations, such as the amazing *Bayeux Tapestry*, which includes Halley's Comet (or is it a space ship?) of 1066. Indeed Halley's Comet appears in several medieval paintings; most famously in the Florentine painter Giotto de Bondone's painting *Adoration of the Magi*. This shows the "Three Wise Men" kneeling in adoration of the Christ Child with the conventional Star of Bethlehem replaced by an astronomically correct illustration of Halley's Comet of 1301. The term "magi" derives from the Persian word *magoi*, meaning "astrologers" or simply men who study the stars.[86]

Modern Space Art and Illustrations[87]

With the invention and dissemination of the telescope, more accurate observation of not-Earth became possible. Its depiction often became more "accurate" as well from the 16th and especially 17th centuries onward, and particularly following the invention of chemical photography in the 19th century and electronic imaging in the 20th century.

Space illustrations, as a self-conscious genre, began at the same time as written space fiction, and much early illustrations were created to accompany and make visual ideas in written texts. Illustrations by Emile Bayard (1837-1891) and by Alphonse de Neuville (1835–1885) were found in Jules Verne's stories about travels to space. They may be the first illustrations that attempted to depict not-Earth according to scientific notions of the time. However, the "grandfather" of modern space illustrations was the Frenchman Lucien Rudaux (1874-1947), a professional astronomer who also wrote and illustrated books based on his observations. The International Association of Astronomical Artists created a space art "hall of fame" in 2000, named after Rudaux, with Rudaux as the first person to be so honored.[88]

Depictions of space environments, some intending to be factually while others were fantastically presented, flourished throughout the late 19th and early 20th centuries, reaching their heights on the covers and sometimes pages of the pulp fiction magazines such as *Amazing Stories* and *Astounding Stories*.

Just before the onset of World War II a group of enthusiasts in England formed the British Interplanetary Society (BIS). Due to archaic restrictions enshrined in British law these enthusiasts, unlike their counterparts in the Soviet Union, the United States and Germany, were prohibited from building rockets. They therefore concentrated on the theory and design of spacecraft. The team was headed by an unusual confluence of talent which included astronomer and future science fiction writer Arthur C. Clarke, engineer Harold E. Ross and draftsman and artist Ralph A. Smith. Smith's meticulous attention to practical design, as demonstrated in over 140 drawings and sketches of spacecraft, would fire the imaginations of space engineers around the world. The post-war design for a lunar landing craft, as imagined by the BIS team and rendered by Smith, had a form and function that was to be uncannily similar to the Apollo lunar landing craft of two decades later.

But it was probably the depictions in the large-sized popular picture magazines in the United States, such as *Life*, *Colliers* and *Coronet* in the 1950s and early 1960s, that really brought space illustrations to the eye of the public. They ignited popular support for space exploration as it became technologically possible for the first time.

Figure 2.2. *Saturn as seen from Titan*, 1944 (by Chesley Bonestell, Oil on board, 16 X 20 inches, Published in *Life* Magazine, May 29, 1944.

(Reproduced courtesy of Private Collection).

Some of the best-known images were done by the person who is most often called the "father of space art", Chesley Bonestell (1888-1986). He was an architect and a special-effects artist in Hollywood before World War II where he developed techniques he later used in his space illustrations. One of his best known paintings is of Saturn as imaginatively seen from one of its moons that appeared in *Life* magazine in 1944. He also worked with Willy Ley, Wernher von Braun and Arthur C. Clarke.[89]

Figure 2.3

Left: *Poet on a Mountain Top*, 1496, by Shen Zhou, Album leaf mounted as a handscroll; ink and color on paper, 38.7 x 60.3 cm. (Credit: Nelson-Atkins Museum of Art, Kansas City, Missouri. Purchase: William Rockhill Nelson Trust, 46-51/2, Photograph by Robert Newcombe).

Center: *Kindred Spirits*, 1849, by Asher B. Durand, Oil on canvas, 44 x 36 inches, (Courtesy Crystal Bridges Museum of American Art, Bentonville, Arkansas).

Right: *Mars Exploration*, 1979, by William Hartman, Acrylic on rag board, 10 x 14 inches. (Used by permission of the artist, William K. Hartmann, Planetary Science Institute, Tucson, Arizona).

Bonestell, who was the second person to be inducted into the IAAA Hall of Fame, was influential in establishing a form of space illustrations that dominates the popular mind to this day. His illustrations are heroic and grand in effect (see Figure 2.2), showing breathtaking imaginary vistas of or from planets, moons and other space objects, with human figures generally dwarfed by the immensity of the environment. This style of painting is very reminiscent of that of the Hudson River School that depicted the (rapidly vanishing) American wilderness in the middle 19th Century.[90] These were typically huge hyper-realistic romantic paintings rendered on wall-sized canvases that overwhelm the viewer with awe and longing. They are literally breath-taking, provoking a strong desire to enter the painting and "go" to the places so evocatively depicted. Both the Hudson River School and Bonestell's style of space paintings are similar to the *shanshui* style of scroll painting that was popular in Chinese (and later

Korean and Japanese) art from the 10th century onward. This style still resonates with most naïve viewers as being stereotypically Chinese (or Japanese or Korean) art in which a lonely figure at the bottom of the painting is dwarfed by huge steep mountains, covered with jagged rocks and warped trees, wrapped in mist, and gouged by rapidly flowing water (see Figure 2.3).[91]

At the same time, space illustration and art in Russia were developing, in part influenced by what was happening elsewhere, but perhaps more strongly influenced by the pioneering ideas of Fedorov and Tsiolkovsky.[92] Konstantin Tsiolkovsky was one of a handful of persons in the late 19th to early 20th centuries who envisioned and created what was technically necessary for space travel. But, unlike any of the others, he was inspired by a worldview that declared space exploration and settlement to be a necessary next step in human evolution. His teacher, Nikolai Fedorovic Fedorov, developed "the philosophy of the common task" that influenced many other Russian intellectuals including Dostoevsky and Tolstoy. Fedorov believed that everything in the Universe was alive. Only humanity had obtained the highest consciousness, and so it was our duty to introduce order and purpose into the chaos of nature. Tsiolkovsky expressed this compellingly in his famous saying, "The planet is the cradle of intelligence, but it is impossible to live in the cradle for ever." Humans must grow up, leave their cradle, Earth, and expand through the solar system and out into the cosmos. Tsiolkovsky believed it was his life task to see that humans did so as soon as possible. Whatever else can be said to be the reasons why the Soviet Union was so powerfully motivated to develop a space program, the philosophy of Fedorov and the genius and work of Tsiolkovsky must be among the most compelling. No other space program can make a similar claim. That may be one reason many national space programs languish while that of Russia remains comparatively vigorous: theirs is inspired by a philosophy of cosmic dimensions while most of the others are based on economic, utilitarian, or nationalistic motivations. As a consequence, most Russian space art and illustrations, in comparison with that of Europeans and Americans, tends to be much more interpretive, subjective, and reflective — philosophical.[93]

Astrophotography as Space Art

Astrophotographs — photographs of the sky, or, more recently of space through telescopes — are often treated as works of art. Indeed, as presented to the general public, astrophotographs, such as those taken using the Hubble Space Telescope (see Figure 2.4), must be viewed *pri-*

marily as works of art since they are framed and color-enhanced for maximum aesthetic effect. Otherwise, most of the photographs would be too dull to attract much attention. Elizabeth Kessler observed that "the aesthetic choices made result in a sense of majesty and wonder about nature and how spectacular it can be, just as the paintings of the American West did. The Hubble images are part of the romantic landscape tradition. They fit that popular, familiar model of what the natural world should look like."[94]

Figure 2.4. Pillars of Creation in a Star-Forming Region (1995). Photograph by Hubble Telescope http://hubblesite.org/gallery/album/entire_collection/pr1995044a/.

(Courtesy of NASA).

Didier Ottinger makes the point more broadly: "Artists of the modern sublime often use scientific materials. Such a revelation occurs when an object transferred to a museum raises issues and evokes symbolic values. By displacing images from the observatory to the museum, the artists of the modern sublime identify the cosmos as a place that cannot be reduced to a useful function or a rational definition. In contrast to technical or political utilitarianism, there is space, as beautiful and useless as a work of art."[95]

Art Taken into Space

Perhaps the first bit of consciously "art" in space was a small ceramic tile about the size of a postage stamp grandly titled, *Moon Museum.* It was carried on Apollo 12 (1969). American artists Robert Rauschenberg drew a straight line, Andy Warhol a penis, Claus Oldenberg sketched Mickey Mouse, and John Chamberlain, Forrest Myers and David Novros all drew geometric designs.[96] In 1971, a small figurine, titled *The Fallen Astronaut* (to commemorate all cosmonauts and astronauts who had died so far), by Belgian Artist Paul Van Hoeydonk was left on the Moon by Apollo 15 astronauts, while sculpture by Joseph McShane, titled *S.P.A.C.E.*, flew as *Payload G38* on the second mission of the Challenger (1984). Lowry Burgess's *Boundless Cubic Lunar Aperture* flew on the Space Shuttle in 1989 as a self-contained "non-scientific payload". This conceptual artwork included holograms and cubes made from all of the elements known to science and water samples from the world's rivers.[97] As part of the AustroMir mission in 1992, Austrian artist, Richard Kriesche, transmitted a video signal to the cosmonaut crew on board the Mir. They returned the altered signals after they had interacted with various devices on board.[98] The West cigarette company commissioned German artist Andora to paint the outside of a Russian Proton rocket (1992) with examples of his art and an advertisement for the cigarette company.[99]

Arthur Woods' *Cosmic Dancer* sculpture was launched in 1993 to the Mir space station to investigate the properties of sculpture in weightlessness and the effects of integrating art into the living and working environment of the cosmonauts.[100] *Ars Ad Astra: The 1st Art Exhibition in Earth Orbit* was organized by Arthur Woods and The OURS Foundation in cooperation with the European Space Agency during their EUROMIR'95 mission. This was the most comprehensive exhibit of art in space so far. Twenty paintings and a laptop computer with 81 digitized art works accompanied German cosmonaut Thomas Reiter on his six-month mission.[101]

Space Art in Space

The space illustrators and artists mentioned so far have been Earth-locked and therefore are not producing true space art. However, there have been a few astronauts and cosmonauts who painted in space. Alexi Leonov, co-commander of the Apollo-Soyuz Test Flight of 1975 was one. He was a prolific space artist in general. He also worked closely with Andrei Sokolov, "the dean of Soviet space art", sometimes taking into space with him work that Sokolov had done on Earth, comparing it with what Leonov actually saw in space.[102] Vladimir Dzhanibekov, commander on five flights, including the Salyut 7 space station in 1985, composed many works about space after he returned to Earth. Dzhanibekov said he tried in his paintings "to show the philosophical side of this not-always-easy work."[103]

Alan Bean, who flew to the Moon on Apollo 12, has described in some detail his aesthetic reactions to the experience, including how he chose the colors, values, shading and other features of his space paintings. He said there is no right or wrong choice: "Any painting will show only how that artist wants to portray a subject, how he or she feels about it. An observer of the painting can either connect with that and like it, or not connect with it and not like it. That is how art is different from science."[104]

Evolution of Space Art

Didier Ottinger says that there have been three historical periods of space art. The first was from the beginning of history through the Renaissance, "when the artist was not yet distinct from the scholar". Copernicus, Tycho Brahe and "Kepler's astronomy evokes a world imbued with magic and as yet untouched by scientific rationalism." "During its scientific revolution, astronomy was still dependent on a system of thought according to which the world was penetrated through and through by correspondences and analogies."[105]

Art during the second, Romantic, period reflects "the time of the schism between intuition and objective knowledge". "Sublime is indeed the proper term for Romantic painting, in its obsession with grandiose, terrifying spectacles: nature in the grip of raging storms, unscalable mountains, fathomless seas and infinite skies".[106]

"The final group of artists sees the conquest of space as the ultimate manifestation of enthusiasm for technology. For them, space seems merely a new field for political struggle, the last refuge of science viewed as an epic. They feel it their responsibility to slide the banana peel of skepticism and irony under the feet of the dreamers and manipulators who, for the benefit of the eternally imbecilic, point a finger towards the waxing or waning moon". In 1998, Joan Fontcuberta presented what appeared to be spectacular images of mammoth constellations. Only later were they revealed to be "photographs of his car windshield spotted with crushed insects." "Haidar published in *Vogue* and *Cosmopolitan* a series of so-called photographs of the Universe, which he created in his studio by setting alcohol-soaked cotton on fire."[107] This is art in the 'pataphysics mode once again.

Or, as Marshall McLuhan once put it, "Art is anything you can get away with".[108]

Space Dance

If "dance is the only art where we ourselves are the stuff from which it is made", there has arguably been a lot of dancing in space. As Chiaki Mukai's space *tanka* about weightlessness suggested, few astronauts or cosmonauts have been able to resist the freedom that zero (or substantially reduced) gravity affords. Whether it is Buzz Aldrin bunny-hopping on the Moon or almost everyone turning somersaults in shuttles and spaceships, spacefarers are clearly blessed with Happy Feet.

This is one instance where life has influenced art: Earthbound dancers envy the spacefarers' freedom. Dancers are gravity haters by definition (see Figure 2.5). To dance free of the bonds of gravity would be heaven to them all. But so far no dancers have gone into space. Consequently, more and more dancers have done the next best thing: they have danced momentarily during parabolic airplane flights.[109]

Figure 2.5. Biomechanics Noordung, 1999, by Dragan Zivadinov, weightless in an Ilyushin 76 aircraft, from Star City, Russia.

(Photo by Miha Fras. Used by permission of the photographer)

Space Music

"I could bear it a little better if there were some music." —
Yuri Gagarin, shortly before liftoff, April 12, 1961
<http://www.yurisnight.net/music/>

Music of the Spheres[110]

"It is almost a truism to say that there is a considerable interaction between the musical and scientific worlds…." "Nowhere is this cross-fertilization more evident than in the astronomical sphere."[111] In the West this interrelation between science and music goes back at least to the sixth century BC philosopher, Pythagoras. His interest in geometry (namely, in the numerical values of lengths, angles, and other properties of lines and spheres), on the one hand, and music (actually, the comparative lengths of vibrating strings that make differing sounds), on the other, coupled with the assumption that the Universe is harmonious, balanced, and perfect, led him to develop a scheme of geometrical and tonal harmony that pervaded the Universe.

A century later, Plato elaborated the perspective of Pythagoras from his Idealist philosophy, while several hundred years after Plato the Alexandrian astronomer, Ptolemy, modified these ideas into a form that remained dominant in the West until Johannes Kepler revisited the issue in the 16th Century, almost 1500 years later. To some extent from the Greeks, but especially through the Middle Ages, the so-called "*quadrivium*" was the basis of formal education. The four elements are typically labeled arithmetic, geometry, astronomy, and music. But it is important to realize that "arithmetic" might better be understood as "numerology" — the intrinsic power of certain numbers and sequences per se — while "astronomy" was the result of careful observations of the heavenly objects as seen with the unaugmented human eye. "Music" was "monodic" — a sequence of individual tones. Even the term "harmony", which is always used to characterize the "music of the spheres", did not mean what it means today. Most of us now assume that "harmony" results when several different tones are sounded together in a way that is considered "pleasing", whereas "disharmony" results from the simultaneous sounding of tones that are not "pleasing." That is to say, "harmony" is polyphonic to us, whereas it was monodic to Greek and Middle Age philosophers.

Johannes Kepler changed all this. He was "a transition figure with one foot in the symbolic and spiritual cosmos of Pythagoras and Plato, and the other embedded firmly in the new heliocentric Universe of his day."[112] While still captured by the old notions of a perfect, orderly Universe, he also was an empiricist who let "facts" speak to and challenge his received beliefs. One of the most important facts — that contributed to the collapse of the old Ptolemaic system — was his discovery first that Mars (and later that all planets) orbit the Sun not in a perfect circle but in an ellipse. The old "orderly" Universe, revolving perfectly around the Earth, as God created it, collapsed. Kepler, nonetheless, clung to the belief that there was a deeper order to the Universe; eventually, in his *Harmoices Mundi* (1619), he published what he considered to be evidence of the grand musical and geometric harmony of the Universe. It could not be heard by human ears, but it could be appreciated by the human mind.

More importantly for our story here, after many years of study and experimentation, Kepler concluded that the music of the spheres was polyphonic — "harmonious" in the modern sense — and not monodic. This conclusion contributed to the development of the "scales" and tone-intervals characteristic of western "classical" music from the late 17th through the late 19th centuries.

Modern "Classical" Space Music

From this point on, the story of music becomes very much like the story of space fiction, poetry, and art. As science (and technology) changed and deepened humans' understanding of the Universe, so also it provoked and enabled new aesthetic expressions. William Herschel, who discovered the planet Uranus in 1781, actually was known for most of his life as a professional musician.[113] It was only later that he devoted more time to astronomy than to music, but he never neglected either. It thus may be poetic justice that Brian May postponed for over thirty years his work on a PhD dissertation in astrophysics at Imperial College, London, because his job as lead guitarist in the rock group *Queen* took too much time. He finally received his PhD degree in 2007.[114]

Franz Joseph Haydn (who knew Herschel), wrote *Il mondo della luna* (The World in the Moon, 1777), an opera buffo, but "the Moon" in the story was a consciously fake setting. It was not intended to depict actual life on the Moon. On the other hand, Jacques Offenbach's opera *Le Voyage dans la Lune* (1875) was inspired by Jules Vernes' story, *De la Terra a la lune* (1872). Though Vernes' story was fictional, both it and Offenbach's opera were meant to reflect real yearnings for space travel. Josef Strauss's *Sphären-Klänge* (Music of the Spheres, 1868) also is not based on any scientific inspiration, though it may reflect a growing popular interest in not-Earth generally.

As better telescopes and more powerful theories began to tell us different stories about not-Earth, so also more musicians began to reflect these ideas in their compositions. In a recent review, Andrew Fraknoi lists "over a hundred pieces of music that make use of serious astronomy".[115]

Probably the best-known piece of classical space music — *The Plan-*

ets (1916) by Gustav Holst — is often excluded from lists of serious space music because it sought to exhibit the essence of each planet according to how astrology described it, and not according to features understood by current astronomical observation or scientific theory. However, much more recently, the British conductor, Sir Simon Rattle, added five contemporary compositions to a 2006 recording by the Berlin Philharmonic of *The Planets*. One was a piece for *Pluto* (2000) by Colin Matthews — written only shortly before Pluto was declared by the International Astronomical Union not to be a planet after all. The other four new additions refer to asteroids. *Asteroid 4179: Toutatis* (2005) is by the Finnish composer, Kaija Saariaho. Toutatis is "the asteroid whose orbit passes closest to Earth" "with an unusual shape and complex rotation" which the music tries to reflect. Brett Dean's *Komarov's Fall* (2006) was "written in memory of Soviet Cosmonaut, Vladimir Mikhailovic Komarov, who died upon re-entry to the Earth's atmosphere in his Soyuz 1 spacecraft in 1967. I happened across a recording of sound effects including the final conversations between …Komarov and ground control [and his wife]. It was absolutely chilling…. He was the first person to die in space.'" "The asteroid '1836 Komarov', discovered in 1971, was named in his honor." Matthias Pintscher's *towards Osiris* (2005) reflects "the myth of Osiris, who was killed by his brother Seth and torn in pieces. His wife re-collected these pieces, put them together, and by swaying her huge wings, re-animated him. It's a very moving story and has a very strong formal structure, like in music." Mark-Anthony Turnage said of his composition, *Ceres* (2005), "the idea of an asteroid hitting the Earth at any moment really appealed to me because I grew up in a religious family and I was always in fear of the second coming. The piece explodes about five eighths of the way through into a big climax which is the asteroid hitting Earth".[116]

Paul Hindemith wrote an opera (1947) and then a symphonic suite (1951) called *Die Harmonie der Welt*, based on the life of Kepler. Willie Ruff, a professor of music at Yale University teamed up with John Rodgers, professor of geology there, to try to realize Kepler's "Harmony of the World" via electronic synthesizers of the era (late 1970s) for human ears to hear.[117]

Contemporary "Post-Classical" Space Music

During the 20th and early 21st centuries, many serious post-classical composers have written space-related music incorporating contemporary as well as traditional instruments and sounds. Typically, these composers use tones, chords, tempi, and intervals not based on the classical ideas of harmony of the 18th and 19th centuries. Indeed, many of these composers were influenced not only by astronomy and space but also especially by quantum physics, with its emphasis on randomness and complementarity. Their music is "post-Newtonian" in conception and execution.[118]

Music from Space

Some of the pieces above combine sounds recorded from space or spaceships with more or less conventional instruments and modes. But there are a few compositions derived entirely from instruments that recorded various space "sounds". These include compilations by Stephen McGreevy, *Music of the Magnetosphere and Space Weather* (1995-2008); *Symphonies of the Planets 1-5* (NASA Voyager Recordings, 1995); Alexander Kosovichev, *Solar Sounds* (1998); *Space Sounds* (2000-2005); and Donald Gurnett, *Selected Sounds of Space* (2003-05).

Popular Music Inspired by Space Events[119]

Popular music has been inspired by some specific space event, beginning perhaps with *Sputnik (Satellite Girl)*, by Jerry Engler and the Four Ekkos (1957). Yuri Gagarin's pioneering flight in the Vostok provoked at least two songs in the Soviet Union, *The Constellation of Gagarin* (1961) and *Motherland Knows Her Son is Flying in Orbit* (1961). Similarly, *Happy Blues For John Glenn* (1962) by Sam "Lightning" Hopkins and *The Ballad of John Glenn* (1962) by Roy West commemorated the flight of America's first man in space.

One of the best known bits of early space satire was *Wernher von Braun* (1965) by Tom Lehrer with the memorable lyrics, apparently actually uttered by von Braun: "'Once the rockets are up, who cares where they come down? That's not my department,' says Wernher von Braun." But more perspicaciously, Lehrer's song ends, "'In German oder English I know how to count down, Und I'm learning Chinese!' says Wernher von Braun".

When first Neil Armstrong and then Buzz Aldrin walked on the Moon, The Byrds produced *Armstrong, Aldrin and Collins* (1969) celebrating it, while Jethro Tull wrote *For Michael Collins, Jeffrey and Me* (1970) commenting on three people who "almost made it", including Collins who was left on the command module circling the Moon and did not get to walk on it. However, it was *Armstrong* (1969) by John Stewart of the Kingston Trio that caused the greatest stir because his lyrics wondered whether a starving black boy in Chicago or a poor girl in Calcutta knew or cared about the feat as poverty, pollution, war, and hate continued on Earth.

In stark contrast are *The Walk of Ed White* (1969) and *Moon Rider* (1977), based on recollections of Eugene Cernan, the last man to walk on the Moon), by the inspirational song group, Up with People. Similarly, Roy McCall and Southern Gold, (*Blast Off Columbia*, 1981) and Rush, *Countdown*, (1982) dedicated to Astronauts John Young and Robert Crippen, celebrated the flight of STS-1, while Deep Purple (*Contact Lost*, 2003) bemoans its loss 20 years later.

There are two particularly poignant songs about space tragedies. One, *Flying for Me* (1986) by John Denver was about the Challenger explosion. Denver had tried very hard to get on that flight which was featured as being one of the first to include several "ordinary people" among the crew. Similarly, Jean Michel Jarre's *Last Rendezvous* (1986) was written to be played on a saxophone by Astronaut Ron McNair on the Challenger. *Beagle 2* (1999) by the British group, Blur, was intended to be broadcast to Earth from Mars by the Beagle 2 rover that apparently failed to land on Mars in December 2003.

The UK band, The Picture Show, paid tribute to Tsiolkovsky in their song *The Revolutionary* (2007): "Konstantine oh Konstantine/ Wants to build rockets to the stars/ Spends all day looking up at them/ Lying in the long grass/ Inspired he says to leave the cradle soon/ To see what there is outside." *Ride, Sally, Ride!* (1983) by Casse Culver celebrates Sally Ride, the first American woman in space, while Elaine Walker's *X Prize Song* (2004) may be the first song celebrating private spaceflight. We look forward to more space music honoring entrepreneurial as well as transnational space activities.

Popular Music on Space Themes[120]

There has been far too much popular music more or less on space themes to do justice to it all here.[121] However, an entire genre called "Space Rock" was spawned around the time of the Apollo moon landings. The progenitors of this new genre were two British bands formed out of the alternative culture of the late 1960s. The first was The Pink Floyd led by the mercurial singer/songwriter Syd Barrett. Songs on the first two Pink Floyd albums included "Astronomy Dominé", "Interstellar Overdrive" and "Set The Controls for the Heart of the Sun". The second band, Hawkwind, would come to define the genre of "space rock" with their albums "X In Search of Space" and "The Space Ritual". The former was a high-concept album complete with story booklet about a spacecraft heading off into deep space with the crew in suspended animation. Hawkwind followed this up with a concert tour that went across Europe and North America called "The

Space Ritual" in which the same themes were explored in an extended format using dialog as well as music. The band themselves were placed on the stage in a format to coincide with mathematical concepts expounded by Pythagoras and others (i.e. to reflect the music of the spheres). The first concert in the USA was held at the Hayden Planetarium in New York. Media of the time referred to "The Spaceship Hawkwind" and the term "space rock" soon became commonplace. Meanwhile, the birth of David Bowie's career was launched by the twin hit singles "Space Oddity" and "Life on Mars". Both inspired by Bowie's particular interest in space exploration, an interest which has now spanned a generation with Bowie's son Duncan Jones writing and directing the critically acclaimed hard sf film entitled "Moon", about lunar Helium 3 mining.

But the dominant pop music of the present is rap, hip-hop, or indie. These forms too have music with space themes. Their roots are partly in rock, but mainly in soul, funk, and to some extent punk — that is to say, primarily in black culture. In 1970, rhythm-and-blues singer, Gil Scott-Heron, was "an inspiration to legion of 1990s rappers, reciting "Whitey on the Moon" "accompanied by only bongos and congas".[122]

Rap, Hip-Hop and Space

During 2007/2008 a subgenre developed in rap/hip-hop designated "SpaceRap". It was identified by the blog OhWord.[123] In addition, a Philly/Puerto Rican trio "took the name Outer Space because their rhyme style was said to be 'beyond the Earth's atmosphere.'" They said that "after all, space is the final frontier and they wanted to make exploring outer space rap's next adventure. As such they abandoned their government names and took the names Planet, Jedeye and Crypt".[124] Commenting on another artist, Kenna, who came to the US from Ethiopia at age three, an interviewer concluded: "There's a new movement afoot in the hip-hop/R&B universe: future funk. It's music that goes black to the future. Like Kenna, it borrows from unlikely sources like alternative and electronica to heighten its sci-fi vibe, without ever losing sight of the groove."[125]

However, the hip-hop/rap artists most directly related to space are Rakim who, on *Follow the Leader* (1988), imagined traveling "at magnificent speeds around the Universe" so that the Earth and other "planets are small as balls of clay," and Nas who, on *I Want to Talk to You* (1999), stated that while "Niggas play with PlayStations / [The government] buildin' space stations / on Mars plottin' civilizations / Dissin' us, discriminating different races / Taxpayers pay for more jails for black and Latin faces."

Channel Zero (1998) by Canibus is a long rap entirely devoted to the US government's alleged cover up of extraterrestrial intelligence findings. It includes reference to the Drake equation, "various geomagnetic gravitational anomaly areas," and ends, "I hope you become aware of what I'm spittin' in your ear was intended to stimulate your left-brain's hemisphere."[126]

Finally, Nas and KRS-One announced in 2004 that they were going to bring 100 students from Washington, DC, to NASA headquarters in order to "promote the powers of math and science" to the members of the hip-hop community. The event apparently never happened but the OhWord blogger anticipated "I doubt this is the last we will hear about these homeboys and outer space. The future (of space-rap anyway) is looking bright and I can see just it now — future b-boys and girls uprocking zero gravity to a cosmic megamix that includes" Nas' *Star Wars* and KRS-One's *Step into a World.*[127]

Filk Music

Filk is inspired by science fiction and fantasy, and is primarily intended for science fiction fan communities, originating at science fiction meetings in the1930's. Attendees gathering to socialize might sing familiar folk songs. Some participants began setting words relating to science fiction to traditional folk tunes, while new works were also composed and sung. Apparently, when a typo in a magazine article about the practice written by Lee Jacobs referred to "filk music" instead of "folk music", a name for the new genre was born. Of the many pieces of filk worth mentioning, Leslie Fish's *Hope Eyrie* clearly stands out. It is often referred to as the "'national anthem' of the pro-space movement, often sung while standing at attention in a reverent manner."[128]

True Space Music: Performed or Composed in Space

However, as I have said before about other art forms, the only true space music is that which has been composed or performed in not-Earth itself. It seems that the first person to sing in space was the very first person in space: Yuri Gagarin. He is quoted as saying about his descent from his flight on the Vostok: "When I was going down, I sang the song, The Motherland Hears, the Motherland Knows."[129] However, apparently the first music to be performed in space was "Jingle Bells", sung and "played" by Walter Schirra and Thomas Stafford on the Gemini 6 mission, during a broadcast to Earth on Christmas Day, 1965.

Astronaut Ron McNair is believed to be the first person to take a musical instrument — a saxophone — into space and play it, on shuttle flight STS-41B in 1984. As noted above, McNair later died in the Challenger explosion in 1986. He had planned to play a work during that flight that had been composed for him by Jean-Michel Jarre. It would have been the first musical piece whose debut occurred in space. The piece later appeared on Jarre's album *Rendezvous* with the sax played by Pierre Gossez. (*Track 6 - Last Rendez-Vous*)

Cosmonaut Aleksandr Laveikin brought his acoustic guitar to the Mir space station and taught Yuri Romanenko how to play it, leaving it with Romanenko when Laveikin returned to Earth. During his 326 day stay on Mir from February 1987 to December1988, Yuri Romanenko wrote twenty songs. They were "optimistic songs, written by a man who feels good." The French cosmonaut Jean-Loup Chretien brought a keyboard with him to Mir in 1988. He also left it there.

Susan Helms played her keyboard during STS-54 in 1993. She was a member of the NASA astronaut band, *Max Q.* Also in 1993, Ellen Ochoa played her flute on STS-56,[130] while Chris Hadfield brought a collapsible guitar with him on STS-74 in 1995, and gave it to ESA astronaut Thomas Reiter on the Mir station. Ed Lu, science officer for the International Space Station Crew Expedition 7 in 2003, played a portable electronic piano that he took with him.

Music for Extra Terrestrials

Music is often used to wake astronauts in space.[131] *Blast-Off Columbia,* mentioned above, was written by Jerry Rucker who worked as a technician on the shuttle's external tank. The song, performed by Roy McCall and Southern Gold, was transmitted as a wakeup call to astronauts John Young and Robert Crippen on the first flight of Columbia, STS-l, 1981.

Voyager 1 and 2, launched by NASA in 1977 to study Jupiter, Saturn, Uranus, and Neptune, are now in deep space, heading towards the truly unknown. In case some extraterrestrial intelligence encounters them, NASA included a message on a 12-inch gold-plated copper disk containing sounds and images intended to show the diversity of life and culture on Earth. Carl Sagan and others compiled 115 analog images and natural sounds, along with music from different cultures and time periods, as well as spoken greetings from Earth in fifty-five languages. Included with the records are a cartridge and a needle with instructions, in symbolic language, that are intended to explain the origin of the spacecraft and how the record is to be played.[132]

Four songs, *Lalala, Bald James Dean, Hot Time,* and *No Love*, by Julien Civange and Louis Haéri, were included in ESA's Cassini-Huygens probe that landed on Titan, a moon of Saturn, in January 2005. The composers said the four songs were written to reflect various phases of the flight. ESA's intention was "to leave a trace of our own humanity and to build awareness, especially among young people, about this adventure outside the specific scientific arena".[133]

Rock and roll is there to stay.

Coda

It is clear from this whirlwind tour of space art — written, painted, danced, sung, and performed — that, as much as some space managers and funders might want to deny or prevent it, aesthetic expression in many forms is fundamental to the human spirit. It will motivate or demotivate us to leave our cradle. We may or may not freely express it in space. But it must be seen as an integral aspect of all space activities, and celebrated — and funded — as such.

1 *Special thanks to Kerrie Dougherty, Curator of Space Technology, Powerhouse Museum, Sydney, Australia, for guidance in researching all aspects of this chapter.*
Philip Freund, *Myths of creation*. London: Peter Owen, 2003; David Adams Leeming, *An Encyclopedia of creation myths*. Santa Barbara, Calif.: ABC-CLIO, 1994; Marie-Louise von Franz, *Creation myths*. Boston: Shambhala, 1995

2 *Taketori monogatari. The tale of the bamboo cutter.* Modern rewriting by Yasunari Kawabata; translation by Donald Keene. Tokyo: Kodansha International, 1998

3 Fraser Cain, "Japanese Moon Probe Nicknamed KAGUYA," *Universe Today* <http://www.universetoday.com/2007/06/06/Japanese-moon-probe-nicknamed-kaguya/>; "KAGUYA selected as SELENE's nickname" <http://www.jaxa.jp/countdown/f13/special/nickname_e.html>; Selene/Kaguya homepage http://www.jaxa.jp/projects/sat/selene/index_e.html

4 F. Hadland Davis, ed., *Myths and legends of Japan*. Boston: David Nickerson & Company, 1912, Chapter VIII, "The Star Lovers", pp. 126f

5 Engineering Test Satellite #VII (ETS-VII / Orihime & Hikoboshi) Home Page http://robotics.jaxa.jp/project/ets7-HP/index_e.html

6 Ovid, *Metamorphoses*. Translated by Mr. Croxall under the direction of Sir Samuel Garth. New York: The Heritage Press, 1961, p. 249f

7 *The epic of Gilgamesh: A new translation, analogues, criticism*. Translated and edited by Benjamin R. Foster. New York: Norton, 2001; Paul D. Hardman. *Homoaffectionalism: Male bonding from Gilgamesh to the present.* San Francisco: NF Division, GLB Publishers, 1993

8 *The essential Odyssey*. Translated and edited by Stanley Lombardo. Indianapolis: Hackett Pub. Co., 2007; Eva Brann *Homeric moments: Clues to delight in reading the Odyssey and the Iliad.* Philadelphia: Paul Dry Books, 2002

9 Richard Heitman, *Taking her seriously: Penelope & the plot of Homer's Odyssey*. Ann Arbor: University of Michigan Press, 2005

10 Lucian, with an English translation by A. M. Harmon, New York: G. P. Putnam's Sons, 1915, Volume II, *Icaromenippus, or the Sky-Man*, pp. 267-323

11 Lucian, *True history and Lucius or the ass*. Translated from the Greek by Paul Turner. Bloomington: Indiana University Press, 1974; Aristoula Georgiadou and David H. J. Larmour, *Lucian's Science Fiction Novel "True Histories": Interpretation and Commentary*. Leiden: Brill, 1998

12 St. Augustine, *The city of God*. Translated by Marcus Dods. New York: Modern Library, 1950

13 Joachim of Fiore (or Flora), *Catholic Encyclopedia*. http://www.newadvent.org/cathen/08406c.htm

14 bn Khaldûn, *The Muqaddimah: An introduction to history*. Translated from the Arabic by Franz Rosenthal. Princeton, N.J.: Princeton University Press, 1967. Three Volumes

15 Plato, *The republic*. Translated and with an introduction by R.E. Allen. New Haven: Yale University Press, 2006

16 Thomas More, *Utopia*. Edited by George M. Logan and Robert M. Adams. Cambridge: Cambridge University Press, 2002

17 Francis Bacon, *New Atlantis*. Edited by Jerry Weinberger. Wheeling, Ill.: H. Davidson, 1989

18 B. F. Skinner, *Walden Two*. New York, Macmillan Co., 1948

19 Johannes Kepler, *Somnium*. Translated with a commentary by Edward Rosen. Madison: University of Wisconsin Press, 1967

20 Francis Godwin, *The Man in the Moone or, A discourse of a voyage thither, by Domingo Gonsales*. San Marino, CA: The Huntington Library, 1961

21 Cyrano de Bergerac, *Other worlds: The comical history of the states and empires of the moon and the sun*. Translated and introduced by Geoffrey Strachan. New York: Oxford University Press, 1965

22 Bernard le Bovier de Fontenelle, *A plurality of worlds*. John Glanvill's translation, with a prologue by David Garnett. London: The Nonesuch press, 1929

23 Jonathan Swift, *Gulliver's Travels*. Edited by Christopher Fox. Boston: Bedford Books of St. Martin's Press, 1995

24 Ludvig Holberg, *Nicolai Limmi's Underground Journey*. New York, Garland Publishing, 1974

25 Voltaire *Micromegas*. Translated by W. Fleming with an introduction and chronology by Ben Barkow. New York: Hippocrene Books, 1989

26 Nicolas Restif de la Bretonne, *La decouverte australe par un homme volant, ou le Dedale francais*, (Southern Discovery by a Flying Man, or the French Daedalus), Geneve: Slatkine Reprints, 1979

27 Based on Aldiss, Brian. *The Trillion Year Spree: The History of Science Fiction*. New York: Atheneum, 1966; Adam Roberts, *The history of science fiction*. New York: Palgrave Macmillan, 2006; Flo Keyes, *The literature of hope in the Middle Ages and today: Connections in medieval romance, modern fantasy, and science fiction*. Jefferson, N.C. : McFarland & Co., 2006; Franz Rottensteiner, "European Science Fiction," in Patrick Parrinder, ed., *Science Fiction* (Longman, 1979); Franz Rottensteiner, *The science fiction book: An illustrated history*. London: Thames & Hudson, 1975; Sam J. Lundwall, *Science fiction: An illustrated history*. New York: Gros-

set and Dunlap, 1978; Eric S. Rabkin, ed., *Science fiction: A historical anthology*. New York: Oxford University Press, 1983; Patrick Parrinder, ed., *Science fiction: a critical guide*. New York: Longman, 1979; Mark Rose, ed., *Science fiction: A collection of critical essays*. Englewood Cliffs, N.J.: Prentice-Hall, 1976; Mark Hillegas, "The literary background to science fiction," in Patrick Parrinder, ed., *Science fiction: A critical guide*. London: Longman, 1976; The Ultimate Science Fiction Web guide" (1997) <www.magicdragon.com/UltimateSF>; Science Fiction Resource Guide (2001) http://sf.emse.fr/SFRG.sfrge2.htm

28 Susantha Goonatilake, *Toward a global science: Mining civilizational knowledge*. Bloomington: Indiana University Press, 1998

29 Christian Bök,*'Pataphysics: The poetics of an imaginary science*. Evanston, Ill.: Northwestern University Press, 2002; Alfred Jarry, *Adventures in 'pataphysics*. translations by Paul Edwards & Antony Melville. London: Atlas, 2001. See also Novum Organum du Collège de 'Pataphysique http://www.college-de-pataphysique.org/college/accueill.html

30 Jules Verne, *A journey to the center of the Earth*. New York: Penguin, 1986; *From the Earth to the Moon*. Translated by Lowell Bair with an introduction by Gregory Benford. New York: Bantam Books, 1993

31 Konstantin Tsiolkovsky, *Beyond the planet Earth*. New York, Pergamon Press, 1960

32 Alexei Tolstoy, *Aelita, or, The decline of Mars*. Ann Arbor, Michigan: Ardis, 1985

33 Kurd Lasswitz, *Two planets*. Abridged by Erich Lasswitz. Translated by Hans H. Rudnick. Afterword by Mark R. Hillegas. Carbondale: Southern Illinois University Press, 1971

34 Karel Capek, *R.U.R.* Translated by Claudia Novack; introduction by Ivan Klbima. New York: Penguin Books, 2004

35 Yevgeny Zamyatin, *We*. Translated and with an introduction by Clarence Brown. New York: Penguin Books, 1993

36 Ivan Efremov, *Andromeda: A space-age tale*. Translated from the Russian by George Hanna. Moscow: Progress Publishers, 1980

37 K. Rosenberg, *Soviet science fiction: To the present via the future*, MIT, 1987

38 Stanislaw Lem, *Solaris*. Translated by Joanna Kilmartin and Steve Cox; afterword by Darko Suvin. New York: Berkley Books, 1970

39 Based on Debjani Sengupta, "Sadhanabu's Friends: Science Fiction in Bengal from 1882-1961," in *Sarai Reader 2003*: Shaping Technologies, 76; Hevil Shah, "Science Fiction in India," Spring 2005 <http://sciencefictionlab.lcc.gatech.edu/subTopicIndia.html>; Samit Basu, "IWE and genre," July 3, 2006 <http://samitbasu.blogspot.com/2006/07/iwe-and-genre.html>; Samit Basu, "The Trousers of Time: Possible futures of Indian speculative fiction in English," July 4, 2006 <http://samitbasu.blogspot.com/2006/07/trousers-of-time-possible-futures-of.html>; <http://flowtv.org/?p=89>; Cyril Gupta "Science Fiction in India" August 2007 <http://www.cyrilgupta.com/wp/?p=22>

40 Dileep Kumar Kanjilal, *Vimana in ancient India*. Calcutta: Sanskrit Pustak Bhandar, 1985

41 Amardeep Singh, "Where women rule and mirrors are weapons," http://www.lehigh.edu/~amsp/2006/05/where-women-rule-and-mirrors-are.htm

42 Shanti Kumar, "Mixing mythology, science and faction: The sci-fi genre in Indian film and television". <http://flowtv.org/?p=89> December 2006

43 Thanks to Stacey Solomone for providing the initial draft of this section on Chinese space fiction.

44 Wu Dingbo, "Chinese science fiction," in Wu Dingbo and Patrick Murphy, eds., *Handbook of Chinese popular culture*. Greenwood Press, 1994, p. 258

45 *Loc. cit.*

46 *Ibid.*, p. 259

47 Henry Y. H. Zhao, "A fearful symmetry: The novel of the future in twentieth-century China," *Bulletin of the School of Oriental and African Studies*, Vol. 66, 2003, p. 456

48 Wu, *op. cit.*, p. 259

49 *Ibid.*, p. 260

50 *Loc. cit.*

51 *Ibid.*, p. 268

52 For other sources of Chinese space fiction, see, "Twelve Hours Later" (Chinese Science Fiction blog) <www.twelvehourslater.org>; Jeremy Goldkorn, "Chinese science fiction" March 23, 2004 <www.danwei.org/internet/Chinese_science_fiction.php>; Lavie Tidhar, "Science fiction, globalization, and the People's Republic of China," <http://concatenation.org/articles/sf~china.html>.

53 Interview with Stacey Solomone, Beijing, August 8, 2007

54 Based on Robert Matthew, *Japanese Science Fiction: A view of a changing society*. London: Routledge, 1989; Robert Matthew. *Science fiction in Japan: A comparative study of the development of the genre in Japan and in the West*. Brisbane: Dept. of Japanese, University of Queensland, 1978; Robert Matthew, *The origins of Japanese science fiction*. Brisbane: Dept. of Japanese, University of Queensland, 1978; Yoriko Moichi, "Japanese utopian literature from the 1870s to the present and the influence of western utopianism," *Utopian Studies*, March 1999 http://www.encyclopedia.com/printable.aspx?1d=1G1:62086570

55 Kyoko Kurita, "Meiji Japan's Y23 crisis and the discovery of the future: Suehiro Tetcho's Nijusan-nen miraiki," *Harvard Journal of Asiatic Studies*, Vol. 60, No. 1, 2001, pp. 5-43

56 Matthew, *Japanese Science Fiction, op. cit.*, p. 103

57 Kumiko Sato, *Culture of desire and technology: Postwar literatures of science fiction in the United States and Japan*. Dissertation, Pennsylvania State University. 2005, 259 pages; Takayuki Tatsumi, *Full metal apache: transactions between cyberpunk Japan and avant-pop America*. Durham, NC: Duke University Press, 2006

58 H. G. Wells , *The Time Machine*. A critical edition with introduction and notes by Harry M. Geduld. Bloomington: Indiana University Press, 1987; *The War of the Worlds*. Introduction and notes by David Y. Hughes and Harry M. Geduld. Bloomington: Indiana University Press, 1993; *The First Men in the Moon*. Edited with an introduction by David Lake. New York: Oxford University Press, 1995.

Writing to some extent in opposition to Wells and the worldview that Wells, Verne, and others inspired, C. S. Lewis, an Anglican theologian, wrote many stories set in not-Earth that had deeply theological perspectives. The best known are *Out of the Silent Planet* (1938), the first in his space trilogy, and *The Lion, the Witch and the Wardrobe* (1950), the first of the seven *Chronicles of Narnia*. Lewis was a colleague of J.R.R. Tolkien whose books (and films) about *The Hobbit* (1937) and the *Lord of the Rings* series (1954) have been enormously popular recently, and set the stage for the *Harry Potter* rage recently.

There is a subfield of science fiction devoted to Christian themes. Some of the work of Philip Dick is of this type. Though totally different in style and theology from either C. S. Lewis or Philip Dick, the stories of Hal Lindsey, beginning with *The Late Great Planet Earth* (1979) and Tim LaHaye and Jerry Jenkin's twelve volume, *Left Behind* series (1995) have had extremely wide readership among Christians waiting expectantly for Armageddon, the Rapture, and the End of the World.

59 Based on Mike Ashley, *The Time Machines: The Story of the Science-Fiction Pulp Magazines from the beginning to 1950*. Liverpool: Liverpool University Press, 2000; Mike Ashley. *Transformations: The Story of the Science Fiction Magazines from 1950 to 1970*. Liverpool: Liverpool University Press, 2005; Thomas D. Clareson, *Understanding Contemporary American Science Fiction: The Formative Period, 1926-70*. University of South Carolina Press, 1992; James Gunn, *Inside science fiction*. Lanham, Md.: Scarecrow Press, 2006; David Hartwell, *Age of wonders: Exploring the world of science fiction*. New York: Walker, 1984; David Hartwell and Kathryn Cramer, eds., *The ascent of wonder: The evolution of hard SF*. New York: TOR, 1994

60 Such as Aldous Huxley, *Brave New World* (1932), [and] *Brave new world revisited*. With a forward by the author; introduction by Martin Green. New York: Harper & Row, 1965; George Orwell, *Animal Farm* (1945). With a preface by Russell Baker and an introduction by C.M. Woodhouse. New York, NY: Signet Classics, 1996; and *Nineteen eighty-four*. (1948), with a foreword by Thomas Pynchon and an afterword by Erich Fromm. New York: Plume, 2003; Ray Bradbury, *Fahrenheit 451* (1953), New York, Simon and Schuster, 1967; Harold Bloom, ed., *Ray Bradbury's Fahrenheit 451*. Philadelphia: Chelsea House, 2003; Kurt Vonnegut, *Slaughterhouse-five, or, The children's crusade: A duty-dance with death*. New York: Dell 1969; Harold Bloom, editor, *Kurt Vonnegut's Slaughterhouse-five*. Edited and with an introduction. New York, NY: Chelsea House Publishers, 2007; Doris Lessing, *Memoirs of a Survivor.* New York: Knopf, 1974; Ursula K. Le Guin, *The dispossessed*. New York: Avon Books, 1974; Margaret Atwood, *The Handmaid's Tale* Boston: Houghton Mifflin, 1986; Octavia E. Butler, *Parable of the Talents* New York: Warner Books, 2000

61 Larry Owens, "Sci-fi and the mobilization of youth in the cold war," *Quest*, Vol. 14, No. 3, 2007, pp. 52-57. Even now, science fiction in the US is still called upon to do more than merely entertain or even simply to educate. Several science fiction writers attended a conference on science and technology sponsored by the US Department of Homeland Security. The motto of the group (which had been formed fifteen years earlier by Arlan Andrews) is said to be "Science Fiction in the National Interest." [Mimi Hall, "Sci-fi writers join war on terror" *USA Today*, May 29, 2007 <http://www.usatoday.com/tech/science/2007-05-29-deviant-thinkers-security_N.htm>]

62 William Gibson's *Neuromancer*. New York: Ace Science Fiction Books, 1984; McCaffery, Larry, ed. *Storming the Reality Studio: A Casebook of Cyberpunk and Postmodern Science Fiction*. Duke University Press, 1991

63 See, the science fiction game reviews on weekly Sci.Fi.com <www.scifi.com/sfw>; "Is there hope for a science fiction MMORPG?" posted by Wilhelm2451 < http://tagn.wordpress.com/2008/01/17/is-there-hope-for-a-science-fiction-mmorpg/>; "Top science fiction MMORPGs," <http://internetgames.about.com/b/2007/12/12/top-science-fiction-mmorpgs.htm?iam=metaresults&terms=top+science+fiction+mmorpgs >. On the positive role of electronic games in general, see Steven Johnson, *Everything Bad is Good for You: How Today's Popular Culture is Actually Making Us Smarter*. New York: Riverhead Books, 2005; David Gibson, Clark Aldrich, Marc Prensky (editors), *Games and Simulations in Online Learning*. Hershey, PA: Information Sci-

ence Publishing, 2007; and "Serious Games" <http://www.serious-games.org/index2.html>. "Grand Theft Auto IV" was widely acclaimed as the first popular game that had excellent alternative plot lines and character development as well outstanding graphics and sound effects. See Seth Schiesel, "Grand Theft Auto Takes on New York" http://www.nytimes.com/2008/04/28/arts/28auto.html?_r=1&oref=slogin

64 Lucie Armitt, *Contemporary women's fiction and the fantastic*. New York: St. Martin's Press, 2000; Marleen S. Barr. *Feminist fabulation: Space/postmodern fiction*. Iowa City: University of Iowa Press, 1992; Marleen Barr, *Lost in space: Probing feminist science fiction and beyond*. Chapel Hill: University of North Carolina Press, 2003; Marleen Barr, ed., *Envisioning the future: Science fiction and the next millennium*. Middletown, CT: Wesleyan University Press, 2003; Patricia Melzer, *Alien constructions: Science fiction and feminist thought*. Austin: University of Texas Press, 2006; Patricia Monk, *Alien theory: The alien as archetype in the science fiction short story*. Lanham, Md.: Scarecrow Press, 2006; Jenny Wolmark, *Aliens and others: Science fiction, feminism and post modernism*. University of Iowa Press, 1994; Jane Donawerth, "Utopian Science: Contemporary Feminist Science Theory and Science Fiction by Women", *NWSA Journal*, Volume 2, No. 4, Autumn 1990, pp. 535-557; Jane Donawerth and Carol Komerten, eds., *Utopian and science fiction by women: Worlds of difference*. Syracuse University Press, 1994; Jane Donawerth , *Frankenstein's daughters: Women writing science fiction*. Syracuse, N.Y.: Syracuse University Press, 1997; Veronica Hollinger, "Feminist theory and science fiction"; Andrew M. Butler, "Postmodernism and science fiction"; and Wendy Pearson, "Science fiction and queer theory", in Edward James and Farah Mendlesohn, editors, *The Cambridge companion to science fiction*. New York: Cambridge University Press, 2003; Camilla Decarnin, Eric Garber, and Lyn Paleo, editors, *Worlds apart: An anthology of lesbian and gay science fiction and fantasy*. Boston: AlyCat Books, 1994. Paul Youngquist, "The space machine: Baraka and science fiction," *African American Review*, Vol. 37, Nos.2-3, Summer-Fall, 2003, pp. 333-343; Carl Freedman, *Critical theory and science fiction*. Hanover, New Hampshire: Wesleyan University Press, 2000; Ivana Milojevic and Sohail Inayatullah, "Futures dreaming outside and on the margins of the western world," *Futures*, Vol. 35, No. 5, June 2003, pp. 483-507

65 See "Space Poetry," *Encyclopedia Astronautica* http://www.astronautix.com/poems/index.htm>; Jonathan Vos Post, "70 major poets who minored in science fiction and fantasy," *The Ultimate Science Fiction Poetry Guide* <http://www.magicdragon.com/UltimateSF/sfpo-7pt0.html>; space poetry <http://www.learnenglish.org.uk/storis/poem_act/space_poetry.html>; John Fairfax, ed., *Frontier of Going: Anthology of Space Poetry*. London: Panther, 1969; John Foster, ed., *Spaceways: An Anthology of Space Poetry*. Oxford University Press, 1986; David Levy, *Starry night: Astronomers and poets read the sky*. Amherst NY: Promethus Books, 2001

66 Li Po, "Drinking Alone Under the Moon", *Crossing the Yellow River: 300 Poems from the Chinese*, translated by Sam Hamill. Rochester, New York: BOA Editions, 2000, p. 83f

67 William Drummond, "The Shadow of Judgement, " in *Flowers of Sion*, in *The poetical works of William Drummond of Hawthornden*, edited by L. E. Kastner, Manchester: The University Press, 1913, Vol. 2, p. 59f

68 Henry More, *Democritus Platonissans, or an Essay Upon the Infinity of Worlds*, Introduction by P. G. Stanwood. Los Angeles: William Andrews Clark Memorial Library, University of California, 1968, "The Argument," Stanza 26. Np

69 Richard Blackmore, "The Creation: A Philosophical Poem in Seven Books," in *The poems of Smith, and Blackmore*. Chiswick: Press of C. Whittingham, 1822, Volume XXVIII, pp. 118, 137, 138

70 Walt Whitman, "On the Beach at Night Alone", in *Leaves of Grass*. New York: Doubleday, 1926, p. 220f

71 Alfred Noyes, excerpt from "Tycho Brahe", in *The Torch-Bearers: Watchers of the sky*. New York: Fredrick A. Stokes Company, 1922, p. 84f

72 T. S. Eliot, excerpt from "Little Gidding", in *Four Quartets*. London: Faber & Faber, 1944. p. 43

73 *Selected Poetry of Robinson Jeffers*, Tim Hunt editor, Stanford University Press, 2001, reviewed by Christopher Cokinos as "Images of Inhumanism" in *Science*, 9 November 2001, Vol. 294, No. 5545, pp 128f

74 Jocelyn Bell Burnell, "Astronomy and Poetry," in Robert Crawford, ed., *Contemporary poetry and contemporary science*. Oxford: Oxford University Press, 2006

75 Gwyneth Lewis, *Zero gravity*. Newcastle upon Tyne: Bloodaxe Books, 1998

76 Alfred Worden, *Hello Earth: Greetings from Endeavor*. Los Angeles: Nash Publishing, 1974

77 Story Musgrave, "Cosmic Fireflies," *Story Musgrave NASA Astronaut recites poetry*, <http://www.spacestory.com/poetry.htm>

78 "100 Tanka Poems Selected from 145,000 Entries. Chiaki Mukai presents awards for completion of poetic verses", *Japanese Ministry of Education, Culture, Sports, Science and Technology* <http://www.mext.go.jp/english/news/1999/01/990106.htm>. See also, "Discovery Astronaut Writes Tanka Poem," *Space.com*, October 16, 2000 http://www.space.com/missionlaunches/missions/sts92_wakata_poem.html

79 *Space Poem Chain* http://iss.jaxa.jp/utiliz/renshi/index_e.html

80 Personal communication from Tsutomu Yamanaka, January 26, 2008

81 Marshall Sahlins, *Stone Age economics*. Chicago, Aldine-Atherton, 1972

82 Jack Goody, *The domestication of the savage mind*. New York: Cambridge University Press, 1977

83 Jack Goody, *The logic of writing and the organization of society*. New York: Cambridge University Press, 1986; Jack Goody, *The power of the written tradition*. Washington: Smithsonian Institution Press. 2000; Eric Havelock, *The muse learns to write: reflections on orality and literacy from antiquity to the present*. New Haven: Yale University Press, 1986

84 John Morieson, "The Astronomy of the Boorong". World Archaeological Congress, Washington D.C. June 2003 <http://bdas.fastmail.fm/astronomers/JohnMorieson/documents/World_Archaeological_Congress.pdf>; Roslynn D. Haynes, "Astronomy and the Dreaming: The Astronomy of the Aboriginal Australians," Helaine Selin, editor. *Astronomy Across Cultures: The history of non-western astronomy*. Boston: Kluwer Academic Publishers, 2000, pp. 53-90

85 Francesca Rochberg, *The heavenly writing: Divination, horoscopy, and astronomy in Mesopotamian culture*. New York: Cambridge University Press, 2004; Clive Ruggles, *Ancient astronomy: An encyclopedia of cosmologies and myth*. Santa Barbara, Calif.: ABC-CLIO, 2005

86 Roberta J. M. Olson, "...And They Saw Stars: Renaissance Representations of Comets and Pretelescopic Astronomy". *Art Journal*, Vol. 44, No. 3 (Autumn, 1984), pp. 216-224

87 Based on Ronald Brashear and Daniel Lewis, *Star Struck: One thousand years of art and science of astronomy*. University of Washington Press, 2001; Eileen Reeves, *Painting the heavens: Art and science in the age of Galileo*. Princeton University Press, 1997; David A. Hardy, ed., *Visions of space: Artists journey through the cosmos*. Limpsfield ; New York: Paper Tiger, 1989; International Association of Astronomical Artists <http://www.iaaa.org>; History of space art <http:www.artscatalyst.org/projects/space/spacearthistory.html>; Space Art <http://www.hobbyspace.com/Art/art2.html>; Ars Astronautica <http://www.arsastronautica.com/>; Art Technologies <http://www.arttechnologies.com/site-2005/links. html>; Leonardo Space Art Project Working group <http://spaceart.org/leonardo/vision.html>; Astrona—Space and Astronomical Art Journal <http://astrona.blogspot.com>; Spacearts—The Space Art Database <http://www/spacearts.info/en/info/index.php>; Ayako Ono's Space Art <http://www.geocities.jp/cosmo21art/aosa/whatsnew_e.html>; Arthur Woods, "From Cyberspace to Outerspace" (2001) <http://space-dreams.com>;Frank Pietronigro: Space Artist (2008) <http://www.pietronigro.com>; Roger F. Malina, "Space Art Definition" Space art projects at Art Technologies <http://www.arttechnologies.com/site-2005/space-art.html>; Roger F. Malina, "In defense of space art: The role of the artist in space exploration," in D.L. Crawford, ed., *IAU Colloquium* Vol. 17,112, 1991, p. 145

88 Space Art Hall of Fame <http://www.iaaa.org/gallery/rudaux/>

89 Some other early modern space artists were Jack Coggins, Dan Beard, Scriven Bolten, Paul Hardy, Mel Hunter, Theophile Moreau, and Charles Schneeman.

90 William K. Hartmann, Andrei Sokilov, Ron Miller and Vitaly Myagkov, eds., *In the stream of the stars: The Soviet/American space art book*. New York: Workman Publishing, 1990, p. 36, 37, 41; Mayo Graham, "Nineteenth-century America: the New Frontiers," in Jean Clair, ed., *Cosmos: From Romanticism to the Avant-garde*. Montreal Museum of Fine Arts, 1999, pp. 68-84; Jean Clair, "From Humboldt to Hubble," in Jean Clair, ed., *Cosmos: From Romanticism to the Avant-garde*. Montreal Museum of Fine Arts, 1999, pp. 25-27

91 Many people are associated with this kind of space art and illustrations, among the best known being Ralph Andrew, Beth Avary, Michael Carroll, Geoffrey Chandler, Jack Coggins, Vincent Di Fate, Paul DiMare, Don Davis, Don Dixon, Frederick Durant III, Bob Eggleton, Frank Germain, Joel Hagen, David Hardy, William Hartmann, Paul Hudson, Mel Hunter, John Foster, MariLynn Flynn, Robert Kline, Jon Lomberg, Pamela Lee, Robert McCall, Ron Miller, Larry Ortiz, Ludek Pesek, and Kim Poor

92 Konstantin E. Tsiolkovsky, State Museum of the History of Cosmonautics <http://www.informatics.org/museum/>; Ben Finney, "The Cosmicization of Humanity" in A. Houston and M. Rycroft, Editors, *Keys to space: An interdisciplinary approach to space studies*. New York: McGraw-Hill, 1999. Chapter 19.3, pp. 19-18 to 19-26

93 Igor Kazus, "The idea of Cosmic Architecture and the Russian Avant-garde of the early Twentieth Century," in Jean Clair, ed., *Cosmos: From Romanticism to the Avant-garde*. Montreal Museum of Fine Arts, 1999, pp. 194-217; also Vitaly Myagkov, "Soviet space art," in William K. Hartmann, Andrei Sokilov, Ron Miller and Vitaly Myagkov, eds., *In the stream of the stars: The Soviet/American space art book*. New York: Workman Publishing, 1990, pp. 54-77

94 Paul Rincon, "Hubble pics 'like romantic art'", news.bbc.co.uk/2/hi/science/nature/4278949>. Also, Jerry Lodriguss, "Catching the Light: Astrophotography" <www.astropix.com>; Gabriel Gache, "Hubble space telescope: Science Meets Art," news.softpedia.com/news/Hubble-Space-Telescope-Science-Meets_art-80553.shtml>; HubbleSite <hubblesite.org>

95 Didier Ottinger, "Contemporary Cosmologies," in Jean Clair, ed., *Cosmos: From Romanticism to the Avant-garde*." Montreal Museum of Fine Arts, 1999, p, 285

96 Space Place, "Moon Museum" < http://www.orbit.zkm.de/?q=node/148>

97 Robert Horvitz, "Art into Space," *Whole Earth Review*, Issue 48, Fall 1985, pp. 26-31 and <http://www.volny. cz/rhorvitz/Artspace.html>

98 Space Place, "Richard Kriesche < http://www.orbit.zkm.de/?q=node/268>

99 Space Place, "Andora, Photon Rockets" < http://www.orbit.zkm.de/?q=node/404>

100 Arthur Woods, "Cosmic dancer: A space art intervention on the Mir space station," (1993-2008) <http://www.cosmicdancer.com>

101 Richard Garriott, a highly successful game developer and son of Owen Garriott, a former NASA astronaut on both Skylab and Spacelab-1, flew on the Soyuz TMA-12 spacecraft from Star City, Russia, as a private customer of Space Adventures in partnership with Zero Gravity Art Inc., to the International Space Station in October 2008. There, he both created art while in space and exhibited specially-commissioned art from Steve Jenson, Drue Kataoka, Greg Mort, John Matthew Riva, Melinda Fager, and Stanley Goldstein. <www.richardinspace.com> and Press release, October 21, 2008, Challenger Center for Space Science Education <www.spaceref.com/news/viewpr.html?pid=26757>

102 Hartman, *op. cit*., 182

103 *Ibid*., 176. Among the Russians best known for space illustrations and art are Vladimir Arepiev, Nadezhda Devisheva, Victor Dubrovin, Rafik Karaev, Peter Kovalev and Olga Kovaleva, Alexei Leonov, Josef Minsky, Vitaly Myagkov, Boris Okorokov, Galina Pisarevskaya, Yuri Pochodaev, Andrei Sokolov, Andrei Surovtsev, Anatoly Veselov.

104 Alan Bean, "An Artist on the Moon" William K. Hartmann, Andrei Sokilov, Ron Miller and Vitaly Myagkov, eds., *In the stream of the stars: The Soviet/American space art book*. New York: Workman Publishing, 1990, p. 116

105 Ottinger, *op. cit*., 282f

106 *Ibid*., 284

107 *Ibid*., 286

108 Roger Malina offers a seven point definition of space art in Roger F. Malina, "Space Art Definition" Space art projects at Art Technologies <http://www.arttechnologies.com/site-2005/space-art.html>, while William Hartmann says that contemporary space art plays four roles: By inspiration; by conveying a sense of history by recording actual events and scientific knowledge of the past; by directing society towards the future; and by bridging the alleged gap between "the two cultures" of art and science. "Or, rather, they see no gap" (Hartmann, *op cit*., pp. 134-139).

109 The Arts Catalyst, "History of Space Art" <http://www.artscatalyst.org/projects/space/spacearthistory.html>; Dragan Zivadinov "Zero Gravity Biomechanical Theater", 1999 <http://www.orbit.zkm.de/?q=node/271>; Colin Fries, "Sports milestones in space," *Quest*, Vol. 10, No. 2, April 2003; Eduardo Kac, "Against Gravitropism: Art and the joys of levitation," 2005 <http://www.ekac.org/levitation.html>

110 Based on Andrew Fraknoi, "The music of the spheres: Astronomical sources of musical inspirations," *Mercury*, Vol. 6, No. 3, May-June 1977, pp. 15-19; Andrew Fraknoi, " More music of the spheres: Further astronomical sources of musical inspiration," *Mercury*, Vol. 7, No. 6, November-December 1979, pp. 129-132; Andrew Fraknoi, "The music of the spheres in education: Using astronomically inspired music in education," *Astronomy Education Review*, Vol. 5, No. 1, April 2006-November 2007 <http://aer.noao.edu/cgi-bin/article.pl?id=193>; Space Music <hpttp://www.hobbyspace.com/Music/; Space Music, "Yuri's night radio!" http://www.yurisnight.net/music/

111 Colin Ronan, "Music of the spheres," *Interdisciplinary Science Reviews*, Vol. 1, No. 2, 1976, p. 156; Bruce Stephenson, *The music of the heavens: Kepler's harmonic astronomy*. Princeton University Press, 1994; Philip Ball, "Science and Music," *Nature*. Vol. 453, May 8, 2008, pp. 160-162

112 *Ibid*., p. 151

113 Colin Ronan, "William Herschel and his music," *Sky and Telescope*, May 1981, 195-204

114 Andrew Fraknoi, "More music…," p. 132; Felix Lowe, "Brian May, Queen legend, hands in star thesis," <http://www.telegraph.co.uk/news/main.jhtml?xml=/news/2007/08/03 /nmay103.xml>

115 Andrew Fraknoi, "The music of the spheres in education…," pp. 2 and 6-10

116 Gustav Holst, The Planets, Simon Rattle <http://ecards.emiclassics.co.uk/planets/rattle_planets1.swf>

117 John Rodgers and Willie Ruff, "Kepler's Harmony of the World: A realization for the ear," *American Scientist* Vol. 67, No. 3, 1979 pp. 286-292

118 In addition to Hindemith, some of the most famous and influential composers of the 20th century produced pieces inspired by astronomy, space, scientists, and scientific discoveries, including Oliver Messiaen, *Visions de l'Amen* (1943) and *Des Canyons aux Etoiles* (1974); John Cage, *Atlas Ecipticalis* (1961); Toru Takemitsu, *Corona* (1962), *Orion and Pleiades* (1964), *Asterism* (1967) and *Cassiopeia* (1971); David Bedford, *A Dream of the Seven Lost Stars* (1965), *Tentacles Of The Dark Nebula* (1969), *Music For Albion Moonlight* (1970), *The Sword of Orion* (1970), *Some Stars Above Magnitude 2.9* (1971), *Star Clusters Nebulae And Places In Devon (1974), Ocean Star a Dreaming Song* (1981), *Of Stars Dreams and Cymbals* (1982), and *An Island in the Moon* (1986); Alan Hovhannes, *Symphony No. 19, Vishnu* (1966), *Saturn* (1971), *Celestial Canticle* (1977), *Star Dawn* (1983); George Crumb, *Night of the Four Moons* (1969), *Makrokosmos I, II, III, IV* (1971-2), and *Star-child* (1977); Karlheinz Stockhausen, *Ylem* (1972) and *Sirius* (1976); Leo Smit and Fred Hoyle, *Copernicus* (1973); and Philip Glass, *Einstein on the Beach* (1976), *Orion* (1983), and *Galileo Galilei* (2002).

At least one person who has written about contemporary serious space music expressed very openly his personal dislike for it, saying it is "not my cup of tea" and disparaging what he terms the "stunts" of John Cage (Andrew Fraknoi, "The music of the spheres in education," p. 2). Cage was arguably the most influential serious musician of the mid 20th century. Fraknoi's judgments seem strange for a scholar who should know full well that most musical forms—as well as most scientific ideas—were roundly rejected when they were first presented, and only later were embraced as "beautiful" or "obvious."

119 Colin Fries, "Flying for us: Space age milestones celebrated in music," *Quest*, Vol. 9., No. 3, 2002, pp. 30-36; Colin Fries, "Space Pioneers remembered in song," *Space.com* August 21, 2000 <http://www.space.com/news/spacehistory/space_songs_side_000822.html>

120 Space Music <hpttp://www.hobbyspace.com/Music/>; Space Music, "Yuri's night radio!" <http://www.yurisnight.net/music/>; U.S. Centennial of Flight Commission, "Aviation and Space Music" <http://www.centennialofflight.gov/essay/Social/music/SH16.htm>

121 Some favorites produced by major popular artists include *Countdown Time in Outer Space* (1961), The Dave Brubeck Quartet; *Space Cowboy* (1969), Steve Miller Band; *Space Oddity* (1969), *Ashes to Ashes* (1980), and *I Took a Trip on a Gemini Spacecraft* (2002), David Bowie; *Moondance* (1970), Van Morrison; *Blows against the Empire* (1970), Jefferson Starship [This is my personal favorite and was nominated for a science fiction Hugo Award]; *Rocket Man* (1970), Pearls Before Swine, and *Rocket Man* (1972), Elton John, were both inspired by a Ray Bradbury short story; *Astronaut Food,* from "The Miraculous Hump Returns from the Moon (1971), Sopwith Camel; *Dark Side of the Moon* (1973) [one of the most popular albums of any kind ever produced], Pink Floyd; *Revenge of Vera Gemini* and *E. T. I.* (1976), Blue Oyster Cult; *Moon Maiden* (1977), Duke Ellington; *Cygnus X-1* (1977) and *Cygnus X-1, Book II* (1978), Rush; *Hello Earth* (1985), Kate Bush; *Standing on the Moon* (1989), The Grateful Dead; *Space Patrol, UFO,* and *Southern Cross* (1992), Country Joe McDonald; *C. T. A. 102* (1996), The Byrds.

122 Apollo Vignette (aka Colin Fries), "A hit or a myth: Critiques of the space race in popular recordings," *Quest*, Vol. 13, No. 4, November 2006, pp. 15-24

123 Oh Word: Hip-hop in its purist form <www.ohword.com> and supported by the blog of Isamu Jordon <www.spokane7.com>

124 Outer Space <www.wordsound.com/outerspace.html>

125 Kenna: Urban music goes to outer space <http:findarticles.com/p/articles/mi_m1285/is_3_32/ai_84237662>

126 Jamaal Abdul-Alim, "Outer space as a muse: Rappers such as Rakim lace lyrics with intergalactic themes, " *Milwaukee Journal Sentinel* May 7, 2001 <http://www2.jsonline.com:80/news/metro/may01/space-jam08050701a.asp>

127 BEAUTIFULLEST MUTHASHIP <ohwordspacerap.blogspot.com/search/label/nasa>

128 Roger Launius, "Got Filk?" *Quest*, Vol. 12, No. 4, October 2005, pp. 6-14

129 "The Cruise of the Vostok," *Time*, Friday, Apr. 21, 1961

130 Sandy Schwoebel, "First flutist in space; An interview with Ellen Ochoa," *The Flutist Quarterly*, Vol 19, No. 1, October 1993, p. 14

131 Colin Fries, "Traditions of the space age," *Quest*, Vol. 11, No. 1, January 2004, pp. 31-39; Karen Miller, "Space Station Music" <http://science.nasa.gov/headlines/y2003/04sep_music.htm?list1003452>

132 Jet Propulsion Laboratory, Voyager: The Interstellar Mission, "Golden Record" <http://voyager.jpl.nasa.gov/spacecraft/goldenrec.html>

133 ESA, Cassini-Huygens, "Rock 'n' Roll is out of this world!" <http://www.esa.int/esaMI/Cassini-Huygens/SEMJBDXEM4E_0.html>

Chapter Three

Space Stories

"Humanity visualized the Earth ... 'as it truly is, bright and blue and beautiful in that eternal silence where it floats'..."

— Archibald MacLeish

Only about 500 people have flown into space since the space age started a half century ago. Although the advent of space tourism may change those numbers dramatically in coming years, going into space today is still a truly special and rare event. Most of this book is organized to explain and inform. In particular we have sought to explain how we can use space to explore, to help Planet Earth, or to discover the amazing physics of the Universe. We have also sought to cover the incredible ways that space systems can help people to communicate, to navigate, to farm, to fish, to combat climate change, or to protect ourselves against hostile weather or natural disasters. We will even spend a good deal of time examining how to design and build satellites and launcher systems. But this chapter is different.

This chapter contains a series of stories about astronauts who have gone into space or about "space-consumed" people, who have either designed systems that have successfully gone into space or about those planning totally new ways of going into space, including the pioneers planning and creating the fantastic new space tourism industry. These are stories about special people who have done, or are doing, amazing things. Astronaut Rusty Schweickart tells us about how remarkable his Apollo 9 experiences were. Astronaut and MIT Professor Jeffrey Hoffman relates to us the thrills and challenges of repairing the Hubble Telescope the first time around. We hear from International Space University graduates from Canada, the U.S.A. and Finland who worked on the Phoenix robot explorer, that amazing device which helped us reveal the latest new information about Mars. They tell us what working on this program was like. They also tell us not only of their part in the "Mars Adventure", but also how important international cooperation in space is to achieving success in complex missions such as Phoenix. We also hear about space tourism and the "space billionaires" who are redefining what space travel might mean in future decades. In time, the successful pursuit of new space tourism ventures could transform not only our planet, but help us to unlock a whole new future for humanity. We start with the space story of astronaut Rusty Schweickart.

Space Story One:

No Frames, No Boundaries—The Apollo 9 Flight in 1969

Apollo Astronaut Russell L. Schweickart

In early 1969 I flew on Apollo 9. I'd like now to take all of you on that trip with me, through that experience, because the experience itself has very little meaning if, in fact, it is an experience only for an individual or a small group of individuals isolated from the rest of humanity. As Neil Armstrong said, the Apollo Program represented a key step forward for all of humankind.

Apollo 9 was to be the first flight of the lunar module, the first time we would take that spacecraft off the ground and expose it to that strange environment to see whether it was ready to do the job. The setting was interesting. In December of 1968, just three months before our mission, Frank Borman, Jim Lovell and Bill Anders had circled the moon on Christmas Eve and had read from Genesis and other parts of the Bible, in a sense to sacramentalize that experience and to transmit somehow what they were experiencing to everyone back on Earth, "the good, green Earth," as Frank called it. And then the next day after Borman and company's readings, came one of those incredible insights. In the New York Times Magazine, Archibald MacLeish wrote an essay about the step that humanity had now taken. He wrote that somehow things rather suddenly have changed, and man no longer perceived himself in the same way that he had seen himself before. Humanity visualized "the Earth now as it truly is, bright and blue and beautiful in that eternal silence where it floats," and "men as riders on the Earth together, on that bright loveliness in the eternal cold, brothers who know now they are truly brothers." To read those moving words as you're preparing to go up into space yourself you are gripped by these thoughts. It's a heavy trip, you realize that it's not just a physical thing you're doing. You realize there's a good deal more to it. So in all the other preparations you make you somehow incorporate these thoughts as well.

All this formed the background against a very, very busy foreground. The foreground involved simulation after simulation. –As Apollo astronauts you were force fed all those millions of procedures. These were the essential procedures we were required to learn. Procedures that could save your life and the lives of your fellows if you run into this problem or that problem.

You attended an incredible number of meetings, going over procedures and detailed checklists and techniques. You learned to think of everything that could happen or go wrong, and then decided in an instant what you will do in each case. Hour after hour in classrooms, you struggled to keep awake so that you could understand all those systems that go into the spacecraft. You learned the "foreground procedures" that would keep you alive or would kill you if you didn't know what you were doing.

You then take part in testing the spacecraft—not a simulated one now but the real one. And those tests go on and on, until you feel the spacecraft is going to be worn out before it ever gets a chance to perform up there where it was designed to work.

And then finally comes the morning when you get up before dawn. Some people are just starting to come to work. You look out the window, and in the distance to the north there is this brilliant, white object standing on its tail with search lights playing on it - and it somehow becomes a bright symbol sitting there on the beach ready for its trip into space. It's the most awe-inspiring thing you've ever seen - beautiful. And you go down the hall and have the last of what seems like an infinite series of physical examinations and you eat breakfast—so much of being an Astronaut seems to revolve around calorie consumption. Then, you go down the hall in the other direction and you put on your suit with the help of all those technicians. You've done it a hundred times before and it's exactly the same, except somehow this morning is a little bit different. And you go down the elevator with your two friends and you get in a transfer van and you go over to the pad and you go up that tower and you look out across that countryside, the sea in one direction and the rest of the country in the other direction. And you realize that all those years and years of hard work - five years, six years, seven years —condense into this moment. And you are deeply moved.

And then you get into the spacecraft and you jostle around and you joke around in the White Room as you're getting in. You put signs on the back of the guy who's helping you get in, so that everybody watching on TV sees these ridiculous signs. It is a montage of all those things large and small. Then you lie there on your back and they close the door and you're right back in a simulator - you've done it a hundred times. And you lie there. During the countdown you may doze off and catch some sleep, waking up when you're called on to take a reading or something. Then they count backwards down to zero and off you go.

Somehow it's anti-climactic. It's much more exciting from the beach, watching it and seeing all that smoke and fire and feeling the power and the concentration of energy that's taking those three people up into space. From the beach you feel that, and it causes your whole soul to oscillate with the throb of that incredibly powerful sound. But you're inside now, you're going up, and everything looks very much like it does in a simulation and you've done this a hundred times. The only difference, at least in most cases, is that it's all working correctly. I mean things aren't going wrong now. The dials read what they should read instead of what some joker outside the simulator throws in as a problem.

And so you go into space. You're lying on your back, and you can't really see out until the launch escape tower gets jettisoned part way up. Then your window is clear, and as you pitch over, getting near horizontal, you catch the first glimpse out the window of the Earth from space. And it's a beautiful sight. So you make some comment - everybody has to make a comment when he sees the Earth for the first time - and you make your comment and it's duly noted. And then it's to work, because you don't have time to lollygag. The work schedule doesn't include time to look out the window and sightsee. You're up there in March of 1969 and the goal is to put a human on the moon and return safety to Earth before the end of the decade. That deadline looms large. So, on with the job.

You get up there in orbit, you separate from the booster, and you turn around to dock with the lunar module. And you have a little problem docking, because a couple of thrusters got shut off inadvertently during launch. You can't understand why you can't control the vehicle. So there's a moment of panic. You go madly around checking switches, throwing switches with increasing urgency. You are trying every thing and any thing. Then somebody notices a little flag that's the wrong way. You then throw the right switches and you dock. You extract the lunar module and now you have to change orbit, so you go through all those procedures. You take out the checklist. You read down the list. You leave nothing to memory. And you've done it. You've changed the orbit. You then light the main engine of the command module with the lunar module now connected on its nose for the first time. For a moment you wonder whether maybe it'll break apart, but it doesn't. You were part of the design process. You knew by the technical design it wouldn't, but now you really know. And that first night in orbit you eat. Always you need to eat—to consume those calories. You doff your pressure suits and stow them under the couches. Then you climb into the sleeping bag and go to sleep.

Up the next morning on a timetable the insistent schedule demands. You eat breakfast and sip your Tang. Then you don the suits. And now you've got a full day of checkout again. You're testing the system that held together the first time you lit the engine, but now you're not just going to light the engine; you're going to wiggle it, test it. Your job is to stress and strain that tunnel between the command module and the lunar module. It is critical to make sure it will really hold together. And again you know it will, but after you've done it for real you have proved it. This is no simulation.

So you've had a busy day there, and again it's eat, doff the suit (you had put on the suit because the spacecraft might have broken apart and it's hard to live in a vacuum.) So, just in case, you do it that way. And you go to bed.

And the next morning it's the same process all over again. It is like the movie Groundhog Day where this guy gets to live the same day over and over again. You haven't quite gotten enough sleep, but it's up—and hurry up—because you're late. You eat while you get into the suits. Then you open up that tunnel and go into the lunar module for the first time. It's an amazing sight out those windows because they're much bigger windows. But again don't stop. You don't have time for that. And so out comes the checklist and you follow it relentlessly down through that day. You are checking out all those same systems that you know so well from paper—but now you're there and you're throwing the switch for real rather than pretend. And you check out the guidance and control system and the navigation system and the communication system and the environmental control system and on and on and on. By the end of the day you're ready for the grand finale - you're going to light up the main engine on the bottom of the lunar module, the engine that will take two of your friends down to the surface of the moon if everything goes right. And you have to demonstrate that that engine will work and that it can also push both the lunar module and the command module around, in case one day that has to be done - little knowing that only the next year that will have to be done to save the lives of three of your friends on the Apollo 13 mission. And you light off that engine and it works, just the way it did in the simulator. It's amazing. So you go back into the command module and you're a little behind again and you hurry up and eat (NASA controllers are really into eating and it really does take the edge off of really high adrenaline stuff you have been doing. Then you take off the suits and get to sleep, because, again, the next day is a big day.

It is up the next day and back through the cycle. Today is the day you check out the portable life support system, the backpack that will be used to walk around on the surface of the moon. This is the critical equipment that will allow people to live and operate and work and observe— to be human in that hostile environment. So you put on the suit that morning knowing that you're going to go outside. And you get over in the lunar module and you go through all of those procedures. You check out the portable life support system and everything seems to work, and you strap it on your back and you hook all the hoses and connections and wires and cables and antennae and all those things to your body. And you sever the connection with the spacecraft. Next you switch on this pack you're carrying on your back. You let all of that precious oxygen flow out the door of the lunar module. Now you're living in your own spaceship and you go out the door. And outside on the front porch of the lunar module, you watch the sun rise over the Pacific and it's an incredible sight—a beautiful, beautiful sight. But you don't look at it, because you really don't have time. You've really got to get moving. That ever challenging and demanding flight plan says you're behind again and you've only got forty - five minutes out there to do all those things you have to do. And so you collect the thermal samples and you start taking the photographs. It is then you have a stroke of luck. Across the way in the command module where your friend is standing, also in his space suit, taking pictures of you while you take pictures of him, his camera jams. It is going to take a while to fix that camera. So you have just a moment to think about what it is you're doing. You have some time to take in the grandeur of it all. But then he gets it fixed and off you go again. It seems like a flash and you're back inside the spacecraft. You know you really need to get moving and get everything back together and taken care of and put away and get the food eaten and the suits off and stowed and get to sleep, because the next day is the biggest test of all.

The next day you have to prove that you can rendezvous. This means separating the two spacecraft by a couple of hundred miles (over three hundred kilometers apart) and then bringing them back together again. We had been given just four to five hours to accomplish something that had not been done before. One of the craft—the LEM—doesn't have a heat shield. This has a very precise meaning for us. Two of you can't come back home unless you get back together. So you get into the lunar excursion module. This vehicle has now become a friend. You go through all the preparation for that rendezvous and you separate. Except when you get to the end of the stroke on the docking mechanism, it goes clunk. You say very perceptively: "What was that? That wasn't in the simulation." About the time you're wondering what it was and if maybe discretion is the better part of valor. In short should we go back, start over. Your friend goes clunk and opens up the fingers. And you say, "Well, we'll find out in five hours whether it's all okay."

So off you go. And five hours later everything has worked right again. It's been a long five hours and you've gone through a lot of tests, but everything has worked and here you come. You're coming back together again. There's no reunion like this reunion - not only because it's your heat shield out there— the only way to get back home—but because that's your friend over there. Dave Scott is your next - door neighbor, but he was never a neighbor like he's a neighbor now. And so you dock. You get back together, and you open the tunnel. Believe me there's a reunion that can't be topped. And you get everything done and get back into the command module. And you're tired. You're absolutely exhausted. You haven't had enough sleep. You haven't had a good meal. In fact, you probably haven't eaten that day. And you sit there and you take off your suit.

And now you've got a piece of that lunar module left sticking on the nose of the command module, and you throw a switch and then it's gone. There's a piece of you that just floats off. It's a machine; but so are we. And it goes away. It just floats off into the distance after doing its job faultlessly. And now your thoughts turn to things like a shower and a bed to sleep in. You think of all those things that you realize you haven't been thinking of for those five hectic days that you've just been through. But there are still five days to go, because the flight plan says you are to go for ten days. They want you to show you can do the whole mission, the endurance part. So for the next five days, while you're thinking about a steak and a shower and a bed and all those things, you float around the Earth doing other tests. And now, for the first time, you have a chance to look out that window. And you look out at that incredibly beautiful Earth down below. You reach down into the cabinet alongside the seat and you pull out a world map and play tour guide. You set up the little overlay which has your orbit traces on it on top of the map, and you look ahead to where you're going, what countries you're going to pass over, what sights you're going to see. And while the other guys are busy you say, "Hey, in ten minutes we're going to be over the Mediterranean again and you might want to look out." So you look forward to that. And you go around the world, around and around and around, performing these tests. Every hour and a half you go around the Earth and you look down at it. And finally, after ten days, 151 times around the world, 151 sunrises and sunsets, you turn around and you light the main engine again for the last time, and you slow down just enough to graze that womb of the Earth, the atmosphere.

And down you come into the atmosphere. As you come back in you experience deceleration and it seems as though you're under an incredible pressure. You know that you're experiencing at least four g's, four times the force of gravity, and you say, "Jim, what is it now?" And he says, "Two tenths of a g." You just can't believe this is only one fifth of what is a human's experience on the Earth.

By the time you reach four or five g's you begin to realize the burden that man has lived under for millions of years. As you look out the window you see your heat shield trailing out behind you in little bright particles.

The heat shield is flaking off, and glowing. In fact, the whole atmosphere behind you is now aglow. We see a glowing sheath that is sort of cork-screwing back up toward space. And finally you slow down enough so that all of the bright lights outside the window, the fireball that you've been encapsulated in, has now dissipated. And you cross your fingers because all through the flight you've been throwing switches. These have activated various pyrotechnic devices. Explosive mechanisms have sealed one fluid from another and one portion of the spacecraft from another. These have been going pop or bang or whatever for days now. And you've a couple more of those to go, the ones that control your parachutes. So you throw the next to last switch and it goes pop and the drogue chutes come out. And you slow down to a couple of hundred miles an hour, and then you throw one more switch and pop, out go the main chutes. They all work. And you realize that the last explosive device, the last switch that you've had to throw, the last surge of electrons through all the wiring has worked. Now that whole thing is behind you. Splash! You're on the surface of the Atlantic. There are people circling around in helicopters and ships. You're back in humanity again. It's an incredible feeling.

And what's it all meant? You know it means success. It means humans will be able to set foot on the moon and return to Earth by 1970? Yes. All of those things that had to work and to be proven have worked. You're that much nearer to that incredible goal of putting man on another planet. Have you opened the door to the future? Have you changed the nature of exploration? Yeah. You've done that.

We will not step back through that door and close it, except perhaps for short periods of time. Are there any practical benefits from it? Yeah. Lots of practical benefits, ad infinitum. After a while you get tired of talking about them, but they're there. And they make a big difference in the world; in fact, you're dedicated to them because they will make that difference.

But I think that in some ways there are other benefits that are more significant. I think that you've played a part in changing the concept of humans and the nature of life, by redefining a relationship that you have assumed all these years. Humanity, the whole of history has assumed we have a relationship to a single planet. But that is now changed. And you now know that, because it's a part of your gut, not a part of your head. And you wonder, you marvel that an Archibald MacLeish somehow knew that. How did he know that? That's a miracle.

But up there you go around every hour and a half, time after time after time. And you wake up usually in the mornings, just the way the track of your orbit goes, over the Middle East and over North Africa. As you eat breakfast you look out the window as you're going past, and there's the Mediterranean area, Greece and Rome and North Africa and the Sinai, that whole area. And you realize that in one glance what you're seeing is what was the whole history of man for thousands of years—the cradle of Western Civilization. And you go down across North Africa and out over the Indian Ocean and look up at that great subcontinent of India pointed down toward you as you go past it, Sri Lanka off to the side, then Burma, Southeast Asia and China, out over the Philippines. You marvel at the expanse of Asia where the bulk of humanity lives. And then it is up across that monstrous Pacific Ocean, that vast body of water. You've never realized how big that it is before. And you finally come up across the coast of California, and you look for those friendly things, Los Angeles and Phoenix and on across to El Paso. And there's Houston, there's home, you know, and you look and sure enough there's the Astrodome. You identify with that, it's an attachment. And on across New Orleans and then you look down to the south and there's the whole peninsula of Florida laid out. And all the hundreds of hours you've spent flying across that route down in the atmosphere, all that is friendly again. And you go out across the Atlantic Ocean and back across Africa, and you do it again and again and again.

And you identify with Houston and then you identify with Los Angeles and Phoenix and New Orleans. And the next thing you recognize in yourself is that you're identifying with North Africa and other parts of the world. You look forward to that, you anticipate it, and there it is. And that whole process of what it is you identify with begins to shift. When you go around the Earth in an hour and a half, you begin to recognize that your identity is with that whole thing. And that changes you.

You look down there and you can't imagine how many borders and boundaries you cross, again and again and again, and you don't even see them. There you are observing what you can't see. You know there are hundreds of people in the Mid-East killing each other over some imaginary line- a line that you can't see. And from where you see it, the thing is a whole, and it's so beautiful. You wish you could take one in each hand, one from each side in the various conflicts, and say, "Look. Look at it from this perspective. Look at that. What's really important?"

And a little later on we did the impossible. One of your friends accomplished the goal set by President Kennedy. One of those same neighbors, the person next to you, went out to the moon. And now he looks back and he sees the Earth not as something big, where he can see the beautiful details, but now he sees the Earth as a small thing out there. And the contrast between that bright blue and white Christmas tree ornament and the black sky, that infinite universe, really comes through. The size of it on a cosmic scale, the insignificance of it forces itself on you. It is so small and so fragile and such a precious little spot in that universe that you can block it out with your thumb, and you realize that on that small spot, that little blue and white thing is everything that means anything to you. On that tiny spot is all of history and music and poetry and art and death and birth and love, tears, joy, games, all of it on that little spot out there—a spot that you can cover with your thumb. And you realize from that perspective that you've changed, that there's something new there, that the relationship is no longer what it was.

And then you look back on the time you were outside on that EVA. You treasure those few moments that you could take, because a camera malfunctioned. You savor that flash of time when you could think about what was happening. And you recall staring out there at the spectacle that went before your eyes, because now you're no longer inside something with a window looking out at a picture. Now you're out there and there are no frames, there are no limits, there are no boundaries. You're really out there, going at unimaginable speeds, ripping through space in a vacuum. And there's not a sound. There's a silence the depth of which you've never experienced before. That very elegant silence contrasts so markedly with the scenery you're seeing and with the speed with which you know you're moving.

And you think about what you're experiencing and why. Do you deserve this, this fantastic experience? Have you earned this in some way? Are you worthy to be separated out to be touched by God, to have some special experience that others cannot have? And you know the answer to that is no. There's nothing that you've done to deserve that to earn that extraordinary privilege. Then you recognize it's not a special thing for YOU. You know very well at that moment, and it comes through to you so powerfully, that you're the sensing element for all of humanity. You look down and see the surface of that globe that you've lived on all this time, and you know all those people down there and they are like you, they are you. Somehow you represent them. You are up there as the sensing element and that's a humbling feeling. It's a feeling that says you have a responsibility. It's not for yourself. The eye that doesn't see doesn't do justice to the body. That's why it's there; that's why you are out there. And somehow you recognize that you're a piece of this total life. And you're out there on that forefront and you have to bring that back somehow. And that becomes a rather special responsibility and it tells you something about your relationship with this thing we call "life." So that's a fundamental change. That's something new. And when you come back there's a difference in that world for you now. There's a difference in that relationship between you and that planet and you and all those other forms of life on that planet, because you've had that kind of experience. It's a difference and it's so precious.

All through this narrative I've used the word "you" because it's not me, it's not Dave Scott, it's not Dick Gordon, Pete Conrad, or even John Glenn - it's you, it's we. It's life that's had that experience.

I'd like to share with you in closing a poem by e. e. cummings. It's just become a part of me somehow out of all this and I'm not really sure how. He says:

i thank you God for this most amazing day:

for the leaping greenly spirits of trees and a blue true dream of sky;

and for everything which is natural which is infinite which is yes

Editorial Note: Rusty Schweickart presented a version of this "story" at the 1974 Lindisfarne Conference. Printed with the permission of Rusty Schweickart.)

Space Story Two:

Rescuing and Repairing the Hubble Space Telescope (HST)

MIT Professor and Astronaut Jeffrey Hoffman

I had the good fortune to be one of the crewmembers on Space Shuttle mission STS-61 to repair the Hubble Space Telescope (HST). Our mission lasted from 2 to 13 December 1993, and it managed to restore this remarkable tool to its full operational and scientific capability. In fact, our efforts to install a set of "eyeglasses" on the telescope finally allowed it to probe the depths of space as never before.

Given the incredible scientific success and enormous public popularity that Hubble has achieved since our rescue mission, it is hard to remember the sense of despair that pervaded NASA and the astronomical community following the discovery, after its 1990 launch, that Hubble suffered from a "spherical aberration" in its primary telescopic mirror. This distortion destroyed the razor-sharp focus (better than 1/10 arc second) necessary to achieve its ambitious goals.

Once the cause of the problem was discovered, optical engineers, astronomers and astronauts came up with an ingenious method of inserting corrective optics to fix the problem. This was often referred to as "putting contact lenses, or glasses, on Hubble". In fact, we really installed extra mirrors rather than lenses. In any case, our mission was the most ambitious repair mission ever planned by NASA. In addition to fixing the optics, we had to replace Hubble's two solar panels, two of its gyroscope modules, two electronic control units, four fuses, and two magnetometers, and install an additional memory module in the main computer and a repair kit on one of the scientific instruments.

In all, we would need five EVAs (or spacewalks) to accomplish all our tasks. Six EVAs were the absolute maximum that could be accomplished in the limited time a Shuttle can stay in space, and we had to hold one in reserve in case of contingencies, so our schedule was packed to the limit.

The repair and upgrade of the HST was the main reason for this mission. It turned out that our efforts also demonstrated the capability and level of maturity of our orbital operations. This was an important secondary objective for NASA. After all, the Hubble Space Telescope was conceived with the idea that the Space Shuttle would allow us to maintain it as a long-term astronomical observatory facility, like ground-based observatories. At the outset we planned to revisit it, as necessary, to make repairs and to replace older scientific instruments with new, state-of-the-art technology.

This was the dream and the promise of Hubble. Now we had to show that we could do it! Both as a former professional astronomer and as an astronaut, I could not imagine a more exciting space flight. I had many astronomer friends who had invested years of their lives with the Hubble Space Telescope, so our rescue mission had not only scientific and institutional significance, but also quite personal interests.

So what was it like preparing for and carrying out this mission? By the time our crew went into medical quarantine one week before launch, we had spent hundreds of hours in flight simulators, practicing both flying skills (ascent and entry) and orbital operations. Our commander and pilot made hundreds of simulated Shuttle approaches and landings in our special Shuttle Training Aircraft. Our robotic manipulator arm operators worked both with electronic simulators and with a full-size mechanical arm, which moved around an HST-shaped helium balloon. The four EVA crewmembers, myself included, spent almost 400 hours underwater, practicing every step of the procedures we were planning to perform.

We looked at these precise activities and motions as "choreography". Useful time outside the spacecraft was the limiting consumable on this flight, and we did not want to make any mistakes. This could cost us valuable time or, even worse, damage the telescope. We also made extensive use of a virtual reality (VR) simulator, where we did a lot of initial mission planning. (Time on the VR computer is a lot less expensive than underwater training time.) We had several mission dress rehearsals, some of which lasted as long as 54 hours. These efforts were carried out in our Shuttle Mission Simulator, which was electronically tied together with both the Mission Control Center in Houston and the Space Telescope Operations Control Center at NASA's Goddard Space Flight Center. The objective was to practice coordination between the crew, Shuttle controllers and HST controllers and to get it as close to perfect as possible.

All this was behind us when we entered quarantine. In addition to protecting us from germs, this was a nice time to "decompress" from our grueling training schedule and relax a little before the flight. Three days before launch we traveled to Florida for our final briefings on late-developing problems and changes to the Shuttle or the payload. Then there were our final equipment checks and a little time with our families (although children under eighteen were not allowed into quarantine, which annoyed my youngest son no end, since his older brother had just turned eighteen!) We had already made several trips to the Kennedy Space Center during the four months before launch. The purpose of these trips was to watch the HST replacement parts being installed into the Shuttle, to check the payload bay for sharp edges that might hurt our spacesuits, and for a dress rehearsal countdown. Each of these trips ended with our flying West, back to Houston. Now, we were finally ready to leave towards the East! Our excitement was very high.

NASA had done everything possible to reduce the risk that our mission would not succeed. For this reason, all seven crewmembers on STS-61 had flown in space before. Among us, we had sixteen flights worth of experience, a record for space flight. Nevertheless, none of us was blasé about the actual launch. Riding a rocket into space is always an overwhelming experience, no matter how many times you have done it! Our original launch date was 1 December 1993. The previous night there had been a total eclipse of the Moon, and I had enjoyed a lovely bicycle ride on the beach watching the Moon disappear and then reappear. However, the weather on December 1st did not cooperate. Strong crosswinds across the Kennedy Space Center runway exceeded limits, and we had to scrub.

December 2 was a much better day for launch, and for me it had a special significance, since my second space flight had launched on 2 December 1990. That mission was also devoted to astronomy, so the date seemed propitious. We had a perfect launch! Once in orbit, we spent a few hours reconfiguring the Shuttle cabin from its launch to its orbit configuration. Unlike many crews with seven people, we were all on the same shift, so we all went to sleep several hours after getting into space. Our commander was strict about our getting enough sleep every night, so we covered our windows with shades and, unlike on most of my other flights, I did not stay up late watching the Earth drift by beneath me!

On the second day, our main tasks were to check out our four space suits and the Shuttle's remote manipulator system (RMS), which would grab and berth Hubble and move EVA crewmembers and equipment around the Shuttle's payload bay. Our lead RMS operator was European Space Agency astronaut Claude Nicollier, backed up by Pilot Ken Bowersox. Several times during flight day two, Ken and Commander Dick Covey fired the Shuttle's engines to make slight corrections to our orbit as we carried out our chase of the Hubble Space Telescope. Finally, halfway through our third day in orbit, we got close enough to catch our first glimpse of the HST. It was just a bright dot, but this was an emotional moment for all of us onboard. Every orbit we got closer and closer, and Hubble got bigger and bigger. Eventually, I could make out the solar arrays through my binoculars and reported to the ground that one of the two arrays was badly deformed. We knew that thermal stresses on the metallic array extenders had been causing the arrays to flex. This flexing caused oscillations lasting several minutes every time the telescope went from night to day or day to night. We were concerned that damage to the arrays might prevent them from being rolled up and brought home, which in fact turned out to be the case.

Once we closed within twenty-five meters of the telescope we switched our Ku-band system from rendezvous radar to communications mode, allowing TV pictures to be sent to the ground, so that the thousands of people who have worked on the HST and on this repair mission could share the excitement of seeing Hubble close up for the first time since its 1990 deployment. As we approached the HST, Claude had the arm ready in grapple position. Dick slowed the Shuttle's closing rate until the HST was hovering motionless over our payload bay (even though we were both moving around the Earth at 5 miles per second). Claude then maneuvered the arm's end-effector onto one of the two grapple fixtures on the HST, closed the snares to capture the telescope, and Dick announced to the world that "We have a firm handshake with Mr. Hubble's telescope." It was a great feeling! Claude then lowered the telescope onto the Flight Servicing Structure (FSS) at the back end of the payload bay until three hooks on the bottom of the telescope were within capture range of the FSS latches.

I sent a command to close the latches, and Claude released the arm from the HST. I then sent another command to attach an orbiter umbilical to Hubble so that it could be powered by the Shuttle rather than by its own batteries during the week of repairs. It was a long day, but as we went to sleep that night we knew that the first critical part of our mission had been a success. Tomorrow we would carry out our first spacewalk!

The unprecedented number of spacewalks planned for this mission led to the selection of four EVA-trained crewmembers. Each of us had actually done at least one spacewalk on a previous mission. We were divided into two teams, Story Musgrave and I were on one team and Tom Akers and Kathy Thornton were on the other. The plan was for each team to go out on alternate days, to prevent anyone from getting too tired. We cross-trained, so that we all knew how to perform each other's tasks just in case one team could not go outside for some reason or tasks had to be rescheduled. Each day, the team staying inside would read procedures to the outside team and help make sure that all the goals of that EVA were accomplished correctly. Quality control was critical, since we did not want to break anything that was not already broken! Story and I had the initial EVA, and there was a certain amount of overhead the first time we went outside the Shuttle. These basic tasks included getting tools in place, putting the foot restraint platform onto the manipulator arm, attaching tools to it, and extending translation aids that allowed us to move from the payload bay up to the bottom of the telescope. The first thing we did was to put protective covers over the low gain antenna and two umbilical plugs on the bottom of the telescope. After all these preliminaries we were ready to begin the repair work.

Our EVA timeline had changed frequently during the past year. We tried to organize the repair tasks roughly according to their priority. Another critical factor that influenced our timeline was the length of time estimated to do the various tasks. Also, we wanted to finish every spacewalk with the telescope in a configuration such that, if we had to come home early, the telescope could be redeployed in working condition. For example, changing out the solar arrays was at the top of our priority list, but we had determined that there was probably not enough time to complete this on the first day. Therefore, we left this for the second day and replaced two gyroscope packages on the first. We also had to allow for the possibility that some of our planned repair tasks would not succeed, so we designed each task and wrote our procedures to be modular. This modularity, plus our crew cross-training, allowed the order of tasks to be changed relatively easily in response to unexpected problems.

Changing out the gyros required one EVA crewmember to go inside the telescope and squeeze under the star tracker light shades in order to reach the electrical connectors. This is a delicate task to accomplish in a bulky space suit. In most EVA tasks, being tall and having a long reach is a plus, but in this case Story could fit into this tight space much easier than me. Thus I rode on the manipulator arm and used a long socket wrench to undo the bolts on the old gyros. Story could then install them as I tightened the new bolts while he held the new gyros in place.

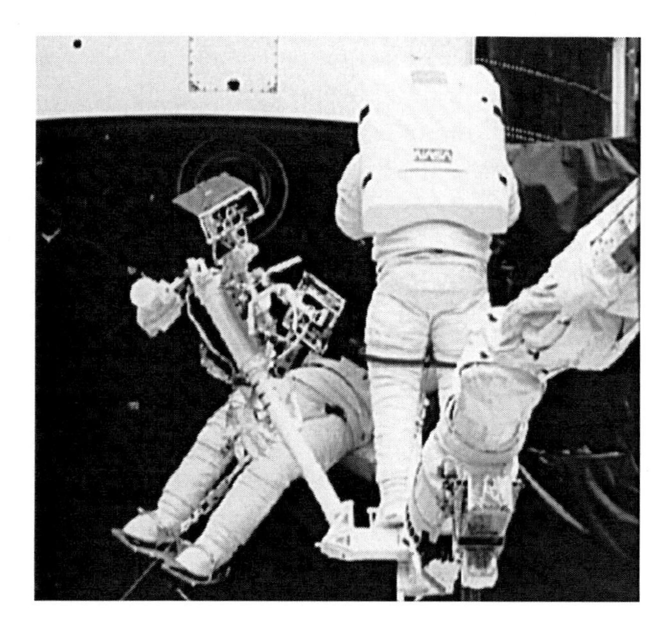

Figure 3.1. Astronauts at work: Jeff Hoffman on the manipulator arm with Story Musgrave wedged into the Hubble's gyro compartment (Courtesy of NASA).

Story reconnected the electrical connectors and extricated himself from his working position, without damaging either the star tracker shades or the scientific instruments that were directly behind him. The original plan given to us when we started training had required us to remove the star tracker light shades to clear the work area before replacing the gyros, but this took a lot of time. We came up with the idea of sliding in under the shades and tried it out in the water. This was the plan finally agreed by all those involved in planning the mission. This team included people from Goddard Space Flight Center, the Space Telescope Science Institute, the Lockheed Corporation, which was the HST prime contractor, and the EVA flight controllers at the Johnson Space Center. All of these team members had a role in planning and executing our mission. We had to convince a lot of people that our plan would not only save valuable time, but also that it did not pose a risk to the telescope. After much practice, we staged a "final exam" to prove to everyone we could do it. We passed!

Our working plan was typical of how we organized ourselves for most of our EVA tasks. One crew member was a "free floater", moving hand over hand around the payload bay and doing most tasks with one hand while holding on to some support with the other or, if two hands were necessary, with both feet locked into a foot restraint, which provided stability. The other crewmember rode on the robotic manipulator arm. This person transported most of the large equipment around the payload bay, since it is difficult to move yourself around hand over hand carrying large pieces of equipment without bumping into things. We practiced our EVA tasks without the robot manipulator, just to be prepared in case it should break, but it is not clear we could have accomplished the whole mission without this valuable robotic tool. Certainly our tasks were accomplished much more easily, quickly and with far less risk. This was an excellent example of how human and robotic systems can work together symbiotically to accomplish tasks better than either could do independently.

The heavy workload for this mission gave rise to a mantra "Keep doing useful work." Useful EVA time was our limiting consumable. Thus, if one person could do a job, we decided that the other person should not wait around for it to be completed. Instead he should go off and start to work on something else. Therefore, after the gyros were replaced, Story went off to get a head start on some of the work that Tom and Kathy would have to do the next day as part of the solar array replacement, leaving me to close the large doors to the gyroscope/star tracker compartment, a simple task we had done many times in the simulators. Like most of the telescope, the door latches were specially designed to be "EVA friendly", meaning "easy to do" when wearing bulky EVA gloves. All I had to do was turn a handle to engage latches on the top and bottom of the door, then flip a few levers and tighten a few bolts with my electric socket wrench. I turned the handle and saw the upper latch engage, but when I looked down at the lower latch, I saw that the bottom of the door was still open by an inch or so and that the latch had not engaged. I opened the door and tried again, with the same result. While we had anticipated that we would encounter problems in the course of our mission, I had hoped that they would at least wait until after the very first task! Was this the first of many? Could it really be that we might not be able to complete our mission? Those were the first thoughts that ran through my mind, but I couldn't let myself get distracted, so I started to analyze the cause of the problem and see if I could fix it.

I asked Claude to lower the arm so that I could push on the lower part of the door while I closed the handle yet again. I was happy to see that this time the lower latch closed properly, but then I saw that the upper latch had not engaged! The door was warped, and the normal closing procedure did not work. I reported the problem on my radio so that the crew inside the Shuttle and the flight controllers and Hubble team on the ground could also start thinking about possible solutions. I eventually came to the realization that the warped shape of the door had turned this from a one-man to a two-man task, so I called Story to come over to help.

If I could push on the top and he could push on the bottom, we might be able to get the latches to close. There was a problem, though. I had my feet fixed to the robotic manipulator and could therefore use both my arms to push and manipulate the door and the latches. Story was free-floating and had to use one arm to stabilize himself and pull himself towards the door while pushing the door in with the other arm. However, pushing was not enough. He also had to flip a lever over a bolt to get the latch to hold. I could push the door with one hand and flip the lever with the other, but with one hand required for stabilization, Story didn't have enough hands!

This led to a lot of discussion with ground personnel. They gave us several suggested techniques, which we tried. None of these worked. Finally, we came up with the idea of using one of our standard Shuttle tools in a unique way. It was a ratcheting strapping device, similar to what truckers use to hold down heavy objects. We envisioned wrapping it around a knob on the door and another attachment point on the telescope. Story could use one hand to stabilize himself and the other hand to tighten the strap. The ratchet would then keep the door closed, and he could move his hand to flip the lever. We felt confident that the technique would work, but ground personnel were concerned that we might exert too much force with the ratchet and damage the telescope. This led to a lot of discussion, which was finally ended when Milt Heflin, our Flight Director, made a command decision, saying "We trained these guys and sent them up to do the job, and they can see the problem better than we can. They'll be careful with the telescope, so let's let them get on with it." And, sure enough, it worked!

The entire EVA had been extended from its original plan by over an hour, and one or two jobs we had originally planned for the first day would have to be rescheduled, but after slightly over eight hours, we finished cleaning up the payload bay and came back in the airlock, having accomplished our main task for the day. An EVA colleague inside the Shuttle helped us to make an inventory of all the tools we had used, to make sure we did not leave anything in the payload bay or (heaven forbid!) inside the telescope. This was typical of the extreme quality control we tried to impose on ourselves during the entire mission. After each spacewalk there were many things we brought back inside with us: our EVA cameras, to reload film, and all our power tools, to prevent their batteries from becoming too cold. Once inside, we got out of our suits, cleaned them up for storage, and reviewed what had happened during the day's activities that might require changes in the next day's plans. This was important work, but I have to admit that what I was looking forward to more than anything else after a full day in a space suit with nothing to drink but a bag of water and nothing to eat but a fruit bar was a nice, big dinner!

We always knew that our spacewalks would result in long workdays. It takes several hours after getting up in the morning and having breakfast before we finished preparing and checking out our space suits and got into them. Once in the suits, we needed to spend forty minutes breathing pure oxygen before depressurizing the airlock, to prevent nitrogen bubbles from forming in our blood (and thus giving us "the bends"). Our spacewalks themselves were scheduled to last about six and a half hours (although our suit consumables allowed us to stay outside for several hours more. Clearly this turned out to be necessary for our first EVA). Cleanup, dinner, and getting the Shuttle ready for us to sleep took another few hours. If we only had to do one spacewalk on the mission, we could perhaps afford to squeeze this schedule and perhaps skimp on sleep; but on such a long flight, with so many spacewalks, we could not afford to let ourselves become fatigued. Before the end of the first EVA day, however, we had one more task, to roll up the old solar arrays in preparation for their replacement the next day. The first one rolled up just fine, but the second, which we could already see was deformed, could only be partially rolled up. This meant that we would not be able to pack it in the payload bay to bring home. Instead, it would have to be jettisoned and left in orbit.

The next day was Tom and Kathy's first spacewalk, and we had spent considerable time going over the revised procedures they would need to use given the problem with the solar array. As planned, Kathy would ride on the robot arm and attach a handle to the array while Tom, the free floater, would disconnect the electrical and mechanical connectors joining it to the telescope. Holding on to the handle, Kathy was now in complete control of the five-meter (16.5 ft.) long solar array. The handle was near the center of mass of the array, so pushing and pulling to translate it did not induce too much rotation. Claude would now move the arm to place her and the array as far above the Shuttle as the arm could reach. From this point she could release the array into orbit. The secret of handling large objects like this in space is not to get them moving too fast. This is one area where you can get fooled by underwater training. The water's viscosity forces you to push hard to get a large object moving, but objects stop on their own as soon as you stop pushing them. In space, if you use the same amount of force to get an object moving it will end up going much too fast, and in a vacuum there is no viscosity to help stop it! To help prepare us for this, we had all spent several days dressed in space suits pushing large masses around on Johnson Space Center's precision air-bearing floor. This was sort of like a giant air hockey game with frictionless pucks weighing hundreds of pounds.

This training had its limitations, since the objects we moved around were not free to move along all six axes as they are in space. This is typical of training for spaceflight: there is no one simulation that can duplicate all aspects of being in space. We use many techniques, each of which gives us partial training, and then we have to integrate the experiences inside our heads, carrying the total training experience into the flight. Having all done spacewalks before, we had experience with this and could appreciate the limitations of our many excellent simulators and take advantage of what each had to offer.

The procedure worked perfectly. Just after sunrise, Kathy released the array, which continued to float motionless over the payload bay. Then, our pilots fired the Shuttle's maneuvering rockets to move us away from the array. The rocket exhaust hit the array, causing it to start tumbling as it slowly moved away from us. This was a breathtaking sight, with the array's two halves flapping like the wings of a giant, prehistoric flying reptile. We were mesmerized, almost forgetting for a few minutes that we still had a busy spacewalk waiting to be completed. But we soon got back to work. The robotic arm took Kathy over to a carrier in the front of the payload bay, where she and Tom removed one of the new arrays, which Kathy carried over to the empty spot on the telescope. Tom moved hand-over-hand to get back to the telescope foot restraint, ready to help gently fit the array into place and close the primary latch. Now they had to connect the three electrical connectors from the new array to the telescope. This was a critical task, because if the attachment bolt went in cross-threaded and the receptacle socket was damaged, we would not have been able to connect the new solar array, and the telescope would not have enough power to operate. Rather than use a power tool for this task, Kathy used a manual wrench, giving her fingertip control and a good feel for any possible binding as the bolt went in.

Tom dropped out of his foot restraint and was floating with his eyes at the level of the union between the plugs and sockets, so that he could make sure that the connector bracket was coming down evenly. With Tom's eyes and Kathy's fingers, we were confident that we could catch any misalignment before it did harm. We had installed mockups of these brackets many times underwater and had installed the real brackets into high fidelity sockets in clean rooms and in a thermal vacuum chamber, always successfully. This would be the first time the brackets had ever seen the real telescope, however, so we all breathed a sigh of relief once they were successfully attached. Now, Tom and Kathy had to pull back while the telescope was rotated 180 degrees to give access to the other solar array. This array was properly rolled up, so Kathy was able to carry it over to the payload bay carrier after she and Tom removed it from the telescope. After that, they installed the second new array, packed away the old array for return to

Earth, and the task was complete! Following the solar array replacement, Tom and Kathy returned to the airlock, their first spacewalk successfully accomplished. Luckily for us all, this EVA was completed in good time, allowing us to get back on schedule after the time crunch caused by our problems with the recalcitrant door the day before.

Day three was dedicated to fixing Hubble's optics, to allow it to peer accurately into the depths of space. The first step was to remove the original Wide Field/Planetary Camera (WF/PC I) and replace it with a new camera, called in NASA-speak WF/PC II. Corrective optics had been installed inside this unit to correct Hubble's vision. We installed handholds on both WF/PCs, to protect the delicate radiator surface on the instruments, which might release contaminating dust if touched. Avoiding contamination was a critical concern throughout our entire mission, but nowhere was it as important as when we handled the telescope's optics. At issue was not just dirt, but any sort of organic material, since even monolayers of organic molecules can poison the ultraviolet reflectivity of Hubble's mirrors. Extraordinary measures had been taken to ensure cleanliness during instrument manufacture and checkout, and we had to continue to preserve this cleanliness in orbit. Almost everything coming near the telescope had been subjected to vacuum bakeout, even the rubberized material on the boot soles and fingertips of our space suits. To remove the unit, we attached the handhold, removed a grounding strap, undid the electrical connectors and opened the latch that held it in place. With Story in a foot restraint at the side of the new unit to assist, I stood on the end of the manipulator arm and gently slid the old unit out of its enclosure. It was roughly the size and shape of a grand piano, so I had to be extremely careful when moving it around not to bump anything.

Figure 3.2. Jeff Hoffman removes the old gyro unit — the size of a grand piano

(Courtesy of NASA).

When it was completely clear, Story looked inside the enclosure to make sure that no insulation had torn loose. Having to repair torn insulation is an example of the sort of surprise that could have added a lot of time to our EVA task. Next, we reinserted WF/PC I part way, as a rehearsal for the critical installation of the new camera and optics. We had inserted WF/PC II many times in training: under water, on the air-bearing floor, and in our virtual reality simulator. But no Earth-based simulators totally duplicate the reality of spaceflight. Here was a great chance for one final simulation, this time in space!

After this rehearsal, I hung WF/PC I on a temporary restraining bracket out over the port side of the shuttle. Meanwhile, Story opened the WF/PC II protective enclosure, and I moved over and attached the handhold, released the latch and pulled out WF/PC II. Story now moved up to a foot restraint on the telescope, ready for what was perhaps the most critical move of the entire repair mission. WF/PC has a pickoff mirror, about the size of a human hand, which sticks out into the main light path of the telescope and diverts some of the light into the WF/PC. WF/PC II was launched with a protective cover over this pickoff mirror, which had to be removed before the instrument could be installed. I swung WF/PC II up towards Story, held it as tightly as possible, and Story gently removed the mirror cover. As with so many of our activities, this was not intrinsically difficult, but it was absolutely critical. One slight touch on the mirror could push it out of alignment beyond the capability of its tilt motors to correct. We moved extremely slowly and deliberately at this point. We had never had trouble removing the mirror during practice sessions, but this was for real!

Once the mirror cover was removed, Story stowed it safely out of the way. Claude moved the arm to position me and WF/PC II in front of the opening from which we had removed WF/PC I. Story positioned himself so that he could help guide WF/PC II into the enclosure rails as I slowly pushed it in. These rails are what allow us to position the instrument to the sub-millimeter alignment accuracy required for successful operation. Initial rail engagement requires a placement accuracy of only a centimeter, but as the track gradually narrows, the tolerance tightens until the instrument engages in latches, which provide the ultimate accuracy. As soon as WF/PC II was fully inserted, I closed the latch, attached the electrical connectors and ground strap, and removed the handhold. At this point, ground controllers carried out a short aliveness test on WF/PC II. If there had been any problem, we would have re-extracted the instrument, checked the connectors, and tried again. In case of a total instrument failure, we could have reinstalled WF/PC I. Luckily, WF/PC II checked out fine. My next step was thus to remove WF/PC I from its temporary stowage bracket, install it into the protective enclosure, latch it down and close the enclosure door. WF/PC I was now ready for its trip back to Earth, where it would be put on display in the Smithsonian Air and Space Museum. We re-stowed both of the temporary handholds we used during the instrument transfers, and the job was complete. Whew!

We had enough time after the WF/PC replacement to unlatch two replacement magnetometers from the payload bay and carry them all the way up to the top of the telescope to install them on top of the old magnetometers. These original magnetometers were never expected to fail and were not designed for replacement. However, both of them failed, so we had to install the new ones right on top of the old ones. This was a fairly straightforward task, but it was easier with two people, so Story "hitched a ride" with me on the end of our trusty robot arm. The arm was working at the limit of its reach, since the magnetometers are near the front (aperture end) of the telescope. Riding fifteen meters above the payload bay gave us an absolutely spectacular view. Sometimes I had to work hard to ignore the magnificence of the environment I was in and concentrate on the job at hand. While attaching the new magnetometers, I noticed that some paint was peeling off the old ones and reported this to the ground. The concern was that paint chips might float into the telescope and contaminate the optics. Ground controllers started coming up with a plan to deal with this unexpected problem while Story and I returned to the airlock, our EVA once again successfully completed.

The fourth EVA day started with the installation of what we called COSTAR (Corrective Optics Space Telescope Axial Replacement). This was an incredibly clever system that deployed mirrors inside the telescope to intercept the aberrated light beam, correct the focus, and then redirect it into the other instruments. To make room for the COSTAR, Tom and Kathy would have to remove the High Speed Photometer (HSP) instrument, which was the most expendable and least used of the first-generation of HST instruments. After Kathy opened the doors on the HSP enclosure, she disconnected the HSP electrical connectors and opened its latches. She grabbed the handles and, with Tom closely monitoring the motion, slid the HSP along its rails and out of the shroud. HSP, like COSTAR, was the size and shape of an old-fashioned telephone booth (the kind you could get into). Although the geometry was different from WF/PC, the flow of activities in the change-out was similar. This meant that first you stowed the old instrument in a temporary fixture. Then you opened the protective enclosure in the payload bay and extracted the new instrument. The next step was to remove the protective cover over the critical optics and then install the new instrument. Finally, you closed the latches, attached the electrical connectors, closed the doors, and installed the old instrument into the protective enclosure for its trip back to Earth. Meanwhile the ground controllers carried out aliveness checks. The main problem in this procedure was that COSTAR almost completely blocked Kathy's view of the enclosure. There was only about one centimeter clearance for COSTAR to engage with its alignment rails, so Tom had to actually get inside the telescope to ensure that COSTAR was properly positioned.

The Hubble Space Telescope's optics were now, at least in principal, capable of performing as originally planned. The astronomical community let out a collective sigh of relief, as did the whole NASA/HST team, but our tasks in orbit were not yet complete. There was enough time left on the fourth EVA day for Tom and Kathy to install some additional memory into the HST computer. (For the record, this upgraded it from the capability of an old 286 processor to that of a 386! Computers in space have to be radiation hardened, which takes years of development, so computers in space are usually at least a generation or so behind those on the ground.) Tom also cut some thermal insulation from one of the instrument carriers inside the payload bay. Ground personnel had come up with a plan for us to fashion from this insulation a cover for the magnetometers to prevent paint chips from floating away. This would be a job for Claude and Ken that evening.

Some of the tasks we planned to do on our fifth day were not particularly designed with EVA repairs in mind. Thus they were quite tricky. While most of the HST is extremely EVA friendly, budget constraints forced some equipment to be built along more traditional lines. For example, our first job on the fifth EVA day, replacing one of the solar array drive electronics (SADE) modules, was the least EVA-friendly of all our tasks. "EVA friendly" electrical connectors have large wing tabs for us to hold, but the seven connectors on the back of the SADE were just like the long, flat connectors on the back of older generations of computers, whose removal requires unscrewing several tiny, two-millimeter screws. The workspace was tightly confined, making this an awkward task to perform wearing spacesuit gloves. As with all our other tasks, we had practiced this many times successfully under water. However, the water tended to damp out motion of the connector cables, and gravity held the screws in place in the connectors after we had unscrewed them. In orbit, the cables bounced around, making the screws spin around inside the threads. Half the screws spun clockwise, staying put, but half rotated counter-clockwise and gradually worked their way out of their holes, a behavior we had not anticipated. We soon started to see little screws floating around in front of us.

Fortunately we carried trash bags on the tool caddy mounted on our chests, so Story started grabbing the screws as they floated by and stuffed them into the trash bag. Eventually, however, there were so many screws floating around inside the bag that when he opened it to insert another one, screws started to float out. The trash bags have since been redesigned, but that didn't help us on that day. Perched up against the telescope, we had little room to maneuver. In an attempt to capture a screw floating out of the bag, Story accidentally bumped it, and it started floating down towards the payload bay. It's not a good idea to have floating debris in the bay, since it might jam up some critical mechanism. Story was standing on the arm that day, and I was free floating, so I had more mobility, and I lunged for the screw, just missing it. I heard Claude on the radio telling me to hold on, and he would drive the arm so that I could get the screw. So I held onto the arm with one hand and strained with my other to grab the screw as the arm started moving. I felt like I was a child again, riding a merry-go-round, reaching out for the brass ring! Unfortunately, the maximum speed of the arm was the same as the speed of the screw, so it remained tantalizingly just out of my reach. It turns out that the maximum speed of the arm is greater when it is moving free than when it is carrying a load, so Ken, the backup arm operator, floated over to the computer and changed a critical parameter. This tricked the arm into thinking that it was not carrying any load. Immediately it speeded up, and I was able to grab the wayward screw. The whole episode became known as "the great screw chase" and was a fine demonstration of how good training allowed us all to work as a team and rapidly improvise a solution to a problem.

The rest of the EVA seemed tame by comparison, but we accomplished several tasks, including installing a repair kit on the Goddard High Resolution Spectrograph in order to provide a redundant power path in case of future failures. When our tasks were complete, it was time to start preparing the telescope to be put back into orbit the next day. The solar arrays had to be lowered and extended. The motor on the first one for some reason did not work properly, and Story had to push the array into place by hand, another example of EVA crewmembers backing up automated systems. The ground personnel asked us to remain outside until all the telescope systems were deployed.

Figure 3.3. Mission accomplished: time to consider the awe of it all (Courtesy of NASA).

There were no more problems, but this gave me some precious extra time to float lazily above the payload bay and watch the Earth go by. We were now finished with all twelve of our original repair tasks plus an extra one to put our improvised covers on the old magnetometers! The only thing left to do was to remove the protective antenna cover from the HST, put away all our tools, return the payload bay to its reentry configuration, and come back inside. We were a happy crew when Story and I floated out of the airlock into the Shuttle Middeck for the final time.

The next day, Claude grappled the HST with the arm, Kathy commanded the latches and umbilicals to release, Claude lifted the HST out of the payload bay, and as soon as the controllers gave the word that the telescope was "go for release" Claude commanded the arm to let go. Dick and Ken eased the Shuttle away from the HST, taking care that our maneuvering rockets did not contaminate the telescope, and we bade farewell to this magnificent astronomical instrument that had been the center of our lives for the past week. Hubble drifted away towards the West, getting a little farther away from us on every orbit. As the Sun rose in the East every ninety-five minutes, we could see Hubble lit up towards our West, a magnificent "morning star" visible for the rest of our flight right up to our final deorbit burn a couple of days later.

We had successfully completed the most complex spacewalking mission in NASA's history, but we would not have proof that Hubble had been restored to its full operating potential until scientists at the Space Telescope Science Institute turned on the new instruments and took some pictures. A couple of weeks were needed while engineers waited for atmospheric gases to escape from sensitive electronics. Sparks could damage the sensitive equipment we had installed if high voltage was turned on too soon. I well remember getting a phone call during the wee small hours of 1 January 1994. An astronomer friend working at the Institute asked me if we had any champagne left over from New Year's Eve. I told him that I still had half a bottle in the refrigerator, and he said, "Well open it and drink a toast, because we just got the first pictures back, and Hubble works!"

The rest, as they say, is history. I will never forget the standing ovation given by the American Astronomical Society after a presentation on our mission. To me this represented one of the most gratifying parts of the entire Hubble enterprise: uniting the unmanned world of space astronomy with the world of human spaceflight. Since our initial rescue mission, astronauts have revisited Hubble several times. They have continued the tradition we started of making repairs to HST systems that were not originally designed for on-orbit servicing, showing over and over the adaptability of humans to deal with unanticipated problems in complex systems. What is even more important about Hubble servicing is that not only did each crew fix the existing problems with HST equipment, but also they installed new, state-of-the-art detectors into Hubble's focal plane. Essentially, every time a crew visited Hubble, they left it as a new, up-to-date telescope, far more sensitive than its predecessor. Hubble has become one of NASA's most popular missions. Its discoveries have rewritten astronomy textbooks, and many of its images have become public icons. I have had many extraordinary experiences during my five space flights, but fixing Hubble surely had the most lasting long-term significance, and, every time I look at one of the magnificent images of the Universe taken by Hubble, I still get a warm feeling thinking back on how our crew played a pivotal role in making that possible.

Space Story Three:

The Phoenix Mission

Eric Choi and His Friends From the Phoenix Mission

In the field of space applications and exploration, there are many instances where astronauts and prominent scientists capture headlines and stir the imagination in a very public way. In the nitty-gritty world of major space programs, such as those associated with constructing the International Space Station, launching a new commercial satellite or developing an exploratory satellite, there are literally hundreds to even thousands of "space heroes" that operate "under the radar". These dedicated people without a great fanfare bring complex and extremely technically demanding space programs to life. The highly successful Phoenix exploratory program that has recently verified the existence of water on Mars is one such program. This particular undertaking is commendable for many reasons. First, it was a very difficult and complex program that took many years to plan and execute. Yet, this device has accomplished its many tasks flawlessly, including a cushioned yet "hard landing" on the surface of Mars. Second, it gave us incredible new information about the physical and chemical make up of the Martian terrain. Third, it represents a very compelling example of effective international planning and cooperation that allowed us to draw on scientific talents from seven countries, namely the United States, Canada, Finland, Denmark, Germany, Switzerland and the United Kingdom.

In this "story" we explore not only the technical and scientific success of the Phoenix mission, but also examine the subtle fabric of international cooperation in the world of space. A network of international space engineers and scientists that helped to develop and execute the Phoenix Mars program had previously learned about the importance of international and interdisciplinary cooperation from their participation in International Space University programs years earlier. Thus in this story we learn about the interlinked careers of three ISU graduates from Canada, the United States and Finland, Isabelle Tremblay, Rob Grover and Jouni Polkko, as they collaborated on Phoenix.

In the mythology of many cultures, the phoenix was a fabulous bird that was consumed in fire and later reborn from its ashes. It was an appropriate name for the Phoenix mission to Mars, because this program would reuse the 2001 Mars Surveyor Lander that had been grounded following the loss of the Mars Polar Lander in 1999.

Figure 3.4. The Phoenix Mars Lander (Courtesy of NASA/JPL).

The reborn Phoenix has turned out to be a remarkably versatile exploratory tool equipped with an impressive suite of scientific instruments. This amazingly capable device that survived a "crash landing" on Mars was an autonomous robotic explorer designed to discover the secrets of the uncharted northern polar regions of Mars. It was particularly optimized to dig within the Martian tundra-like soil in search of water and the chemicals of life. This spacecraft contained a unique payload that included a robotic arm, a stereo camera, a mass spectrometer, a wet chemistry laboratory and an oven to heat recovered samples, optical and atomic force microscopes, a conductivity probe, and a meteorology station. Led by the United States, the Phoenix mission included instruments and science team contributions from Canada, Finland, Denmark, Germany, Switzerland and the United Kingdom.

Three scientists and engineers, all graduates of the International Space University, relate their various forms of participation in this complex international program. First we hear from Isabelle Tremblay, of the Canadian Space Agency (CSA), who served as Senior Design Engineer. Next we hear from Rob Grover, of the Jet Propulsion Laboratory (JPL), who served as one of the lead mission designers, and finally we learn about the participation of Jouni Polkko, of the Finnish Meteorological Institute. He was the scientist who designed the Phoenix's atmospheric pressure sensor. Each of these ISU alumni has, in his or her own way, served to make the Phoenix program a remarkable success.

The Phoenix Story of Isabelle Tremblay

Just four years after the go-ahead, the Phoenix spacecraft was launched atop a Delta II rocket that would send this unusual mechanical explorer to Mars. Isabelle Tremblay (an alumna of the 1998 ISU Summer Session Program and shown in Figure 3.5) not only played a crucial role in the design and development of the meteorology station but, as an accomplished artist, she also prepared the official logo of the Phoenix mission. She says: "Creating a logo was a perfect opportunity to express, and share, how I felt about the mission. My design juxtaposes the search for water on Mars with the phoenix mythology; in the background is an image of Mars prominently showing the northern polar region where the Phoenix mission was to land."

Figure 3.5. Isabelle Tremblay (SSP 1998) and the Phoenix mission logo

(Courtesy of CSA/NASA/University of Arizona).

"At 09:26 UTC on 4 August 2007, the Delta II rocket carrying Phoenix soared into the heavens on a column of fire. The rising Sun was just nearing the horizon, and its emerging light illuminated the rocket's exhaust plume against the still dark sky." As Isabelle and the other spectators watched, the winds began to twist the plume into a shape resembling wings and a long tail, looking very much like the majestic bird in her logo. "It was a wonderful omen – the Phoenix had arisen."

The Phoenix Story of Rob Grover

Rob Grover, who participated in the ISU Summer Session Program of 1997 and shown in Figure 3.6, now works at the JPL on NASA planetary programs. After a nine month, 679-million kilometer voyage, Phoenix arrived at Mars and began the perilous entry, descent and landing phase of the mission; Rob monitored the landing from his console at JPL. Rob knew the spacecraft particularly well since his first job after graduating from the University of Washington was working on the mothballed 2001 Mars Surveyor Lander that would be reborn as Phoenix. His next assignment was to work as an attitude control engineer on the Mars Odyssey orbiter, and later he had the chance to be a systems engineer on the Spirit and Opportunity rovers. For the Phoenix mission, he was one of the lead engineers - you might say that, after being involved in all these programs, Rob has Mars in his blood.

Figure 3.6. Rob Grover with a Mars Exploration Rover engineering model

(Courtesy of NASA/JPL).

Over five years of work by hundreds of engineers and scientists from seven nations came to fruition with Phoenix. This number does not include the many others who were involved in building the Delta II launcher, or those who were involved with the tracking, telemetry, command and control operations.

When Rob and the operations team at JPL received confirmation of a successful landing at 23:53 UTC on 25 May 2008 everyone shouted with joy. The successful "crash landing" was a critical milestone in the mission's overall plan. Phoenix had arrived intact and still functional at its landing site in the unexplored northern plains of Vastitas Borealis. Rob explained: "We had rehearsed and gone over it for years - it was terrific that everything could go so perfectly. At the moment of touchdown there was a lot of joy and excitement, but some caution as well. You always think of the little things that could still go wrong, but when we got confirmation that everything was fully deployed and operational it was truly an exciting moment."

The first picture from Phoenix was received later that evening. "It was spectacular. It was at that point that the mission became very real. In the engineering world, a lot of the mission exists only as numbers and models. But once the first pictures came down there was the excitement of seeing a new place on Mars and discovering what it looked like. My spacecraft was sitting on another planet - there was my baby."

Following the successful landing, control of the Phoenix mission was transferred to the Science Operations Center at the University of Arizona. From there, the international science teams prepared and uplinked commands to their various instruments aboard Phoenix, and then received and analyzed the downlinked data. The opportunity to resurrect the Mars Lander program, convert it into a more capable facility and achieve success with the new Phoenix spacecraft was a once in a lifetime opportunity.

The Phoenix Story of Jouni Polkko.

Jouni Polkko (SSP 1991, see Figure 3.7) of the Finnish Meteorological Institute (FMI) was one of the scientists at the Science Operations Center at the University of Arizona. Jouni's institute designed and provided the pressure sensor element of the Phoenix meteorological station. Jouni worked on this project himself from 2004. The Phoenix pressure sensor was actually the fourth that FMI had built for Mars missions. Three earlier units were lost due to the failures of the Mars 96, Mars Polar Lander and Beagle 2 missions. In January 2005, a similar sensor device recorded the pressure profile of Titan's atmosphere during the descent of the Huygens probe.

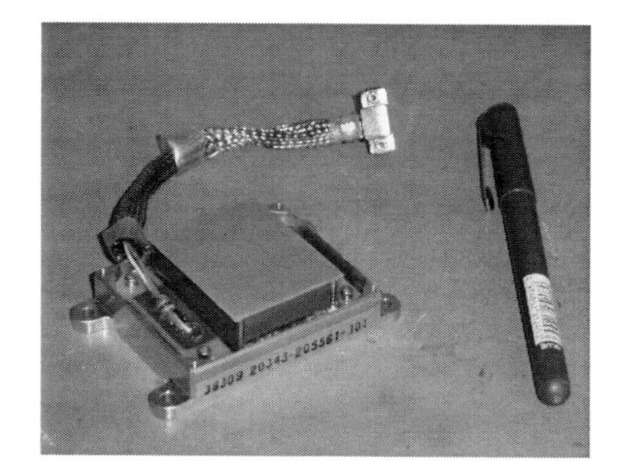

Figure 3.7. Jouni Polkko and the Phoenix atmospheric pressure sensor

(Courtesy of the Finnish Meteorological Institute (FMI)).

"The first scientific data arrived in the afternoon of 26 May," recalls Jouni. "Our instrument worked perfectly, measuring a pressure on Mars at the landing site of 8.55 hPA. In the following days, the first dust devils were observed by the pressure instrument." Isabelle was with Jouni in Arizona, providing engineering support for the operation of the meteorology station. She remembers the experience with great fondness: "The community and team spirit were so strong. It was great to see people from all over the world working together for common goals. I wish the world would be more like that in other areas."

Originally designed for a surface mission of 90 Martian sols (92 Earth days), Phoenix far exceeded its planned lifetime, returning data for months before succumbing to the onset of Martian winter in late 2008. Its legacy is the wealth of scientific data returned, and the path it has blazed for further robotic and eventual human missions to Mars. For all the ISU alumni who were privileged to be involved, it was the journey of a lifetime.

Acknowledgement

The author of these stories, Eric Choi, gratefully acknowledges the assistance of Dr Leslie K. Tamppari, Phoenix project scientist of the NASA Jet Propulsion Laboratory, during the preparation of this article.

Space Story Four:

The Rise of Space Tourism—Up Close and Personal

Emeline Paat-Dahlstrom and Eric Dahlstrom

When SpaceShipOne (SS1) touched down on the dusty Mojave desert on 4 October 2004 it was greeted by thousands of people lining and cheering the runway. This remarkable vehicle had just broken the X-15 high altitude flight record held since 1963. It became the first fully reusable spacecraft in history by flying into "outer space" and then doing so again a second time within days of the initial launch. This commercial spaceship had also broken the invisible "wall", that space was the sole domain of governments and their contractors. A handmade sign proclaimed this new era as follows: "SpaceShip One - NASA Zero."

A new era had finally begun where dreams of going into space seem to be much more possible. It is a new age where a handful of young and brilliant engineers can develop a reusable rocket and fly multiple flights. Perhaps most impressively they will do so at a cost of less than a tiny fraction of a single launch of the Space Shuttle. In a year when NASA was still reeling from Columbia's accident, which effectively grounded its whole Shuttle fleet, the only other human spaceflights launched in 2004 were from two Soyuz vehicles. Three of the five human spaceflight launches in 2008 were commercial – thanks to SS1. It gave the average person the hope to dream and indeed to begin to believe that space tourism may not be too far in the future after all. Suddenly the idea began to spread that this generation may have the chance to see spaceflight become a mass-market commodity.

Burt Rutan, world famous master designer and aerospace engineer, started developing SS1 back in 2001. His company Scaled Composites was financed by Microsoft co-founder billionaire Paul Allen. He had designed and built dozens of innovative airplanes, including the Voyager that flew around the world nonstop. But he had never taken on a spacecraft project before. Scaled Composites at the time was just one of twenty-six spacecraft teams that were competing for the Ansari X Prize. This was a $10 million purse conceived and engineered by Peter Diamandis, one of the founders of the International Space University. The prize was to be awarded to the first of the registered teams to build a successful space plane that could fly into space (i.e. to an altitude of over 100 kilometers) and then repeat the feat within eight days.

The X-Prize competition was co-sponsored by the Ansari family, First USA Bank and partly financed through a clever "hole-in-one" insurance deal. Peter Diamandis' initial purpose, of course, was much more than to have an exciting competition. His aim was to help accelerate development of the human spaceflight industry through the commercial competitive process. Teams registered from around the world using both off-the-shelf and innovative technologies with different launch and landing configurations. A number of developers like Rocketplane and Armadillo Aerospace in the U.S. and the Da Vinci Project in Canada gave Burt Rutan a run for his money. Eventually, their lack of financing and the tight schedules took them out of the running. But, as Peter Diamandis had predicted, the competition ignited the birth of the suborbital space industry and encouraged hundreds of people to put down deposits for a chance to fly in space - hardware sight unseen. Now, the regular X Cup event held in Las Cruces, New Mexico, showcases every known commercial space vehicle developer around the globe and is attended by thousands of space enthusiasts, further affirming that the industry is about to go mainstream.

Brian Binnie, the pilot of SS1, successfully flew a flawless flight that day - giving confidence to SS1's investors and would-be passengers that the design and technology could be safe enough to carry commercial passengers. He carried enough ballast for two more passengers inside the spacecraft, a requirement to win the X Prize. The implications of those two empty seats played out as multi-billionaire Sir Richard Branson took a gamble on the potential of the market and announced that his newly formed company Virgin Galactic would order five more second generation, bigger capacity spaceships from Scaled Composites for commercial use. Not shy to market his brand, Virgin's logo took center stage at Branson's command. Thus the logo was prominently placed right in the middle of both of SS1's tail booms for the world media to see. The event's webcast broke records, sandwiched between Victoria Secret's fashion show and Britney Spear's concert.

The event was fully licensed as an "experimental launch" by the U.S. Federal Aviation Administration (FAA) and the office of Commercial Space Transportation, under the leadership of Patricia Grace Smith, awarding civilian astronaut wings to Brian Binnie and pilot Mike Melvill who flew the two preceding SS1 missions. Later that year, the U.S. Congress passed the Commercial Space Launch Amendment Act, effectively giving the green light to commercial space tourism by laying out the mechanism to fly private citizens at their own risk. The terms also restricted FAA regulations to third party liability and public interest for eight years, giving the equivalent leeway to vehicle developers and tour operators during the "barn storming" years of general aviation. Congress is expected to consider during 2009 whether it will change the regulatory structure for space tourism safety regulation in the post-2012 period.

The success of the X Prize launches got the small town of Mojave, California, and its airport a spot on the world map. Now lauded as the first truly commercial spaceport, other aspiring ports were determined to follow suit. In 2007, Spaceport America in New Mexico, backed by Virgin Galactic, unveiled its plans for an ultramodern spaceport facility and entertainment center – the future homeport for Virgin's commercial flights. Other current and future airports like Kennedy Space Center, Oklahoma and Wallops in the U.S., and Dubai, Singapore, and Kiruna abroad were determined to follow suit. These spaceports are essential elements to the growth of the industry and to the future possibility of point-to-point suborbital flights, extending spaceflights for pleasure into the realm of rapid business transit and package delivery worldwide.

Now as Scaled Composites continues to build its second generation mother ship White Knight Two and SpaceShipTwo, Virgin Galactic continues to sign up would-be passengers at U.S.$200,000 a ticket. It has set up a network of informed space agents recruiting space cadets around the world with deep pockets. Celebrities, pilots, adventurers, and millionaire space enthusiasts are high on the list. Training of the first 100 founding flyers and agents has commenced at the U.S. NASTAR Training Center, and interior spaceship and spacesuit designs are in progress. Nor is Virgin Galactic alone; others are signing up potential space tourists as well. On a variation on the theme Jeff Greason of XCOR has promised to develop a two-person jet that would fly to an altitude of 37 miles (or 60 kilometers). That would take a pilot and a co-pilot passenger on a ride into the "dark sky", or high enough to see the curvature of the Earth. This would not be a ride into space (since outer space is defined as beginning at an altitude of 100 kilometers), but the offering would be at a "bargain basement" fare of $100,000. It might even be offered $99,995 for the thrifty minded thrill seeker. Indeed the true, budget-constrained, would-be pseudo-space tourist can sign up through Space Adventures for a ride on a supersonic Russian Foxbat for under $20,000. For years we have worked in this incipient "industry" and remain truly optimistic that space for the masses is now a reasonable prospect in the not too distant future.

In the same year as the Ansari X Prize was won, Robert Bigelow, Budget Suite's multi-billionaire and CEO of Bigelow Aerospace, announced the $50 million America's Challenge Prize. Bigelow ups the ante by soliciting competition for a space vehicle that could transport seven passengers to an orbital station by 2010. This would provide the transport mechanism for his fleet of inflatable space modules, dubbed "space hotels", which he has been progressively building for a human-rated private space station. At the moment, PayPal billionaire Elon Musk's Space X is the only private company potentially capable of building such a transportation system to fit Bigelow's needs. Elon Musk is working on his Falcon series of launch vehicles intending to service the future orbital business with an ultimate goal of sending humans to Mars and beyond. And then, of course, there is the ever-reliable Russian Soyuz vehicle that, since 2001, has been sending private citizens to the International Space Station via the U.S. company Space Adventures. But the unpredictable and potentially volatile political climate in Russia and the apparent cost increase per ticket since it started could eliminate this opportunity at any time. Without a transportation system, the space hotel business has no future.

So, as space tourism continue its journey from science fiction to reality, the essential pieces for making spaceflight for the masses possible begin to fall into place. But we are still a long way from evolving into the grand vision of a space faring species. The industry needs more mainstream investors to fund the research and development of these transportation systems. It needs more young and enthusiastic engineers and designers to come up with innovative solutions or use existing technologies to build the vehicles. It needs the support of policy makers and legislators who will create the laws and regulations in a timely manner in conjunction with the industry's timeline and schedule. It needs entrepreneurs to grow the business and marketers to sell the spaceflight programs to the general public. Each component is interrelated and needs each other's support to thrive. There is still a lot to be done, but the future belongs to those who have the vision, the resources, and the drive to make it happen. As Peter Diamandis says: "The best way to predict the future is to create it yourself."

Space Story Five:

The Visionary Space Billionaires

Joseph N. Pelton

One of the most special clubs in the world is the Commercial Spaceflight Federation-formerly the Private Spaceflight Federation. This group, now some four years old, promotes the new personal commercial space industry. It's a rather special club with a rather special mission. Its typically super wealthy members are the people who believe that space commercialization can really happen. This is because they are also the people who are actually in the process of "making it so". These new "space billionaires" grew up watching Star Trek. They actually believe that some day humans can explore other worlds and escape the bonds of Earth's gravity well. This is a short story that augments the Dahlstroms' recollections about the birth of the space tourism industry, and profiles some of the key people who are trying to change the world.

This time the story is told with the focus, not on space scientists and engineers, but on the "world changing billionaires" who know in their very being that yesterday's constraints do not represent the boundaries of tomorrow. They are people who, like Peter Diamandis, believe that if you want to see a new reality you start by doing something about it yourself.

When I, along with Jeff Hoffman and some colleagues from the Aerospace Corporation, participated in 2008 on a study on commercial space safety for the FAA as commissioned by Congress, I got to know several of the key participants and their top executives a little. Then, when I wrote a book entitled License to Orbit: The Future of Commercial Space Travel, I got to meet a few more of this remarkable group of billionaire world changers and their very bright minions. Here are a few of my impressions.

First of all: "What makes many of the members of this exclusive club so special?" For one thing, they have already conquered other worlds and are looking for a fresh and tougher fight.

Who, exactly, are these people? The current batch of space billionaires includes Sir Richard Branson who heads the Virgin empire. He owns his own island, his own airline, his own entertainment company and much, much more. In his usual, rather brash way of doing things, he calculates that the best way to get into space - other than by riding a balloon into the stratosphere – something which he has done, with some very scary results

- is to start his own space tourism business and fly there himself. He also plans to take his family along too - including his ninety-year old parents. Others of this strange new breed of billionaire spacefarers include Elon Musk, who made his first bundle by starting PayPal and now is not only building rockets but is also Chairman of his own all-electric sports car company as well. Then there is Paul Allen, who co-founded Microsoft along with Bill Gates, and we must not leave out John Carmack, who designed video shooter games like Doom and Quake to make his own multi-millions. Then, as the Dahlstroms have mentioned, there is Las Vegas-based Robert Bigelow, who owns Budget Suite Hotels and wants to open a new chain of habitats in space. We also mustn't forget Jeff Bezos, who created Amazon.com and who now has his own private spaceport in the most remote parts of Texas, not too far from the border with New Mexico. These guys - not to overstate the case - are real risk takers. Not only do they have Type A+ personalities but they also have enough money to have super triple platinum credit cards that can buy into advanced space technology in a serious way.

Perhaps the most famous, headline grabbing, financially committed of the visionary space billionaire gang is Sir Richard Branson. Having made his estimated $6 to $7 billion dollars in music, transportation, and other ventures, he has now moved into "dream fulfillment mode". His Virgin Galactic company will begin flying SpaceShipTwo vehicles on sub-orbital flights via Virgin Galactic in late 2010. Some people ask: will Virgin Galactic flights be safe? Sir Richard's answer is that he will be going up with his family on the first flight, along with his 93 year-old Dad and 90 year-old Mom. Richard Branson is clearly not one to do things by halves!

Jeff Bezos is another guy in the club who is not averse to taking some big risks. He founded Amazon.com on a dream and the thought that an Internet-based world could redefine retailing. His enterprise survived the "dot com bubble" when it burst and, today, he is willing to take an even bigger gamble with his Blue Origin space tourism business that flies its experimental New Shepherd launcher out of a remote West Texas location. He figures that, if he can beat the odds in Internet retailing, he can do it again in the commercial space business.

Elon Musk made over a billion dollars by the time he was reaching "cruising altitude" in his thirties, when he sold his PayPal enterprise. Rather than taking the cash to live on a tropical isle for the rest of his life, he challenged himself to make money out of the spaceflight industry. He said: "Oh yes, maybe I will also show conventional aerospace companies and NASA what they are doing wrong along the way". Thus he founded Space X (full name—Space Exploration Technologies). He started to design and build his Falcon class launch vehicles with just a few dozen employees. He rather quickly convinced the U.S. Air Force to buy one of his cut-rate launchers (for under $10 million). Then not too much later he landed a NASA R& D contract, valued at over $200 million, to build a vehicle that could shuttle cargo back and forth to the International Space Station on a commercial basis. After four tries he has finally managed to get it up - I mean, the Falcon 1, of course. We can only imagine what Elon Musk plans to do in his forties. Perhaps he will start a "time travel" business.

When Elon testified before the U.S. Congress, this is what he had to say:

> *"The past few decades have been a dark age for development of a new human space transportation system. One multi-billion dollar Government program after another has failed...*
> *... When America landed on the Moon, I believe we made a promise and gave people a dream. It seemed then that... someone who was not a billionaire, not an astronaut made of "The Right Stuff", but just a normal person, might one day see Earth from space. That dream is nothing but broken disappointment today. If we do not now take action different from the past, it will remain that way." (Elon Musk, giving evidence to the U.S. Senate Hearing on Space Shuttle and the Future of Space Launch Vehicles, in April 2004.)*

Paul Allen is the old guy in the space billionaires' club. He is the fellow who made his fortune way back in the 1980s and early 1990s by co-founding Microsoft. He has since bankrolled professional sports teams, a science fiction museum and given boatloads of money to charity. He still has his entrepreneurial juices flowing, and believes in the future of commercial space as well. He was the fellow that bankrolled Burt Rutan of Scaled Composites when he built the White Knight carrier aircraft and SpaceShipOne that successfully claimed the X-Prize in 2004.

After Paul Allen claimed the X-Prize, he formed a partnership with Burt Rutan and Richard Branson to start the SpaceShip Corporation. So far this enterprise is building spaceplanes for Virgin Galactic and Jeff Bezos's own Blue Origin's space tourism venture. The business plan for the Space-

Ship Corporation is to build dozens of the new SpaceShipTwo craft. At least this is what Burt Rutan says is the plan.

And let's not forget the incredible vision of Robert Bigelow, the hotel czar, who had the "crazy" idea that people would someday like to travel to hotels in space - not just to show in sci-fi movies, but for real. Budget Suites' magnate Robert Bigelow has gone after this goal with the only speed he knows - full steam ahead. Bigelow Aerospace launched a 33%-scale model prototype of his inflatable Genesis I habitat on 12 July 2006. The final version of this habitat is designed to be some 27 times larger in volume (i.e. 3x3x3). The second flight of Genesis II took place in June 2007; it, like Genesis 1, was launched on a Russian Dnepr rocket. And Robert Bigelow, who does not do things in a small way, is offering a $50 million prize for a commercial space transportation company that can demonstrate safe transport to, and BACK DOWN FROM, his space habitats. A successful return of his hotel guests to Earth seems to be a key part of his business plan.

The deployment of the first Bigelow Aerospace habitable commercial space structure, tentatively called the CSS (Commercial Space Station) Skywalker, is targeted for 2010. He will sell commercial experimental space on this structure that will have a larger volume than the International Space Station to pay the initial investment costs.

Finally, we come to John Carmack. He, like Elon Musk, became a billionaire while still in his thirties. He also has the space bug. His Armadillo Aerospace company has already designed a number of launch vehicles and computer-optimized craft that have shown amazing versatility as a "lander" vehicle for the Moon or Mars.

If anyone tries to look at this disparate group of space billionaires and find some common ground it is far from being immediately clear. They vary in age, height, weight, handsomeness and ancestry. Despite their differences and the diverse paths they have followed to arrive at this new commercial space industry, their aims are common. This new breed of space leaders represents the future. Their commitment to finding "new ways forward" is vital to a new surge in space innovation. We can only hope they can indeed deliver new ways for humanity to travel "to *The Farthest Shore*", or at least to low Earth orbit.

Chapter Four

The Future of Space[1]

"Earth has an astronomical future as a habitable planet for probably some 6 billion years, enough time, indeed for a new race of thinking creatures to evolve once again from blue-green algae if all advanced terrestrial life forms become extinct. But the achievements that a continuously evolving intelligent species might make in 6 billion years are unimaginable. A species might need only 50 million years to colonize the whole Galaxy...."

— Eric Burgess, Science Editor, Christian Science Monitor

What we can Learn from History

The most common approach that people take in projecting the future is to assume that the past is the prologue. The book Megamistakes suggests that extrapolations that address complex social situations, changing cultural practices, or situations where dramatic new innovations are introduced may prove to be well off base in predicting the future. You would never use extrapolations to predict the next day on the stock market, the demand for a new consumer product, or the success of a new theatrical production. You would certainly be unwise to use trend analysis to project the future of totally new space systems or of exploratory probes to previously unknown regions. Scientist, author and futurist Sir Arthur C. Clarke (ISU Chancellor for 17 years) often pointed out how easy it is to overestimate the future in the short term and underestimate the future in the long term. One of Sir Arthur's "Three Laws" states that, "Any sufficiently advanced technology is indistinguishable from magic." The long-term effects of technological wildcards can be so unpredictable and so profound that even the most vivid imaginations will fail to anticipate the stunning realities of the future.

To discuss the future of space requires imagination and an understanding of the entrepreneurial spirit of the inventive mind. No major industry in the history of the modern technological world has experienced steady, consistent and long-term growth. Spurts of growth are followed by periods of retreat and retrenchment, and then new opportunities emerge. As human society expands and evolves across the planet with unprecedented speed, social, political and technological variables have become more dominant and unpredictable. The only constant over the last century has been change.

Some of the most successful and well-capitalized companies in the world, such as Intel, Google, Microsoft and Orange Telecommunications, did not exist 25 years ago. Some of the most successful corporations of twenty-five years ago no longer exist, or exist only as subsidiaries of others. What is true of the broader commercial world is true for space as well.

From the time of Sputnik in October 1957 through the Apollo 11 landing of astronauts on the Moon in 1969 there was enormous progress in space programs. This was a time that saw not only the development of rockets that could carry astronauts safely to the Moon, but also successful missions to other planets, the development of sophisticated communications satellite technology such as the Applications Technology Satellite 6 (ATS-6), and a host of space firsts. However, following the cold-war inspired space race of the 1960s, the last 40 years has seen a slowing of progress by the various space agencies and the shift of many functions to commercial organizations.

Space Entrepreneurs

As mentioned in the previous chapter there has recently been a great surge in initiative within the commercial spaceflight industries. Suddenly there are plans to fly early participants into space by organizations such as Sir Richard Branson's Virgin Galactic, Jeff Greason's XCOR, and Jeff Bezos's Blue Horizons. There are over twenty different ventures that are seeking to enter the space tourism business.

Robert Bigelow, the billionaire owner of Budget Suites, has successfully launched inflatable Genesis I (see Figure 4.1) and Genesis II space habitats into space, with plans to offer space tourists the opportunity for a stay in outer space aboard a private space station. To this end he has established a major design competition for the commercial development of reliable transport to, and from, low Earth orbit by commercial operators.

The sudden arrival of commercial players into the space arena is often called the "Personal Spaceflight Revolution", or "NewSpace". Many of these new ventures sprang from the Ansari X PRIZE. The winning vehicle, SpaceShipOne, designed by aviation pioneer Burt Rutan and funded by Microsoft Founder Paul Allen, achieved this remarkable feat in 2004. This unique vehicle now hangs in the National Air and Space Museum in Washington, D.C., next to the Spirit of St. Louis in which Charles Lindbergh was the first to fly, solo, and non-stop, across the Atlantic Ocean in May 1927. Building on the success of the Ansari X PRIZE, the X PRIZE Foundation has partnered with Google to offer the US$30M Google Lunar X PRIZE for the first private team to land and operate a robotic spacecraft on the surface of the Moon. The prize is being pursued by a growing number of teams from around the world, including the ISU-bred company Odyssey Moon Limited. Commercial ventures that range from the NASA funded Commercial Orbital Transportation Systems (COTS) program to a burgeoning number of spaceports in the United States and around the world seem to be fueling a new awareness of the potential of space and new innovative thought about what the future might hold.

Figure 4.1. Image of Genesis I in low Earth orbit (Courtesy of Bigelow Aerospace).

The future of space, like many other enterprises, depends on innovation, entrepreneurship, capital, public support and opportunity. It is no accident that innovation and growth comes in spurts. The small percentage of inventive and adventurous minds that have led humanity forward should be more available today than ever before. Ninety percent of all humans that have ever lived will be alive in some part of the 21st century. This will be true if some self-inflicted or external catastrophe doesn't destroy us and if projections of a human population of between 10 and 12 billion people by the end of the century prove to be accurate. Technology continues to deliver information and education to more and more people; there will also be more, and better, schools and universities than ever before. There are enough innovative minds, but is there a sufficient political will? Are there enough economic incentives and entrepreneurs with a sense of adventure to pursue new opportunities in space? Motivation is the key to success in space.

Archytus of ancient Greece discovered the concept of reaction jet energy in the Fourth Century B.C. He used steam to create jet propulsion and he devised a clever wooden pigeon that flew around on a string. It was a brilliant invention, but the idea was converted into a toy and a mere conversation piece. In the seventeenth century Sir Isaac Newton provided the world with the basic concepts associated with the Laws of Motion to show, with careful illustrations, how a projectile fired from a high mountain at a sufficiently high speed could go into Earth orbit. Creative minds such as that of Edward Everett Hale in his book The Brick Moon (1869) suggested the plausibility of launching a satellite into polar orbit and using it for navigation, Earth observation and communications. Thus over the years inventive minds conceived of ways to use rocket propulsion, to understand the laws of gravity, and to put outer space to practical use. Despite these admirable intellectual achievements it was not until there was a political will to go into outer space, plus the capital and entrepreneurial skills to make it happen, that space travel ultimately occurred.

At the outset of the Space Age, the drive to launch things off our planet's surface was linked to military objectives, and closely tied to missile launch capability. Also, because it was understood that the research and engineering to design and build rockets would be quite expensive, these efforts were essentially left to governments—either civilian space or defense agencies. The major accomplishments in space, as represented by the U.S.S.R.'s Sputnik launches and the United States Mercury, Gemini and Apollo programs in the early days, were all publicly financed. More recently the space programs of China, India, Canada, Europe, Japan, and a growing number of other countries, as well as the US$100 billion International Space Station (ISS), all were derived from public treasuries; they were implemented through public space agencies and aerospace contractors working under their supervision. These activities are what might be called "big" space programs that have been accomplished by governments and giant aerospace industries. These organizations were, and still are, staffed by thousands of scientists and engineers. Seldom do these big public space programs cost under a billion dollars; often they cost much, much more.

As we anticipate the second decade of the 21st century, winds of change are blowing. The old models of who carries out space programs, or who is entitled to go into space, and even commercial business models surrounding space activities are changing. These changes hinge on several key questions: Why do we want, or need, to go into outer space? Are space programs best carried out by public space agencies, public defense organizations, private industry, or a creative combination of these entities? If commercial space activities are to stand alone, what are the markets that will drive the needed investment? What is the best way forward in order to ensure that the greatest benefits are derived from space? Do we actually need new space systems and capabilities to save the planet and the human race against catastrophic climate change? How does one achieve the maximum return on investments in space? And are new ways of doing things in space possible, viable and desirable? Almost everyone has his or her own answers to these questions. Indeed, there is good reason to support the proposition that new answers are needed to these questions in the light of the modest progress made in space programs over the past 40 years.

The commercialization of space is spreading into many sectors, including satellite communications, remote sensing, space navigation, and "geomatics". These are now successful and innovative billion dollar businesses. Questions related to the militarization of space have also reached a critical threshold with the use of satellites for military communications, observation, detection and even targeting of weapons. The question that hangs in the air is: what is next?

Perhaps the most powerful proposition concerning the future of space is the role to be played by the new entrepreneurial companies that are frequently bankrolled by unconventional billionaires such as Paul Allen, Jeff Bezos, Robert Bigelow, Sir Richard Branson, and Elon Musk. Even more modest but persistent investments in space technology by people like John Carmack, a self-proclaimed "low level millionaire", are making headway with innovative new capabilities. These are people who are impatient to see things happen and happen quickly. When they think of something that they wish to see happen, they apply their own capital and business sensibility to speed things along.

It is for a variety of reasons that we seem to be on the verge of producing capable and cost-effective new space vehicles and totally new commercial space industries that launch paying customers into the high frontier of space. These entities are focused on launching space tourists for short trips into space, creating viable space habitats, deploying and operating commercial space stations, developing new and environmentally friendly hypersonic jets, and creating totally new space industries. Such new enterprises in space are as exotic as mining asteroids for rare mineral resources or using solar power satellites to beam clean sources of energy to the inhabitants of our planet. Space entrepreneurs are setting their sights on the vast resources of space waiting to be exploited.

Why do we Want, or Need, to go into Outer Space?

> *"We have an incredible challenge facing us right now because we are reaching potentially catastrophic times. In the coming decades, technological change and the pressures on our civilization will reach the breaking point if we don't embrace prudent decisions here on Earth – of conservation and rational resource utilization, together with the expansion of our civilization and economic sphere into space. If we can't do that responsibly as a species, we'll be doomed. Creating an off-Earth economy and multi-planet civilization will help secure the long term prospects of humanity."* - Dr. Robert Richards, Founder, International Space University, and CEO, Odyssey Moon Ltd.

The answers to this question are almost limitless. Responses include:

- Because it is our destiny.

- Because space systems provide key applications vital to business, government and the Earth's ecology. (These include satellites for communications, remote sensing, navigational, meteorological, emergency services and climate monitoring.)

- Because space systems may be critical to saving the Earth from catastrophic collision with Near Earth Objects (NEOs), a comet or an errant large-scale asteroid, or perhaps even building a heat irradiator that transfers the heat trapped by "greenhouse" gases in the atmosphere out into the cosmos.

- Because space telescopes and sensing systems are critical to understanding the nature of the Universe, the Big Bang, and the possibility of other intelligent life in the cosmos.

- Because space tourism could allow a large number of people to witness outer space and to see the fragility of our planet alone in outer space. Such a new type of human experience on a vast new scale may be critical to significant reforms to save the Earth's ecosystem, and to combat global warming and ozone depletion. (Non-polluting space planes could also provide a practical solution for global long distance travel.)

- Because, as observed by astrophysicist Stephen Hawking, it is critical for people to explore space and establish colonies beyond Earth to allow the ultimate preservation of the human race as a multi-planet species.

- Because space is the high frontier that holds a good deal of humanity's future opportunity and the potential for clean energy, new materials, new pharmaceuticals, climate change mitigation, and yet-to-be-discovered wonders that could completely reshape the world.

However, space is not guaranteed to be free from conflict and aggression. There are militarists who want to put weapons in space and turn it into a battlefield. These armed force planners, from various parts of the world, in the worst traditions of strategic defense concepts such as MAD (Mutually Assured Destruction), Star Wars (the defunct U.S. strategic defense initiative of the 1980s), and MRVs (Multiple Re-entry Vehicles), could seek to deploy laser and nuclear weapons in space. Space-based military systems, if not controlled and tempered, could serve to extend human warfare into the cosmos. There is also a more moderate concept of "dual-use" space technology, combining space systems for civilian and military uses, such as communications and surveillance satellites that can monitor ice flows and deforestation as well as the movement of military assets and troops. Only time will tell which of these lofty — and not so lofty — goals for space exploration, space applications, space commercialization, space science and space defense will be realized.

Are Space Programs Best Carried out by Public Space Agencies?

> *"The best way to create the future is to create it yourself."*–
> Dr. Peter Diamandis, Chairman and CEO of the XPrize Foundation.

Significant progress has been made in space over the past 50 years. Communications satellites are a US$100 billion a year industry, the Hubble telescope has revealed the history of the Cosmos almost back to the Big Bang, and space probes have explored the Sun and virtually the entire solar system, while astronauts have been to the Moon and dream of soon traveling to Mars. From all our space experience we have learned not only how to explore and apply space, but we have gotten smarter about how to do so. Space agencies generally take on the longer-term projects that require the most research effort and the most far-reaching technologies. Private enterprise evolves commercial applications, develops markets and finds technology spin-offs from space developments in both military and civilian space agency programs. Even in military space programs there is a greater reliance on commercial systems and the so-called "dual use" of commercially designed and deployed systems.

Over the last twenty years the cooperative relationships between large aerospace corporations, smaller and more agile commercial consulting firms, civilian space agencies and military space programs have been sorted out to share research effectively and implement new responsibilities and duties. This is particularly true in the United States, Canada, Japan, and Europe, but even in the more recent nations involved in space, such as India and China, has the division of tasks between government and commercial entities become clearer and more effective in recent years.

The 21st century, with its increasing social and economic pressures exemplified by a global recession in 2009, has placed constraints on new space spending. It has also coincided with the rise of new entrepreneurial space companies with their new capabilities to innovate in space. This has, in part, been stimulated by the advent of the space challenge prizes (such as the Ansari X PRIZE flight of SpaceShipOne in 2004 and the Google X PRIZE designed to inspire a non-governmental and entrepreneurial Moon landing) and now the US$50 million Robert Bigelow funded America's Challenge to develop commercial transport to low Earth orbit to support Bigelow space habitats. Even NASA has caught the spirit of this new entrepreneurial initiative, and has started some very innovative programs. This has included various Prizes or Challenges to develop Moon landers, the new materials and solar-powered climber robots needed to build a space elevator, and other technological advances.

Perhaps the most ambitious of these new commercial initiatives is the US$400 million that NASA is providing on a matching fund basis to develop reliable commercial access services to low Earth orbit. This unconventional NASA funding is to support the R&D needed to create new commercial launchers to shuttle cargo and — possibly in the future — astronauts to the International Space Station. Currently, Orbital Sciences and Space Exploration Technologies (SpaceX) are vying to develop commercial replacement vehicles (if NASA exercises contract options) for the Shuttle. Under this plan commercial vehicles would provide crew and cargo launches to and from the ISS for NASA, and also sell lifts to orbit for commercial purposes for space experiments, material processing and even space tourism.

Despite this progress and the advances in space technology and exploration seen since the start of the X PRIZE Foundation, we still have a very long way to go. As has been said more than once: "The meek shall inherit the Earth. The rest of us will go to Mars."

How can the Greatest Benefits be Derived from Space?

> *"What will the future bring? Perhaps 500 years from now satellites will be relaying three-dimensional holographic images to Earth and even to space colonies. Mining robots on the Moon "plugged into the brain" of a controller on Earth could be producing new wealth and resources a quarter of a million miles away. It is hard to say how far we might go into space using new tools such as telepathy, teleportation, or artificially intelligent Von Neumann machines that can not only reproduce themselves but improve their design on their own. The only ultimate barrier to our progress in space is the limits of our imagination and the political will to do the intelligent and moral thing."* - Dr. Joseph N. Pelton, Former Dean and Chairman of the Board of the International Space University, and Founder of the Arthur C. Clarke Foundation.

There are many intelligent and informed people who say: "Why do we need a space program? There are so many urgent needs right here on planet Earth that it does not make sense to waste money on outer space." There is a probably apocryphal story about a United States Congressman, at a hearing for the National Oceanic and Atmospheric Administration (NOAA), who said: "Why do we need to spend billions of dollars on new weather satellites when I can turn on the weather channel and see everything I need to know about the weather all over the world." The disconnect between what space programs make possible and the key services they provide is, of course, in part due to the fact that satellites and space probes are well out of sight, and their roles in society have become invisible.

Over 12,000 television channels are provided worldwide by communications satellites, along with extensive Internet connections to much of the world. Our knowledge about the critical functions of the ozone layer and the Van Allen belts in protecting humans from extinction only comes from space programs. Knowledge about the climatic conditions on Venus and Mars may help to save us from the worst ravages of global warming or from the next ice age. Today space programs divide their investments between broad categories of space exploration, space transportation systems, space applications, new technology developments, new products and services, "spin-offs", educational development and research, and space sciences.

In most countries, including Canada, Europe, Russia, China, Brazil and India, a great deal of space agency money is spent on developing space applications, vital space education and space sciences to protect the Earth and to study space weather. The United States spends far more of its resources on space exploration and the development of "manned vehicles" than all of the other space agencies. In fact, the United States has outspent the rest of the world in these areas by at least ten to one.

Critics of the United States space policy have proposed reforms that would:

(i) seek a better balance, with more emphasis on space applications and science related to human needs here on Earth;
(ii) give a special focus to environmental concerns, such as monitoring global warming, developing cleaner and greener jet engines and rockets (i.e. hydrogen fueled vehicles), etc.,
(iii) develop capabilities to detect and ward off "killer asteroids" and near-Earth objects,
(iv) let smaller entrepreneurial companies and universities develop more aspects of future space programs, rather than maintaining huge NASA Centers and supporting multi-billion contracts with giant aerospace corporations.

These critics foresee a good deal of future innovation in space coming from new entrepreneurial entities such as Elon Musk's SpaceX, or the SpaceShip Corporation, founded by Sir Richard Branson, Microsoft co-founder Paul Allen, and aerospace guru Burt Rutan. The teams competing for the Google Lunar X PRIZE have come up with much more innovative and diverse ideas than would ever have emerged from a conventional space agency.

Clearly, when looking to the future, what is needed is both better balance and better diversity. Robert Zubrin, noted Mars exploration advocate and critic of NASA, has suggested a program that would send rockets to Mars, generate methane fuel that could power a return trip, reduce the size of the rocket and the cost of sending an astronaut to Mars, and allow a safe trip back to Earth. Zubrin has also said that using the Shuttle to send astronauts into low Earth orbit is as sensible and cost efficient as using an aircraft carrier to pull water skiers. Many critics of current government space programs say that the future of space should be based on institutional and budgetary reform. This means a better balance between the duties and responsibilities of public space and military agencies, universities and research institutes, entrepreneurial companies, and more money used for prizes and challenges to develop new space capabilities.

In short, we might benefit from a shift in how public and private monies are spent on and divided among space exploration, space sciences, space applications and even space education. NASA, at the age of 50, is thought by some to have developed arterial sclerosis — or at least to have gotten a bit stiff in the joints and rusty in its old age. Its efforts to develop a viable space plane produced over a half dozen programs that were started and stopped without success over the span of a decade. The Space Shuttle alone costs US$3 billion a year to operate about six missions a year, much of it due to the standing armies of workers that need to be kept employed by congressional decree. NASA, at the start of the Space Shuttle program, indicated that it would operate a launch a week and reduce the cost of access to low Earth orbit by at least an order of magnitude. Instead, costs for manned access to orbit have skyrocketed. The ISS space station is now twenty years behind schedule and its total cost has ballooned to over US$100 billion. Funds to use the ISS for experiments and prove its value have been cut, simply to build it. Some critics compare the ISS, in terms of its costs and benefits to the United States, to the Vietnam War where it was famously said: "To save the village we had to destroy it."

The Aldridge Commission, which was formed after the Columbia Space Shuttle disaster, reported its findings with some harsh criticisms of NASA. This Commission's findings indicated that NASA was trying to do too many things and not doing any of them particularly well. They suggested that NASA should focus on long term space research, truly advanced technology and convert NASA Centers to Federally Funded Research and Development Centers (FFRDCs). These centers would compete for projects, letting smaller and more entrepreneurial companies play a greater role, while using international cooperation to better advantage. A similar approach could, of course, be applied to other space agencies as well. All too often the budgets of space agencies are spent on the salaries of aerospace corporations and agency employees without the results of these large investments becoming apparent. Most advocates of the "New-Space" paradigm feel that government space programs should return to their roots. They should not be "jobs programs"; rather they should be dedicated to true innovation and meaningful new adventures on the edge of human knowledge and capability.

Such an approach could deliver "more bang for the buck", rapid innovation, and allow a more rapid spin-off of new technology to other fields of endeavor. A further reform would be to restore high level and strategic oversight of space programs at the national or regional level. In the case of the United States, this could mean restoring a National Space Council within the Office of the Vice President. In the case of Europe, Canada, Japan, China, India - as well as the United States - actions along this line could not only mean providing strategic oversight at the highest levels but also suggest the wisdom of creating an International Space Council. Its aim would be to coordinate meaningful new global space initiatives, particularly focusing on international cooperation in space exploration, while helping humanity cope with climate change, heat dissipation from the Earth's atmosphere, global protection of the Earth from destruction from comets or asteroids, and developing "green rocket and jet propulsion" to reduce atmospheric pollution.

The Maximum Return on Investments in Space

It is easy in most businesses, including high tech ventures, to assess if you are making money and if there are net profits. Space programs, especially those with long lead times, and R&D programs that may take a decade or more to pay off, do not have the same easy metrics available to measure success. NASA Administrator Daniel Goldin became famous—some say infamous—for his "better, cheaper, faster" mantra that he used to guide the United States space agency for longer than any other American space administrator. The results of this initiative were mixed, in that some programs were developed, competed, awarded, constructed and launched faster and at lower costs, but other programs failed, perhaps in part by being rushed and not well managed. The initially unsuccessful billion-dollar Hubble Space Telescope project was first considered a failure. But the retrofitting of the optical resolution system on this remarkable space instrument—as described in chapter 3 by astronaut Jeff Hoffman—turned this project into one of the most successful and celebrated space achievements in the history of space science. One of the Explorer satellites of the 1980s ended up with major schedule delays and budget overruns, but this mission confirmed the residual microwave radio noise left over from the Big Bang. It verified the basic theories about the creation of the Universe almost 14 billion years ago. These two examples, and others that could be supplied, suggest that conventional ways of assessing businesses or even governmental programs may not apply to the exotic world of space programs. Nevertheless, there are systematic management processes that can be applied to space programs to enhance the worth and effectiveness of these activities.

For those that say that space programs are an unnecessary luxury, Arthur C. Clarke quite starkly reminded us of how important a space program can be when he said: "The dinosaurs did not survive because of a lack of a successful space program." The combined wisdom of archeologists and geologists has indeed uncovered evidence that a huge asteroid impact led to the demise of the dinosaurs. The dinosaurs had ruled the world up to that point for hundreds of millions of years, but they had no way to know that the huge asteroid was coming and even less of a program that could have diverted its path and saved them from extinction. Of course, humans are currently unprepared for such an eventuality as well, and it is convenient to say we will worry about such "unlikely events" tomorrow. Meanwhile, in mid November 2008 a meteoroid with the explosive force of 100,000 pounds of TNT crashed down in Northern Canada; this was a microscopic event compared with what could happen if the asteroid Apophis collides with Earth in 2036.

But the biggest threats to Earth right now may be a depleted ozone layer that can result in massive mutations in all animal and plant life, including humans, and rampant global warming that peer-reviewed scientific evidence shows has already elevated the Earth's mean temperature by almost 1 degree Centigrade (approaching 1.5 degrees Fahrenheit) since the 1850s. It should be noted that, as far as animal life is concerned, a global temperature rise of only a few degrees is already threatening our ecosystems and several species. With increasing global warming, the human species will ultimately be threatened as well. It is only space-based systems like remote sensing devices, radar systems, multi-spectral sensors, meteorological imaging and ozone detectors that allow us to monitor the health of the planet. Space systems have shown us that our rain forests have been greatly depleted, thus lessening the amount of oxygen pumped into the atmosphere. Space systems allow us to monitor the pollution of the oceans, the decreased vegetation, and the melting of glaciers and the ice cap.

The latest message from Mother Nature that she is about to have a heat stroke is the fact that the so-called Northwest Passage is now open and ships are steaming from the Atlantic to the Pacific via the North, rather than having to pay passage through the Panama Canal. This is an unmistakable sign that global warming is real and the coming consequences could be serious indeed. But space systems not only alert us to dangers and tell us the speed with which global warming is occurring; atmospheric models based on observations of other planets and the Sun's interactions tell us of longer-term consequences. Finally, if it becomes necessary to create some sort of heat irradiator that allows the effects of excess greenhouse gases to escape into the void of space, it will be space systems that have truly become our saviors. Space systems also allow us to monitor hurricanes and evacuate people from their destructive paths. Weather satellites and Earth Observation satellites have now saved countless lives by assisting with prediction and emergency recovery from a host of natural, and even man-made, disasters. They have aided with search and rescue for downed pilots, lost boats and ships at sea, and hikers and explorers in the remotest parts of the world, atop treacherous mountains or in the most arid of deserts. International cooperation in space is much more important than ever before. International space cooperation that is directed at preserving life on Earth is a particularly good idea, a much better idea, for instance, than putting lethal weapons in outer space.

So how can better balanced, more targeted, more productive and more internationally coordinated space programs be accomplished? To put it another way, how can we eliminate waste and cancel unproductive space programs? The simple answer is that, in the absence of fundamental governmental policy reforms towards space agency priorities, there is no simple answer. However, we can look to the future and suggest some important changes in process, institutional organization, and international cooperation. We can employ different types of incentives on the one hand or penalties on the other. These types of reforms might include the following:

- Imposing increased fines against practices that create space debris, developing more specific and precise due diligence review processes against the creation of space debris, and having better systems for de-orbiting a satellite once its useful life is over.

- Creating financial or tax incentives for commercial organizations to develop important new space capabilities, such as solar power satellites, or a lunar based solar energy plant to produce and relay "green" energy to Earth, or to design and/or deploy a space elevator (see Figure 4.2) to lift cargo to the Geo orbit—at very low cost and with zero pollution.

- Approving a new International Space Council backed project to remove orbital debris from low Earth orbit, or to construct a space heat pipe to cool the biosphere or a greenhouse gas "exhaust" to refresh the Earth's atmosphere.

- Creating of a global commercial consortium to undertake key new space initiatives, such as to repair the ozone layer that protects Earth from stellar and cosmic radiation.

- Agreeing on a new international treaty such as the "Space Preservation Treaty" to de-militarize space.

Figure 4.2 A visualization of a space elevator from space (Courtesy of NASA).

More specific examples of possible longer-term new space programs presented along with the associated economic, social, cultural, business and survival values are presented in Table 4.1. A much more active process of interactive discussion and participation by the public, legislators, businesses, academic researchers, journalists, entrepreneurial enterprises and inventors along with governmental space officials could certainly produce a more coherent and international vision of the future in space, and help to prioritize the goals for the next two to three decades.

Candidate Future Major Space Programs Through 2050		
Longer Term Space Project	**Participants**	**Pay-off or Expected Value**
Improve global space infrastructure to monitor the ozone layer and global ecology. These systems would help detect and cope with global warming.	All space-faring nations. Objectives achieved through international collaborations of government and private sector.	Potentially could help save humans and many other species from extinction.
Develop an extended heat irradiator or heat pipe that could pump excess heat out into the cosmos.	All space-faring nations. Objectives achieved through international collaborations of government and private sector.	Potentially could help save humans and many other species from extinction.
Develop a nuclear propulsion system for the ISS. This could convert the ISS into a "manned" space transport system.	The existing partners for the ISS.	Provide a cost effective habitat for astronauts to go to the Moon and beyond. Radiation shielding would need to be added.
Construct a space elevator that would provide low cost and safe access to Geo orbit.	Could be an international consortium of private companies and several countries.	Could provide low cost and safe access to space for application satellites, manned missions to the Moon, Mars and beyond.
Expand space-based astronomical observatories. Constellation systems, etc.	Could be an international consortium of private companies and several countries.	Ability to understand the Universe and the Big Bang in much greater detail.
Expand search for extraterrestrial intelligence (SETI) that would employ a Van Neumann approach to create an expanding network of intelligent probes to travel the Galaxy.	All space-faring nations. Objectives achieved through international collaborations of government and private sector	Could identify habitable planets, expand our knowledge of the Universe and possibly find other life forms.
Develop system and technology to remove space debris, especially from LEO orbit	All space-faring nations. Objectives achieved through international collaborations of government and private sector	Allow greater safety for space habitats in low Earth orbit. Reduce hazards to commercial and governmental systems.
Space Mining: there are such opportunities as bringing to Earth resource rich asteroids containing precious metals worth billions of dollars	International entrepreneurial commercial space companies.	Replenish key natural resources, new robot systems, develop new space transport capabilities
Expand planetary safeguard program to map the solar system against threats to Earth, and develop systems to divert hazards into the Sun.	All space-faring nations. Objectives achieved through international collaborations of government and private sector	Protect the Earth from catastrophic damage by comets, meteorites or other near Earth objects.
Develop new improved rocket and jet engines that use "green fuels" and oxidizers such as liquid H2 and liquid O2	International entrepreneurial commercial space companies.	Replace highly polluting jet airplanes and rockets with "green" engines.
Expanded life sciences experiments related to humans, animals and plant life to explore new medications, life-time extension, etc.	International entrepreneurial commercial space companies in cooperation with nations.	New cures, new strategies for living in space and possible life extension.
Creation of a full scale space colony that would sustain itself economically by beaming electric power back to Earth	International entrepreneurial commercial space companies in cooperation with nations.	Developing the tools for humans to live in space on a multi-generational basis.

Table 4.1 A Space Prospectus (Prepared and copyrighted by J. Pelton; used with his permission).

The twelve examples shown in Table 4.1 of how we could progress in space over the next thirty years to new heights—both figuratively and literally—are only illustrations of the potential we now hold in our hands. Many of the ideas are certainly not new. Astrophysicist Gerard O'Neill, astronaut Rusty Schweickart, space visionary Arthur C. Clarke, author Michael Benson, space entrepreneurs Brad Edwards, Robert Richards, and Peter Diamandis among others have suggested these or similar ideas previously.

The rather incredible thing is that, if we were to take the money currently being spent around the world on military and civilian space programs, all of these seemingly grandiose things and more are quite possible with billions of dollars to spare. If we were to apply only half of current space budgets around the world to the above twelve objectives we could do them all with money to spare. There are many satellites and launchers that are currently being designed and built to support commercial space programs. These could, and would, continue since there are already established markets that support these businesses with revenues of over US$125 billion/year and growing. Since these commercial applications in communications, broadcasting, remote sensing, and navigation are largely supported by business demands and established revenue streams, they currently need little or no government support beyond long-term research.

The new space tourism business, and even space habitats for "participants", in-orbit experimentation plus the Bigelow America's Challenge, and the X PRIZE initiatives continue within the bounds of normal commercial activities and market based Initial Public Offerings (IPOs). We believe that entrepreneurial talent and prizes will represent a major motive force forward in space, securing our future world with increasing space-based commerce and the use of off-Earth resources. There are areas where governments and consortia of governments must also step up to provide new leadership and vision, particularly in seeding commercial customer relationships that lower the short term barriers to entry to long term space markets that can change the standard of living for everyone on Earth.

At this point one thing is absolutely clear: inspired and focused progress in space has stalled due to a lack of leadership, a lack of vision, and a failure of nerve and initiative. These human limitations have prevented us from realizing our true potential as a space faring species of explorers and innovators.

1 Selected Readings and References:

Bradford Edwards and Eric Westling, *The Space Elevator: A Revolutionary Earth-to-Space System*, (2002) BC Edwards, Houston, Texas.
Gerard O'Neill, *The High Frontier*: Human Colonies in Space, (2000) 3rd Edition, Apogee Books, Burlington, Canada
Joseph N. Pelton, *Global Talk*, (1984) Sijthoff and Noordhoff, Aaphen en den Rhyn, Netherlands
Stephen Schnaars, *Megamistakes*: *Forecasting and the Myth of Rapid Technological Change,* Board book, 1989.

Chapter Five
The Universe and Us

"We shall not cease from exploration
And the end of all our exploring
Will be to arrive where we started
And know the place for the first time."

— T.S.Eliot, from "Four Quartets", Little Gidding, part 5, 1942

Space Sciences[1]

Space science embodies an incredibly broad range of exciting topics, which we now address. Beginning with a discussion of our Universe, we will then explore our solar system, our home planet Earth and the Moon.

The Universe - Where Are We?

We live on planet Earth, one of eight planets orbiting around a star called the Sun. Our Earth, 6371 km in radius, is located at a mean distance of 149.7 million kilometers (1.497×10^8 km) from the Sun, whose mass is a third of a million times greater than the Earth's. Distances in the Universe are so large that we really need more suitable units than kilometers to measure them. The mean distance of the Earth from the Sun, called one Astronomical Unit (AU), is a convenient unit for measuring distances in the solar system. Another important unit of distance is the light year, which is the distance traveled by light in one year, 9.46×10^{12} (almost 10^{13}) km. It takes light 8 minutes to travel from the Sun to the Earth; so, when we look at the Sun we are actually seeing it as it was 8 minutes ago.

After Mercury and Venus, Earth is the third planet from the Sun. Beyond the Earth's orbit are the planets Mars (at 1.5 AU), and the giant, gaseous planets Jupiter (at 5.2 AU), Saturn (9.5 AU), Uranus (19.2 AU) and Neptune (30.0 AU). Thus Neptune is 4.2 light hours from the Sun. Beyond the planet Neptune, at 30 AU to 55 AU from the Sun, there is a band of small objects in orbit around the Sun, called the Kuiper Belt; these are a remnant of solar system formation. It is estimated that this belt contains as many as 70,000 bodies greater than 100 km in size. The dwarf planet Pluto, ~ 40 AU from the Sun, is now classified as a Kuiper Belt object (KBO). Outside the Kuiper Belt is a spherical cloud of billions of comets called the Oort Cloud, at more than a thousand times the distance from the Sun to Pluto. The outer part of the Oort Cloud, the boundary of the solar system, is about 1 light year from the Sun and about a quarter of the distance to the next star, Proxima Centauri, which is part of a triple star system 4.2 light years away. Traveling to Proxima Centauri in a rocket at just above the escape velocity from the Earth (12 km/s) would take 100,000 years. The distance between stars in the vicinity of the Sun is vast - the average distance between stars compared to the diameter of the Sun is a factor of about 250 million. Within 17 light years from the Sun we know of 65 stars.

Stars are clustered into vast stellar groupings, or associations ("islands in the Universe"), in the shape of thin disks called galaxies (see Figure 5.1). Our solar system is located in one such galaxy, which can be seen in the night sky as a band of light called the Milky Way. There are ~ 300 billion stars in our Milky Way galaxy which is about 100,000 light years across. Located about 26,000 light years from its center, our solar system rotates once around the center every 220 million years, at a speed of 220 km/s. Galaxies are usually found in groups, our galaxy being one of 30 in the Local Group which is about a million light years in diameter. Other members of the group include the Andromeda galaxy, which is the most dominant, and the Large and Small Magellanic Clouds, which are dwarf galaxies.

Figure 5.1. Diagram illustrating the position of our Sun in the Milky Way galaxy.

As we look farther out into the Universe, we notice that groups of galaxies form into clusters, which may contain as many as 2000 members. These clusters can be as large as 10 million ly in diameter and have a mass of a trillion Suns (10^{15} M_{sun}). Nearby clusters include the Virgo Cluster, the Hercules Cluster and the Coma Cluster. Clusters can form into an even larger structure called a Supercluster more than a hundred million light years across. On even larger scales we see massive structures that look like a large bottle full of soap bubbles, with the galaxies being confined to the surfaces of the bubbles, at distances beyond 10^{10} light years. Insignificant as we humans are on planet Earth, with our small brains we can contemplate and measure the enormity of the Universe and appreciate the beauty and the complexity of its structure.

The Solar System

Investigating the formation of the solar system is one of the most challenging problems of modern science; it is termed stellar-planetary cosmogony. Based on fundamental theoretical concepts and the variety of observational data available we consider the origin and evolution of stars and planets, trying to find the most plausible scenario.

Figure 5.2. Diagram illustrating how the solar system formed 4.5 billion years ago. The number 1 shows the proto-planetary nebula collapse; 2 shows gas-dust disk formation around the proto Sun; 3 and 4 show the following stages of disk compression and development of solid grains and dust clusters that eventually gave rise to the eight planets orbiting the Sun at present, as shown in 5. (Courtesy of G. Fazio)

It is generally accepted that, like other planetary systems, our solar system was born out of the original proto-planetary nebula and gas-dust disk around the forming star. The disk compressed, its density increased, and as gravity instabilities within the disk developed solid grains and dust clusters formed. After numerous collisions occurring over ~ 10^8 years, planet embryos – i.e. so-called "planetesimals" - became planets and other small bodies in the solar system. These processes are illustrated in Figure 5.2. The evident difference in the composition of the inner and outer planets is explained by the condensation of high temperature and low temperature materials from the proto-planetary disk depending on their respective distance from the Sun. This fractionation is responsible for the rocky inner planets and the gaseous-icy outer planets, several of which have ring systems.

Key mechanical and chemical properties of the solar system place important constraints on scenarios of its origin and evolution. All the planets (except Venus and Uranus) and their satellites orbit the Sun in the same pro-grade (i.e. anticlockwise) direction. This is to say that pro-grade planets move in the same direction as the Sun rotates around its axis. The planets' orbits, although elliptical, are nearly circular and have very small inclinations to the imaginary plane containing the Earth's solar orbit. The satellites of planets are locked in resonance with the planet's intrinsic rotation and therefore their same side faces the planet, as is the case for our Moon. Some of the outermost satellites, captured by the planet's gravity field, have stranger orbits. There is a peculiar mass and angular momentum distribution in the solar system. This is to note that the Sun comprises 99.8% of the solar system mass. But the planets, because they revolve around the Sun at such large distances, carry nearly 98% of its angular momentum.

Chemical constraints involve there being similar abundances of chemical elements in the Sun and the most primitive meteorites, which are considered to be formed of the original material and the remnants of "planetesimals". There is some evidence that the inner planets were formed of matter resembling that of so-called chondritic meteorites, while the gaseous-icy giant planets preserved their original composition for the most part. The asteroids (between Mars and Jupiter) have a composition intermediate between the silicate/metal-rich inner planets and the volatile-rich outer planets. Comets are mainly composed of water ice and other frozen volatiles, and so these bodies encapsulate the most pristine matter from which the solar system was formed.

We believe that all planets form in the general process of stellar origins and evolution. In short planets can be viewed as being by-products of star formation. Planetary formation strongly depends on the mass M of the star and the particular stars' position on the Hertzsprung-Russell (H-R) diagram. For a body to become a star it has to be 8% of the Sun's mass. In essence it has to be this massive in order for nuclear fusion reactions to occur. Bodies with one percent of the Sun's mass are regarded by astrophysicists as planets. (This threshold is ten times greater than the mass of Jupiter). Bodies of intermediate mass are termed brown dwarfs. The stars that have planets are those of late spectral classes (G, K, M). of the Hertzsprung-Russell (H-R) diagram. Chemical elements heavier than the cosmically most abundant elements of hydrogen and helium were produced in the interiors of stars during the course of stellar evolution, a process known as "nucleo-synthesis".

The discoveries of proto-planetary accretion disks and extrasolar planets place some additional constraints on the theory of solar system formation. Numerous accretion disks around young stars, where planets could be born, and more than three hundred extrasolar planets have recently been discovered within distances of several tens of light years away from the Earth. These planets, however, are very different from those in the solar system. They usually have eccentric orbits very close to the central star and generally have a mass several times that of Jupiter. This means, for example, that the effective temperature of such a planet is more than 727° Celsius (1000 K) and the composition of its atmosphere could be very exotic. It is unlikely that life could exist there. No Earth-sized planets have yet been found, but there are still instrumental constraints, which prevent their discovery. Could our solar system be unique? In our solar system, the planet Mars, Jupiter's satellite Europa and Saturn's satellite Titan are thought to be the most probable abodes for prebiotic organics and/or primitive life forms. [2]

The main characteristics of the planets and their satellites are summarized in Beatty, et al., 1999. All of the giant planets have rings populated by particles ranging from microns to meters in size, and also some satellites (called "shepherds") that control the rings' shapes and dimensions.

Since their origin some 4.55 billion years ago, the inner planets have experienced dramatic changes in the course of their evolution. Their heavy bombardment by asteroid-size bodies in their early epoch, about 4.0 billion years ago, and their internal heat source due to the decay of long-lived radioisotopes played significant roles. Impacts scarred the surface and left behind numerous craters while internal heating was responsible for tectonic processes and widespread volcanism. For the outer giant planets, the heat released from the interior exceeds the incident solar flux, mainly because of their continuing contraction. Planetary geology is closely related to the differentiation of the planetary interiors into shells (core, mantle, and crust). Impact structures are most clearly seen on Mercury and the Moon, which have no atmospheres. Heavily cratered terrains are preserved on the Martian surface though, in the presence of an atmosphere, ancient craters have been eroded by the weathering processes. On Mars, there are great ancient shield volcanoes at heights up to 24 km above the mean surface level, which (despite the relatively small size of the planet) are the tallest in the solar system. Another remarkable geologic feature is the Marineris valley, which is hundreds of kilometers wide and 8 km deep and puts the so-called Grand Canyon in the U.S. to shame. This valley extends along the equator for more than 3000 km (see Figure 5.3). On Venus, geological features were only revealed using the radar technique aboard the Venera 15 and 16 and Magellan missions, because of its very thick atmosphere and clouds which fully obscure the surface when observing in the optical part of the spectrum (see Figure 5.4).

Figure 5.3. This is a global view of Mars. Many different geologic features can be distinguished on the planet's surface of mostly red-orange color. Huge volcanoes up to 24 km in height above the mean surface level are seen in the elevated Tharsis region at the left. They are partially obscured by the clouds in the very rarefied atmosphere (white patches). The Valley Marineris, some hundreds of kilometers wide and 8 km deep, extends from just South of the equator for more than 3000 km. Cratered terrain is seen at higher southern latitudes. At the top of the image is the North polar cap, composed of water ice and overlying frozen carbon dioxide ("dry ice" deposits) formed during the Martian winter. (Courtesy of NASA)

The atmospheric properties of our neighboring planets Venus and Mars differ dramatically from that of Earth and from each other. The pressure at the Venusian surface reaches 92 atmospheres and the temperature is 462 degrees Celsius. This would be like trying to live in a very high pressure Volcano. The surface of Mars is dramatically different on the other end of the spectrum of what we consider normal on planet Earth. On Mars the average pressure is only 0.006-atmosphere and the average temperature is about –53 degrees Celsius. The composition of the atmospheres of both planets is mostly carbon dioxide, with relatively small amounts of nitrogen and argon and a tiny mixing ratio of water vapor and oxygen (on Mars). Because the axis angle or "obliquity" of Mars is nearly the same as that of Earth (25.19 degrees, compared to 23.45 degrees for Earth), there are pronounced seasonal variations resulting in temperature contrasts between summer and winter exceeding 100 C. The temperature at the winter pole drops to below the freezing point of CO_2 and so "dry ice" deposits cover the polar caps, though they are mainly formed of water ice. Seasonal variations and the release of carbon dioxide from the polar regions is one of the important drivers of atmospheric circulation on Mars, involving both meridional and zonal wind patterns. By way of contrast, the atmospheric circulation on Venus is mainly characterized by "super-rotation", such that the zonal wind speed at the surface of less than 1 m/s increases up to nearly 100 m/s near the upper cloud level at about 60 km altitude. Venusian clouds consist of quite concentrated sulfuric acid droplets, defining (in addition to the very hot and dense atmosphere) a hostile environment on our closest planet, which until recently, was thought to be Earth's twin.

Figure 5.4. A global view of Venus composed of numerous Magellan images obtained using the synthetic aperture radar (at centimeter wavelengths), which penetrates through the very thick atmosphere and clouds. The surface exhibits complex geological patterns involving widespread volcanic activity, extension-compression processes, faulting, fractures and other powerful deformations of Venus' crust. (Courtesy of NASA)

In terms of its natural environment, Mars is another extreme, though more favorable in its weather and climate and, therefore, much more accessible for exploration and future human expansion throughout the solar system. This planet could harbor life beyond Earth at the microbial level; extant or extinct life could be found. Unlike Venus where an assumed early ocean was lost soon after the "runaway greenhouse effect" responsible for its contemporary, furnace-like conditions developed, ancient Mars had plenty of water until a catastrophic drought happened about 3.5 billion years ago. There is evidence that contemporary Mars still has some water (the depth of its original ocean being estimated to have been nearly 0.5 km) stored as permafrost and in deep-seated ice layers (termed "lenses"). The recent NASA Phoenix lander close to the North Pole of Mars (see Fig. 5.3) revealed features which confirm such a scenario, as well as the earlier NASA Odyssey mission which showed the existence of subsurface water at about 1 m depth, unevenly distributed over the planet. Our general understanding is that ancient Mars had a much more pleasant climate than nowadays, when water coursed its surface until its quite dense atmosphere was lost into space and water ice was buried beneath thick sand-dust sediments.

The gaseous giant planets with their numerous satellites and rings are completely different. These planets also underwent differentiation of their interiors and, as a result, they have rather large rocky cores and extensive gaseous-icy mantles, the outermost shell being referred to as an atmosphere. Their effective temperatures range from -138°C (135 K) (Jupiter) to -235°C (38 K) (Neptune). Because of the continuing contraction of these massive planets, the heat released from their interiors is about two or three times the incident solar flux. This internal heat (which is many orders of magnitude greater than that for the terrestrial planets) is responsible for specific features of their atmospheric circulation. These include, at certain latitudes, zones and belts with very strong shear turbulence, which leads to the formation of eddies. Spectacular examples of long-lived eddies (anticyclones) are the Great Red Spot (GRS) on Jupiter and the Great Dark Spot (GDS) on Neptune. The dimensions of the GRS have varied from 10,000 to 14,000 km in the north-south direction and 24,000 to 40,000 km in the east-west direction, which is nearly three times the diameter of the Earth. The GRS also rotates counter-clockwise with a period of seven days. The atmospheres of the giant planets consist mostly of hydrogen, helium, and hydrogen-bearing compounds, such as water, ammonia, and methane.

Of special interest are the satellites of the giant planets, studied at close quarters by space missions such as Voyager 1 and 2. Of particular interest are the Galilean satellites of Jupiter. The closest to the planet, Io, exhibits widespread volcanism due to heating via the tidal forces exerted on it, in its eccentric orbit, by other satellites orbiting Jupiter. Io's surface is covered with sulfur deposits from volcanic eruptions, which give Io its yellow-orange color. Europa also experiences tidal heating which melts the ice about 10-15 km below its icy shell, giving rise to what is believed to be an electrically conducting salty ocean that might be as deep as 50 km. That might be "warm" enough to be home to complex organic molecules. Therefore, Europa is on the short list of future space missions to search for signs of life in the solar system. Ganymede is less subject to tidal heating; its icy surface is more cratered than Europa's, though it possesses an even stronger magnetic field. In turn, Callisto, having a heavily cratered terrain is regarded as a dead body, with impact features preserved throughout the history of the solar system, though it also has a magnetic field comparable in strength to that of Europa. Its origin is still an enigma.

Saturn's Titan and Neptune's Triton are intriguing. Their bulk density is comparable to that of Ganymede and Callisto, which means that they are formed half of rocks and half of water ice. Titan is unique, with its very dense atmosphere composed of nitrogen and argon at a surface pressure of 1.6 atmospheres and a temperature of -179°C (94 K). Since the acceleration due to gravity on Titan is only one seventh of Earth's, to create such a high surface pressure the mass of gas on Titan is nearly ten times the mass of the Earth's atmosphere. Liquid methane and other hydrocarbons could exist on Titan's surface; the recent Huygens lander of the Cassini mission found some features resembling methane lakes. Titan is thus another target in the search for life elsewhere in the solar system.

Unlike Titan, Triton is deprived of an atmosphere and the temperature of its nitrogen-methane surface is only -235°C (38 K); similar conditions are found on Pluto. A peculiar phenomenon on Triton, discovered during the Voyager fly-by, is the geyser-like eruptions of liquid nitrogen; this is called cryovolcanism (i.e. cold volcanism). Its probable source relates to the dissipation of tidal energy generated by the gravity interactions of Triton with Neptune. Like Pluto, Triton could have been captured from the Kuiper belt by Neptune's gravity.

The most abundant, and dynamic, objects in the solar system are small bodies - asteroids, comets, meteoroids, and meteor dust. Asteroids are mostly located in a main belt between the orbits of Mars and Jupiter (from 2.7 to 3.2 AU). They range in size from about one thousand kilometers (Ceres) to Ida, 52 km long and shown in Figure 5.5, to only several meters; bodies less than l m in size are called meteoroids. The number of asteroids larger than 1 km is estimated to be $\sim 10^5$, though only about half this number has been cataloged. Three special groups of asteroids, which have eccentric orbits approaching, or even crossing, the Earth's orbit (Apollo, Amor, and Aten) are referred to as Near Earth Objects (NEO). If one of these asteroids a few km in size impacted the Earth it would constitute a terrible threat to our home planet. Meteorites are fragments of asteroids, which have fallen onto the Earth's surface. Asteroids and meteorites are commonly classified by their composition as stony, iron, and iron-stony (a mixture). The most primitive stony meteorites (carbonaceous chondrites) come from the outer part of the main asteroid belt. Encapsulating primordial material, they give us crucial information about conditions existing when the solar system was formed.

Figure 5.5. Image of the asteroid Ida, and its tiny moon Dactyl, as viewed from the Galileo spacecraft en route to Jupiter in 1991 (Courtesy of NASA).

Comets are regarded as the most pristine bodies in the solar system. Comets are classified into short period (orbiting the Sun with a period less than 200 years), and long period (having a period more than 200 years). The main family of comets is located in the Oort cloud (at 10^4 to 10^5 AU) and in the Kuiper belt beyond the orbit of Neptune (from 30 to 100 AU from the Sun). Comets are icy bodies (see Figure 5.6) having a nucleus, atmosphere (coma), and tail (always pointing away from the Sun). Their very porous nucleus, in the "dirty snow ball" model of a comet, is composed mainly of water ice with other frozen gases and dust. The coma and tail are produced as the ices sublime when the comet traveling along its very eccentric orbit approaches the Sun; they can be partially ionized. There are two types of tail: one is mainly composed of fine dust pushed outwards by the solar radiation pressure, and the second (consisting mainly of ionized particles) is strongly influenced by the solar wind plasma and the interplanetary magnetic field.

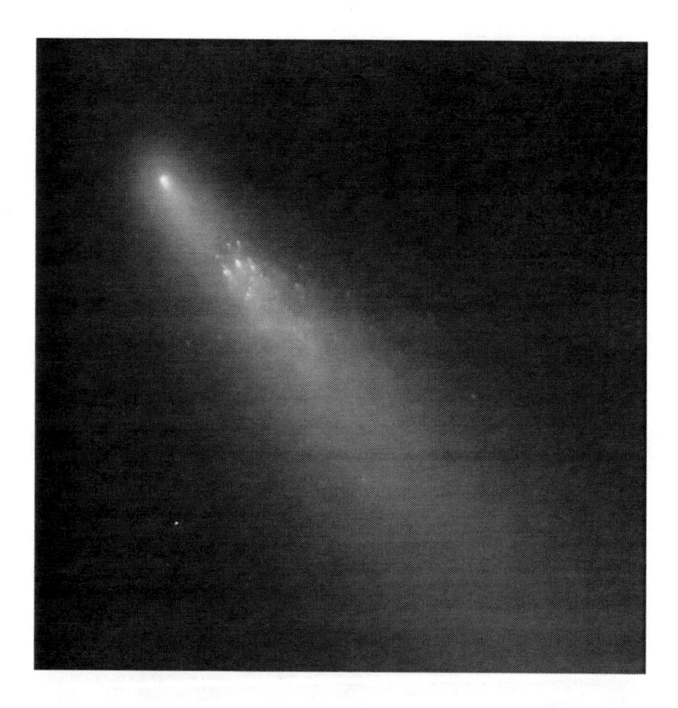

Figure 5.6. A Hubble Space telescope image of the comet Schwassmann-Wachmann 3, breaking up into smaller pieces as it approached the Sun, taken in April 2006

(Courtesy of NASA).

Interplanetary dust particles ranging in size from millimeters to nanometers are either collisional fragments of asteroids/meteoroids or particles formed by sublimation from a cometary nucleus. The former are randomly distributed in interplanetary space, the smallest particles slowly drifting towards the Sun due to the Poynting-Robertson effect, while the latter form around cometary orbits. Periodically the Earth meets some of these dust particles in its orbit around the Sun when this closely intersects the comet's orbit. This phenomenon is well known via the meteor showers named after the direction of the constellation in the skies where the meteors come from (e.g., Aquarids, Perseids, Quadrantids, etc.).

In conclusion, we can state that the study of solar system bodies gives us insight into the problem of its origin and evolution and also serves as a powerful tool to increase our knowledge of the Earth. In other words, the planetary sciences allow us to understand most comprehensively the nature of our home planet and our place in the Universe. Small bodies can carry prebiotic matter and thus help to answer the fundamental question of the origin of life on Earth.

The Earth and the Moon

Earth is the third planet from the Sun and the largest of the group of four terrestrial planets. The Earth's orbit has very small eccentricity

(0.017), which means that its orbit is almost circular. The Earth moves along its orbit at 29.8 km/s, and the period P_{orb} = 365.256 days (1 year). The sidereal period of rotation around its axis (relative to the stars, or 1 sidereal day) is P_{rot} = 23 hr 56 min 4.99 s. The inclination of the Earth's equator to the plane of its circumsolar orbit (the ecliptic plane) is 23°27', which leads to significant seasonal changes on the surface of our planet.

The Moon, as seen in Figure 5.7, is the brightest object in our skies after the Sun. The semi-major axis of the Moon's orbit is 383,398 km, and its eccentricity is 0.055. The inclination of the Moon's orbit to the ecliptic is 5°09'. Its sidereal period of revolution around the Earth is 27 days 7 hr 43 min, while its so-called synodic period (relative to the Sun, and corresponding to the lunar phase changes) is 29 days 12 hr 44 min.

Figure 5.7. A full Moon as seen from the Earth. The dark areas are low land areas (basins or depressions, historically called mare), whereas the grey areas are elevated regions (highlands, historically called continents). Numerous impacts created a heavily cratered terrain on the Moon's surface. The small bright spots indicate fresh impacts.

(Credit: http://upload.wikimedia.org/wikipedia/commons/thumb/d/dd/Full_Moon_Luc_Viatour.jpg/594px-Full_Moon_Luc_Viatour.jpg)

The Earth-Moon system is unique in the solar system, having the smallest planet-to- satellite ratio of 81. This means that the two bodies exert significant mutual gravity forces on one another. On the Earth, these are clearly manifested by the ocean tides, which are larger than those caused by the Sun. In turn, the Earth locks the Moon's orbit and its intrinsic rotation in resonance, such that the period of lunar rotation equals the sidereal period. This means that the Moon orbits the Earth in exactly the same time as it takes to complete one rotation around its axis. This is why the Moon always presents the same aspect to the Earth and why we can only see the "dark side of the Moon" from a Luna-Zond orbiter or an Apollo capsule. The tidal energy lost by the Earth to the Moon is responsible for the Moon moving steadily away from Earth by about 3 cm per year.

The Earth is not a significant body on the cosmic scale. Its equatorial radius is R_e = 6,378 km, its polar radius is R_p = 6,356 km and, hence, its oblateness is 0.0034. The pear-shaped figure of the Earth is called the geoid. Its mass is M_e = 5.974 x 10^{24} kg. (This is more or less six sextillion metric tons which is still very small on a Cosmic scale.) The mean density is 5.515 x 10^3 kg/m^3, and the mean acceleration due to gravity at the surface is g = 9.78 m/s^2 (a bit more at the poles than at the equator). The Earth's surface has continents and oceans, the latter occupying nearly two thirds of the whole surface. The age of the Earth is now estimated with reasonably good accuract to be 4.55±0.07 billion years.

The main geological mechanism operating on Earth is plate tectonics, which means that its outer shell (the lithosphere) is not homogeneous but split into several large plates. This mechanism involves a spreading zone, where hot lava ascends from the upper mantle pushing the lithosphere plates apart and making "cracks" (rifts) between them, and subduction zones where some plates covered with sediments are slowly pushed deep under the continents. The spreading zone coincides with the global system of mid ocean ridges at the bottom of all the oceans. Both the spreading and subduction zones are associated with powerful volcanic activity and earthquakes. Plate tectonics are responsible for the drift of continents, which continuously move away from each other, as was first suggested by the German scientist Alfred Wegener in 1912. Reconstruction of the process back in time led to the conclusion that, ~250 million years ago, there was one super-continent Pangea which then disintegrated into several pieces giving rise to the present continents. In support of this model, we may compare the shapes of the eastern part of South America and the western part of Africa, noting that they fit together like two pieces of a jigsaw. The model has been confirmed by studies of the bottom of the ocean and of the magnetic properties of recent lava flows there.

The oceans comprise nearly 97% of all the water on Earth (the hydrosphere), about 10^{21} kg, and cover 361 million km^2. The remaining 3% is fresh water in rivers, lakes, and glaciers and a small amount in the atmosphere, as well as in the polar icecaps of the Arctic and Antarctica. The mean depth of the ocean is 3,900 m, while the maximum depth is 11,000 m in the Marian trough in the Pacific.

The Earth's interior has a complicated structure revealed by seismic soundings. The speed of propagation of longitudinal and transverse seismic waves depends on the density and elasticity of the rocks. The waves also experience reflection by and diffraction at the boundaries between different layers; transverse waves do not travel through liquids. The main zones are the upper crust, the partially melted upper and lower mantle, the liquid outer core and the inner solid core. The thickness of the crust is ~35 km under the continents and about half that under the oceans. The region between the crust and upper mantle is the lithosphere, to a depth of about 70 km, and above the asthenosphere extending to a depth of 250 km. The boundary between the crust and the upper mantle is the Mohorovicic boundary, or the Moho for short. The dozen large plates of the lithosphere "float" on top of the asthenosphere, therefore ensuring the ever-continuing action of global plate tectonics.

The crust, composed of bedrocks (basalts and granites) having a mean density ~3000 kg/m^3, comprises less than 1% of the mass of the Earth. The mantle takes nearly 65% and the core 34%. The thickness of the mantle is 2900 km; it is composed mainly of silicate rocks, as well as silicate and magnesium oxides modified under high pressure. The outer radius of the liquid core, composed of molten iron and nickel, is 2250 km, and the inner solid core has a radius of 1220 km. The temperature reaches ~1527°C (~1800 K) at a depth of 100 km, ~4727°C (~5000 K) at the mantle-core boundary, and ~7727°C (~8000 K) at the center where the density exceeds 10,000 kg/m^3. Today, the liquid core and partially melted mantle are accounted for by heating due to long-lived radionuclides – uranium, thorium, and potassium – in the Earth's interior. Dynamo action in the liquid core is responsible for the significant magnetic field of the Earth. Its strength is ~31,000 nanoTesla at the magnetic equator and twice that at the poles. Near the North geographic pole is the South (North seeking) magnetic pole (with the angle between the geographic and magnetic axes being ~11.5^o). In the last 10 million years the magnetic field has reversed its polarity 16 times.

The Earth has a unique atmosphere in the solar system, composed mainly of nitrogen and oxygen, and having five stratified layers defined according to their temperature; they are the troposphere, stratosphere, mesosphere, thermosphere, and exosphere. The troposphere stretches from the surface to 12 km (somewhat higher at low latitudes and lower at high latitudes). The mean temperature at the surface is +15°C with variations from -85°C (inner regions of Antarctica) to +70°C (West Sahara). The temperature decreases with increasing height at roughly 6°/km. Small amounts of water vapor, carbon dioxide (now almost 0.04%) and methane are mainly responsible for the so-called greenhouse effect. This raises the Earth's temperature by about 33 °C compared with what it would have been in the absence of these molecules (which absorb infra-red radiation). The mean gas density at the surface when the temperature is 15.0°C, and the relative humidity is zero, is 1.225 kg/m^3 (at standard atmospheric pressure, 1013 hPa).

In the stratosphere (12–50 km), the temperature rises from approximately –50° C at 12 km due to the absorption of solar ultraviolet (UV) radiation between 200 and 300 nm by molecular oxygen and ozone. The stratosphere shields all living fauna and flora on the planet from intense ultraviolet radiation and thereby prevents the destruction of the biosphere. Ozone raises the temperature of the upper stratosphere to nearly 0° C at 50 km. In the mesosphere above the temperature falls, reaching its minimum value (about – 90 ° C) at the mesopause at 85 km. Above this level, in the thermosphere, EUV (extreme UV, at wavelengths shorter than 200 nm) and soft X-ray radiation is absorbed, causing the heating and ionization of the atmospheric molecules and atoms. The temperature steadily increases to 572°C-727°C (800-1000 K) at ~300 km altitude, but that varies from about 227°C (500 K) at solar minimum to 1227°C (1500 K) at the maximum of the Sun's 11-year cycle of activity. This causes atmospheric density variations by almost two orders of magnitude at 400 km altitude, and, hence, dramatically influences the lifetimes of Low Earth Orbiting satellites and the International Space Station. Also at these heights is the ionosphere, an electrically charged gas, with its prominent D, E, F_1 and F_2 layers where the plasma density is high enough to reflect radio waves of short wavelength (at frequencies below ~ 10 MHz) and thus to enable long distance radio communications around the Earth. Above about 500 km altitude is the exosphere, mostly hydrogen and helium, where collisions between atoms are very rare and where upward moving atoms have sufficient thermal velocity to escape into outer space.

The Moon has a radius of 1,737 km and a mass of 7.348 x 10^{22} kg: its bulk density of 3,350 kg/m^3 is comparable with the density of the Earth's mantle. This fact hints at its origin resulting from a collision of a Mars-sized body with the Earth soon after the Earth was formed. The acceleration due to gravity on lunar surface is only 1.63 m/s^2, which makes the Moon a very efficient launch pad for solar system exploration. The Moon's gravity is too small for it to retain an atmosphere, whether or not there was one in the past. The Moon is a rather inhospitable world, with no atmosphere and very marked temperature variations from -170° C at night to +130° C during the day, with both day and night lasting nearly a fortnight. Recent high-resolution infrared observations from spacecraft show a broad absorption at ~2.8-µm due to adsorbed water and hydroxyl (OH) in the upper few millimeters of the sunlit surface on the Moon. The amount of water varies from 50 parts per million (ppm) near the equator to about 1000 ppm near the poles. This is still dryer than the driest desert on Earth. The origin of the water is still controversial and could have been "delivered" by cometary or asteroid impacts or both.

The Moon has high mountains and deep depressions, termed seas (mare). The surface is heavily cratered, the large craters being named after great astronomers such as Tycho, Ptolemy, and Copernicus. The craters were formed long ago by impacts: in the absence of an atmosphere (and so no winds) they experienced little erosion, by meteorites and the solar wind. The range of crater sizes stretches from centimeters to hundreds of km, with the total number of craters on the near side of the Moon bigger than 1 km exceeding 300,000. After an impact, the molten mantle close to the surface is believed to have flowed as lava into the basins. The mountains reach heights of 2 km (Carpathes) to 6 km (Apennines). The rocks brought back to Earth by the Apollo and Luna missions have been estimated to have been more than 4.4 billion years old when they solidified.

An intriguing question is whether the Moon has a liquid core According to modeling partially supported by Apollo seismic data, the thickness of the upper lunar crust ranges from 60 to 100 km and the thicknesses of the upper and middle mantle below are about 400 and 600 km, respectively. Together they constitute a strong solid lithosphere. The lower mantle goes down to ~ 1600 km and thus the radius of the core could be roughly only about 150 km. The very small magnetic field of the Moon (~ 10^{-4} of Earth's) could indicate either a solid core or be due to the lack of a dynamo in the slowly rotating body.

Using the Moon's resources is on the agenda of the space faring nations. Metallic elements and their compounds in rocks could be useful for building lunar habitats. In parallel, oxygen and water could be produced from H_2 and O_2 locked into the rocks. The incident solar radiation ensures an essentially unlimited supply of energy. Some people value the opportunity to extract ^{3}He isotope deposited by the solar wind on the Moon's surface and bring it to Earth for use in controlled "pure" thermonuclear fusion reactors. There are many opportunities for research into planetary science and radio astronomy on the Moon, which could be referred to as investigations both on and from the Moon. In solar system exploration and exploitation, an exciting future lies ahead.

The Space Environment

Nearly all of the known matter in the Universe exists as an ionized gas, termed a plasma; some aspects of the special behavior of plasmas are discussed. The matter in the Sun and all the regions between our planets in the solar system are plasmas. The field of space weather is the study of plasma phenomena and its effect on the Earth and spacecraft.

The Electromagnetic Spectrum

Almost all processes occurring on the Earth are powered by solar energy. Energy radiated by the Sun is transferred to the Earth in the form of light, electromagnetic waves, i.e. oscillating electric and magnetic fields, which travel at the velocity of light, 3.0×10^8 meters per second (also denoted m/s). The frequency of the oscillations, in Hertz (also Hz or cycles/s), and the wavelength of the radiation lambda (λ) given in meters are inversely related to each other.

In the visible part of the electromagnetic spectrum with which we are most familiar, blue light has a wavelength of ~ 400 nanometers, or nm (4×10^{-7} m, or 0.4 micrometers, μm), and red light ~ 700 nm. The spectrum is illustrated in Figure 5.8. (Note that the symbol ~ means "approximately.") This is shown on the right-hand side of the diagram in Figure 5.8. Radiation of even longer wavelengths comes in the form of infrared (IR) radiation, microwave radiation, and radio waves. Commercial radio broadcasts at ~ 100 MHz use Frequency Modulation (FM, see the expanded scale at the bottom left of Figure 1) to convey the information carried by audio frequency (~ kHz) sound waves to the listener. TV transmitters operate at Ultra High Frequency (UHF), almost up to 1000 MHz, or 1 GHz. The longest naturally occurring radio waves have a wavelength ~ 40,000 km, the circumference of the Earth, and a frequency of ~ 8 Hz. Extremely Low Frequency (ELF) radio signals from lightning excite resonances of the spherical cavity between the Earth and the ionosphere at this frequency.

At wavelengths shorter than the visible are ultraviolet radiation, X-rays and Gamma rays. Radiation of shorter wavelengths, i.e. of higher frequencies than the visible, is more energetic. For such radiation we often think of their particle-like behavior; photons are "packets of energy" which travel at the speed of light. The energy of a photon is directly proportional to its wavelength. Although a photon has zero mass, it does have momentum.

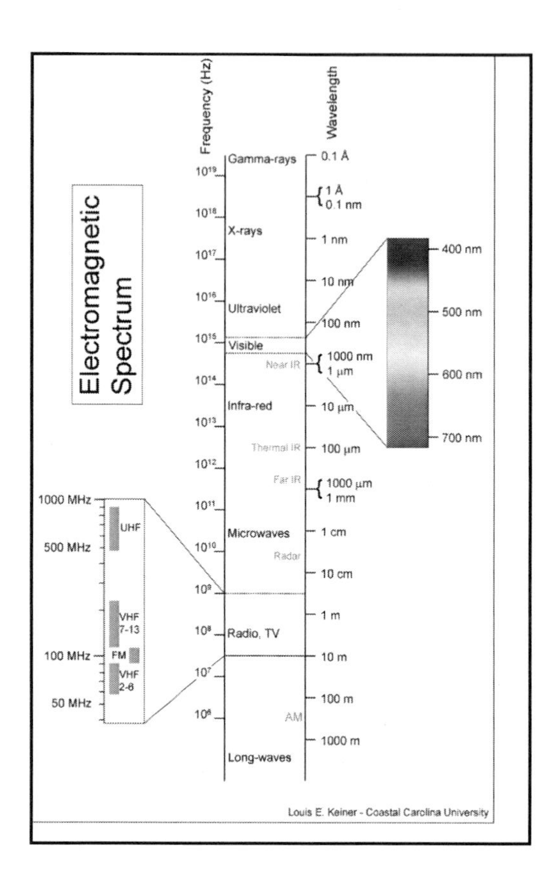

Figure 5.8. Shown here, the electromagnetic spectrum stretches across more than fourteen decades of wavelength, from Gamma rays with a wavelength of 0.01 nm through the visible (400 to 700 nm) to very long wavelength (> 1 km) radio waves. Correspondingly, the frequency of the radiation ranges from >1019 Hz to < 106 Hz (1 MHz), down even to 8 Hz.

(Courtesy of http://kingfish.coastal.edu/marine/Animations/Images/Electromagnetic-Spectrum-3.png)

Electromagnetic waves may be reflected, refracted, absorbed, scattered or polarized by interacting with matter. For example, visible light is reflected by silvered glass (a mirror), refracted through a glass prism (as was first investigated by Isaac Newton around 1670) which splits the light into its constituent colors (red, orange, yellow, green, blue, indigo, and violet), absorbed by a filter, scattered by a rough surface or by particles whose size is comparable with the wavelength, and polarized by passing through a sheet of Polaroid. Radio waves are reflected by the ionosphere where and when their frequency equals the plasma frequency at a particular height in the ionosphere. Radio waves are partially absorbed when propagating through the ionosphere due to collisions between electrons in the ionospheric plasma and atmospheric atoms or molecules.

Atoms of any particular element radiate only at particular wavelengths. Astrophysically speaking, hydrogen is the most important gas, and it has a strong emission line at a wavelength of 121.6 nm. This is in the ultraviolet, and is termed the Lyman alpha line, the first line in the Lyman series of lines. When the Balmer alpha line emitted by hydrogen on the

Sun's surface (the photosphere) passes through the solar atmosphere above it (the chromosphere and the corona), it is partially absorbed, so that a dark absorption line is apparent in the solar spectrum at a wavelength of 656.3 nm (red).

The spectrum radiated by a hot body such as the Sun is a continuous spectrum. As Figure 5.9 shows, the amount of energy radiated as a function of wavelength, termed the spectral radiance and described by Planck's radiation law, varies greatly with temperature T (in K, i.e. degrees C + 273). This figure is drawn for what is called a blackbody, i.e. a body that absorbs all of the energy which is incident upon it and which reradiates it totally. The Sun is a very good approximation to a blackbody with a temperature near 5527°C (5800 K); the maximum of the curve occurs in the yellow part of the visible spectrum to which our eyes are most sensitive. The Earth and its atmosphere also behave like a black-body, with an absolute temperature of 15°C (288 K); the maximum of this curve is at a wavelength of ~ 10 μm, in what is broadly known as the thermal infra-red part of the spectrum. The area under any curve in Figure 5.9 is the total energy radiated by that black-body; the Stefan Boltzmann law states that this is proportional to T^4, the fourth power of the absolute temperature of the radiating body.

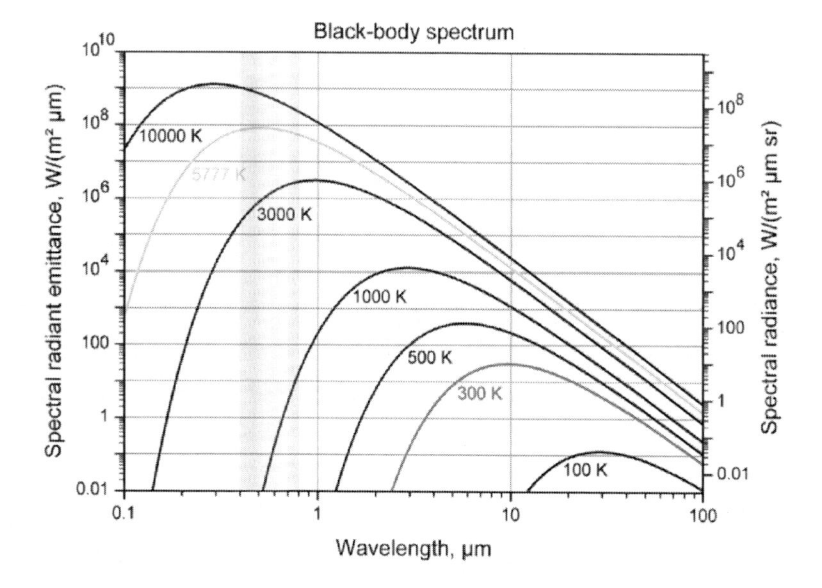

Figure 5.9. The spectrum of blackbody radiation at different absolute temperatures is shown as a function of wavelength λ. The spectral radiance of the Sun plotted at 5777 K peaks in the visible part of the spectrum, here shown as ranging from wavelengths of 0.4 mm (or 400 nm) to 0.8 mm (800 nm). The Earth-atmosphere system behaves as a blackbody with an absolute temperature of just less than 300 K (actually ~ 288 K); the peak of the radiation emitted by the Earth is at wavelengths close to 10 mm.

(Courtesy http://800px-BlackbodySpectrum_loglog_150dpi_en[1].png)

If the source of radiation is moving at some velocity relative to an observer, say an astrophysicist on the Earth looking through a telescope, the wavelength of the light received differs from that emitted (λ) by a small amount, delta λ. This is called the Doppler shift, and is due to the Doppler effect. In everyday life this effect is evident if we listen to the siren of a police car. When the police car is moving towards us, the pitch (or frequency) of the siren increases, delta v is positive and delta λ is negative; both are directly proportional to the velocity of the car v. When the car is moving away from us at a velocity v, delta v is negative and delta λ is positive. Thus, for a source, which is moving away from us, the radiation emitted is red shifted, to longer wavelengths, whereas for a source coming towards us (when v is negative) it is blue shifted. In astrophysics this accounts for the cosmological red shift of distant galaxies radiating at earlier times, when the Universe was smaller than it is now. Measurement of the red shift therefore tells us about the expansion of the Universe.

The Plasma Universe

What is a plasma? Fluorescent lights, lightning, rocket exhausts, thermonuclear fusion reactors, the solar system, stars, and galaxies all contain a substance that simultaneously obeys the laws of electromagnetism and fluid dynamics. In 1928 Irving Langmuir named this substance "plasma"; it is sometimes called the *fourth state of matter*, the other three being solid, liquid and gas. When a gas is heated it is completely ionized, forming a collection of electrically charged particles (both positive ions and a mixture of negative ions and electrons) possessing all the qualities of a fluid but with additional properties caused by the presence of electric and magnetic fields. A plasma is an ionized gas in which the particle dynamics due to long-range electromagnetic forces are more important than those due to short-range collisional interactions. Plasmas are usually very hot, or tenuous, or both. A plasma in space is an assembly of positive particles (protons, and other ions) and negative particles in one volume, in which the dominant interactions are collective, i.e. all the electrons act in unison together, as do all the protons and other ions. In some cases the motion of a plasma can accurately be described by looking at how individual charged particles act.

Plasma physics is essential to explain most astrophysical and space phenomena. In fact, we live in a plasma Universe since more than 99% of the known matter in the Universe is in the plasma state. Nearly all plasmas in the Universe also have neutral atoms that co-exist with the ionized plasma gas. For most of the plasma in the Universe - or in laboratories on Earth - collisions between the individual particles (electrons and ions) and neutrals (atoms and neutrons) occur very rarely. How plasmas behave, and their motions in space, are determined by the most important forces acting; they are gravitational and electromagnetic forces. Understanding charged particles moving in a gravity field, an electric field, and a magnetic field is crucial in both astrophysical and space physics. The force due to gravity depends directly only on the mass of the particle, thereby affecting neutrals as well as all the components of the ionized plasma. For charged particles experiencing an electric field, in the absence of any other force, electrons will move in one direction and the ions will move in the exact opposite direction, along the electric field direction. The motion of a charged particle in a magnetic field is very different. Charged particles moving in magnetic fields, in the absence of other forces, move in a spiral about the magnetic field direction, with electrons and ions spiraling in opposite directions.

The combined forces acting on a charged particle result in rather complicated motions. Figure 5.10 illustrates various motions of a charged particle under the influence of magnetic and electric fields. Panel A shows the motion of an electron with only a perpendicular velocity in a constant magnetic field. Panel B shows charged particle motion when it has both a parallel and perpendicular velocity. Panel C illustrates the bouncing motion that a charged particle will perform when in a magnetic field that changes (either increases or decreases) with time or with position. Panel D shows the drift motion of a charged particle along a shell in a very realistic astrophysical or space magnetic field, namely that due to a dipole. It is these motions of charged particles that have trapped high-energy plasma in the Earth's magnetic field, so forming the Van Allen radiation belts.

Plasmas can also be treated as a collection of charged particles. This approach leads to other important ways of describing how a plasma behaves or interacts with fields and matter collectively. Because a plasma can act collectively, there is a unique way in which it moves when its individual charges are displaced and released to move back towards their original positions. These motions are called *plasma oscillations*. Their characteristic frequency of oscillation, called the plasma frequency, depends only on the density of the plasma. The higher that the density of the plasma is, the higher is the resulting frequency of its oscillation.

Magnetic field *reconnection* (also called *annihilation*) is a way of describing what happens when two different plasmas and their associated magnetic fields collide. In general, plasmas with their separately frozen-in magnetic fields do not mix. But when these magnetic fields are in opposite directions the field lines can connect, allowing the plasmas to mix and some plasma caught up in the process to be accelerated, sometimes to extremely high velocities. Magnetic fields in a plasma behave like stretched rubber bands. The effect of field connection is like that of cutting the bands and retying them to produce shorter but more relaxed segments. Panel A of Figure 5.11 shows the frozen-in magnetic field configuration before reconnection and panel B afterwards. The annihilation of magnetic fields may occur where oppositely directed fields are pushed together by outside forces, such as the solar wind impinging on the Earth's dayside magnetosphere. Energy is released in this interaction as charged particles, which are energized near the "neutral line" between the two opposing fields. The annihilation process is believed to occur in many space plasma situations, such as in solar flares, in coronal mass ejections, at the Earth's dayside magnetopause, and in the center of the Earth's magnetotail. Charged particles accelerated via magnetic annihilation in solar flares are known to be a major hazard for human space travelers.

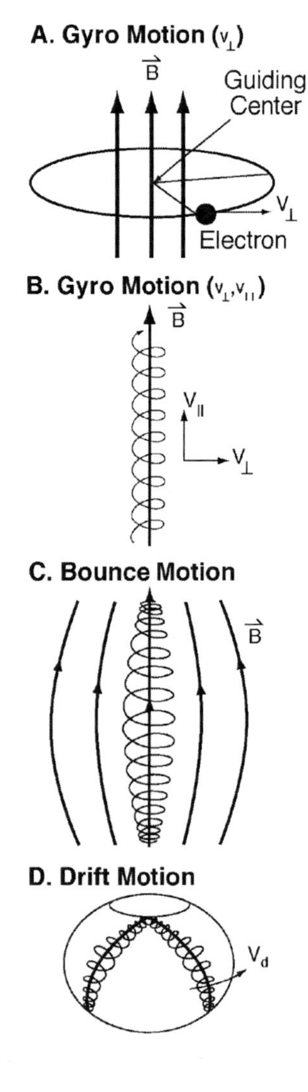

Figure 5.10. Panel A shows the motion of an electron having only a perpendicular velocity in a constant magnetic field. Panel B shows charged particle motion when it has both a parallel and perpendicular velocity. Panel C illustrates the bouncing motion that a particle will perform when it encounters a magnetic field that changes (increases or decreases) with time or position. Panel D shows the drift motion of a charged particle along a shell in the magnetic field due to a dipole, encountered in most planetary and astrophysical objects. (Courtesy NASA).

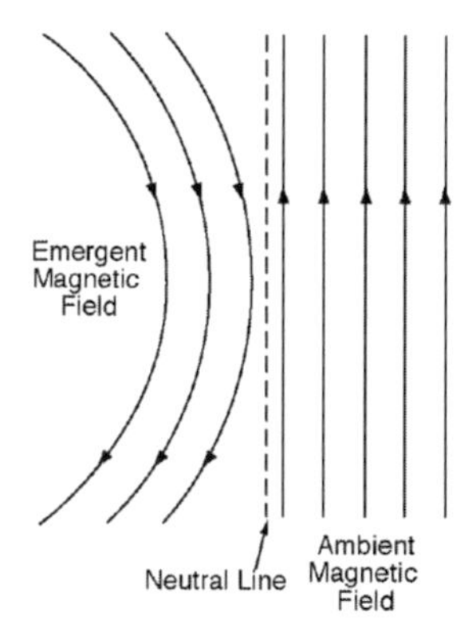

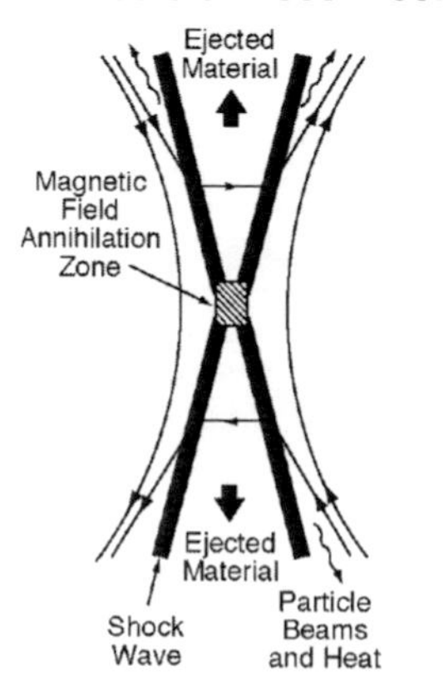

Figure 5.11. Diagram illustrating the process of magnetic annihilation, or reconnection. (Courtesy of NASA)

Our Closest Star, the Sun

As our closest star, the Sun is an object of immediate interest. It is an ordinary star, though special to us; it is not an especially hot star, and one that is middle-aged. The Sun has been a source of wonder, different beliefs and worship since the earliest times. From the God Ré for the ancient Egyptians, Shamash for the Babylonians, Surya for ancient Hindus, Helios for ancient Greeks, and Amaterasu (the Japanese Goddess of the Sun believed to be the ancestress of the royal family), among others, the Sun was both adored and feared by these peoples. Total solar eclipses were also a source of fear in ancient times. The oldest records of eclipses dating from around 800 BC were engraved on tablets found in Syria and China. The first recorded observations of the Sun, by the Chinese, date from 165 BC. Naked eye observations of sunspots were possible through dense fog or dust. In 1612 AD, Galileo and Kepler were the first astronomers to observe sunspots with a telescope.

Nowadays, the Sun is studied in three different ways, to address questions about the three main layers of its structure. The invisible interior is probed by helioseismology techniques, which, as for the Earth's seismology, "listen" to the sound of oscillations inside the Sun. The visible atmosphere is observed with imaging and spectrographic techniques, whereas the extended atmosphere is probed by *in situ* instruments. Wavelengths used to observe the Sun range from Gamma rays to radio waves. There are several ground-based solar observatories around the globe, but space observatories are preferred because the Earth's atmosphere is opaque to many wavelengths.

The Sun's core (25% of the solar radius, see Figure 5.12) is where thermonuclear fusion occurs. The energy produced in the core is transported out to a distance of 0.7 solar radii by radiation. Energy is then transported to the solar surface by convection. The core and the radiative zone rotate as a solid body whereas in the convection zone the (angular) speed of rotation varies with latitude. The temperature is ~ 15 million K in the core, progressively decreasing to ~ 5,800 K at the surface; the density decreases by a factor of 16,000 from core to surface.

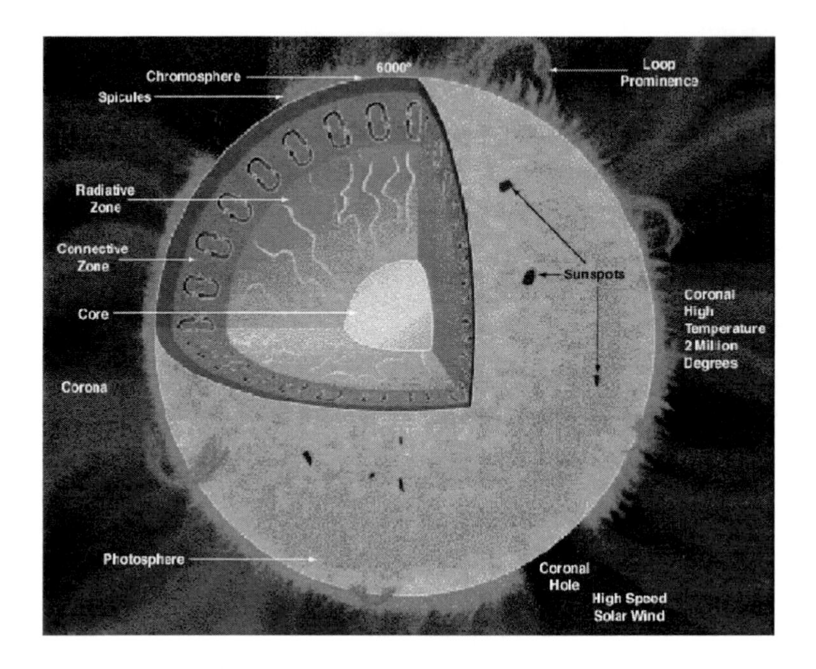

Figure 5.12. The structure of the Sun (Courtesy NASA).

The solar atmosphere is split into four layers. Rotating with the convection zone, the photosphere is the deepest; it is the "visible" surface of the Sun. (However, it is extremely dangerous to look directly at the Sun with the naked eye.) This layer goes out to a height of about 500 km, the temperature decreasing to 3927°C (4,200 K); most of the Sun's light comes from this layer. Solar features such as sunspots and faculae exist here. These are a manifestation of the Sun's magnetic field and of solar activity. Sunspots (see Figure 5.12) are seen as dark areas because they are colder than their surrounding plasma, and are the sites of strong magnetic fields. Faculae are brighter regions where the magnetic field is even stronger than on the rest of the solar disk. The photosphere is observed at visible wavelengths with imagers, magnetographs and spectrographs.

The chromosphere is the layer immediately above the photosphere, with a thickness of about 1,500 km. Here the temperature starts to increase, reaching 10,000 K. It can be observed from the Earth in two different ways - with the naked eye at the very beginning of a total solar eclipse (before the corona becomes visible) and with some instruments at some specific wavelengths. Features appearing on the solar disk are filaments and prominences, two versions of the same phenomenon, i.e. enhanced plasma regions contained by magnetic field lines, the first dark on the solar disk, and the second bright beyond the solar limb (the Sun's apparent edge) due to the contrast with the blackness of space. They are located along the neutral line of the magnetic field, i.e. at the region of opposite magnetic field polarities. This layer is observable from the ground at visible wavelengths where spectral lines of hydrogen and calcium are emitted, as well as at radio wavelengths. Coronographs are instruments built to simulate a total solar eclipse for making observations of this thin, and important, layer, either from space or from the ground.

The transition region is the next layer, where the temperature increases sharply up to 100,000 K. Here other solar features appear. Coronal holes, for example, are less bright areas, more or less extended on the solar disk, with lower temperatures and densities. Here, the magnetic field lines are open into the interplanetary medium, so that charged particles can escape from the solar atmosphere at high speed. This layer can only be observed from space, mainly in the ultraviolet and X-ray parts of the spectrum.

The corona extends to more than a million km (about 10 solar radii) from the Sun. Its temperature can exceed 2 MK, as its density continues to drop. The lower part of the corona can only be seen from Earth when the moon blocks out the light from the solar disk during a total solar eclipse. Violent events that release energetic plasma into interplanetary space can also be observed sporadically; these are solar flares and coronal mass ejections (CMEs). Flares are sudden events always associated with features like sunspots or prominences. They can vary in severity and they evolve through different phases. Since they emit at different wavelengths, they can be observed from the ground (radio) as well from space (ultraviolet). Coronal mass ejections are huge bubbles of plasma and magnetic field lines ejected from the Sun at speeds ranging from 500 to 2500 km/s, higher than the usual solar wind speed. They produce a bow shock ahead of them; if the CME is coming directly towards the Earth, it will cause extensive auroral displays as it interacts with our magnetosphere.

As mentioned earlier, different types of activity occur at different heights in the solar atmosphere. Among these phenomena, one that gave an early indication of the Sun's varying activity was sunspots. Their changing positions on the solar disk were used to find the rotation speed of the photosphere at different latitudes, whereas their changing number demonstrates the cycle of solar activity. The number of sunspots seen on the disk increases and decreases, with an average periodicity of 11 years (see Figure 5.13), and sunspots move towards the solar equator as solar activity increases. Sunspots, arched magnetic fields in prominences, and differential rotation are closely connected, and indicate that the cause of the solar cycle is its magnetic field. The Sun's magnetic field has its origin in its interior, being produced by a dynamo effect. As a consequence of differential rotation and the high electrical conductivity of the plasma in the convection zone, the magnetic field becomes twisted.

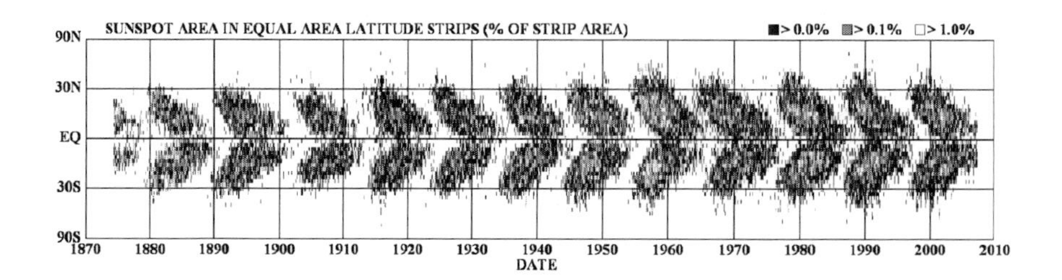

Figure 5.13. Maunder butterfly diagram. (Courtesy of NASA)

The average time for one rotation of the photosphere as seen from Earth is about 27.3 days, ranging from 26.3 days at the equator (whereas its sidereal rotation is about 25.3 days) and up to 30 days near the poles. The rotation rate varies also with altitude. The combination of the differential rotation and the magnetic field leads to the solar cycle, a mechanism that provokes the reversal of the Sun's dipolar magnetic field every 11 years; therefore the true period of the cycle is ~ 22 years. The classical model explains that differential rotation changes the poloidal magnetic field at solar minimum (mainly North-South) into an intense toroidal magnetic field (mainly East-West). This intense field floats up through the photosphere because of its buoyancy, and emerges out of the photosphere as active regions containing sunspots. There is a complex, even chaotic, reorganization of magnetic fields at the maximum of the solar activity, before the poloidal field closes up, with its polarity reversed.

The expansion of the corona into interplanetary space takes the form of a continuous flow called the solar wind. The solar wind carries the solar magnetic field with it, away from the Sun; energetic charged particles move in helical paths along these field lines to interact with the various magnetic environments of the planets. The topical subject of the interaction of the solar wind with planetary magnetic environments is known as space weather.

What happens at the edge of our solar system, at the outer boundary of the zone of influence of the Sun called the heliosphere? The heliosphere is like a bubble that extends far beyond the planets, blown outwards by the solar wind. Eventually it is stopped by the interstellar medium; this happens where the pressures balance, at a rather ragged boundary called the heliopause. Before this, the solar wind travels at an average speed ranging from 200 to 800 km/s until it reaches the termination shock, investigated by the two NASA Voyager spacecraft, where the speed of the solar wind drops abruptly as it begins to feel the effects of the interstellar medium. The heliosheath is the outermost region of the heliosphere, just beyond the termination shock. Beyond the heliopause is another bow shock formed as the heliosphere plows its way through interstellar space.

Sun-Earth Interactions

The Sun affects the Earth's environment in several complex ways that are investigated directly by scientific instruments aboard satellites. In the early days of the space age the first US satellite (Explorer 1) discovered the Van Allen radiation belts around the Earth; Explorer 3 and 4 explored these in more detail.[3] The discovery prompted one young researcher to exclaim – "my God, space is radioactive!" Then the magnetometer aboard a satellite in a highly elliptical orbit around the Earth, IMP 1, the first Interplanetary Monitoring Platform, investigated the detailed shape of the magnetopause.[4] This is the outer boundary of the magnetosphere, the region of the Earth's magnetic field in space, which is illustrated in Fig.5.14.[5]

The geomagnetic field is compressed by the solar wind which flows away from the Sun at supersonic speeds, > 400 km/s, pulling the Sun's magnetic field lines out with it. On the dayside of the Earth the magnetopause is often located ~ 10 Earth radii (R_E) upstream from the Earth's center. When the Sun is active and producing much faster solar wind streams than usual (up to ~ 1000 km/s) the magnetopause lies closer to the Earth, at 7 R_E or occasionally even at 6 R_E. If the interplanetary magnetic field lines have a southward component, they can readily "connect" to the northward pointing geomagnetic field lines at the dayside magnetopause. As the solar wind flows past the magnetosphere, it pulls geomagnetic field lines on the night side out into a comet-like tail. The magnetotail extends beyond the Moon's orbit, at ~ 240 R_E from the Earth, on the night side.

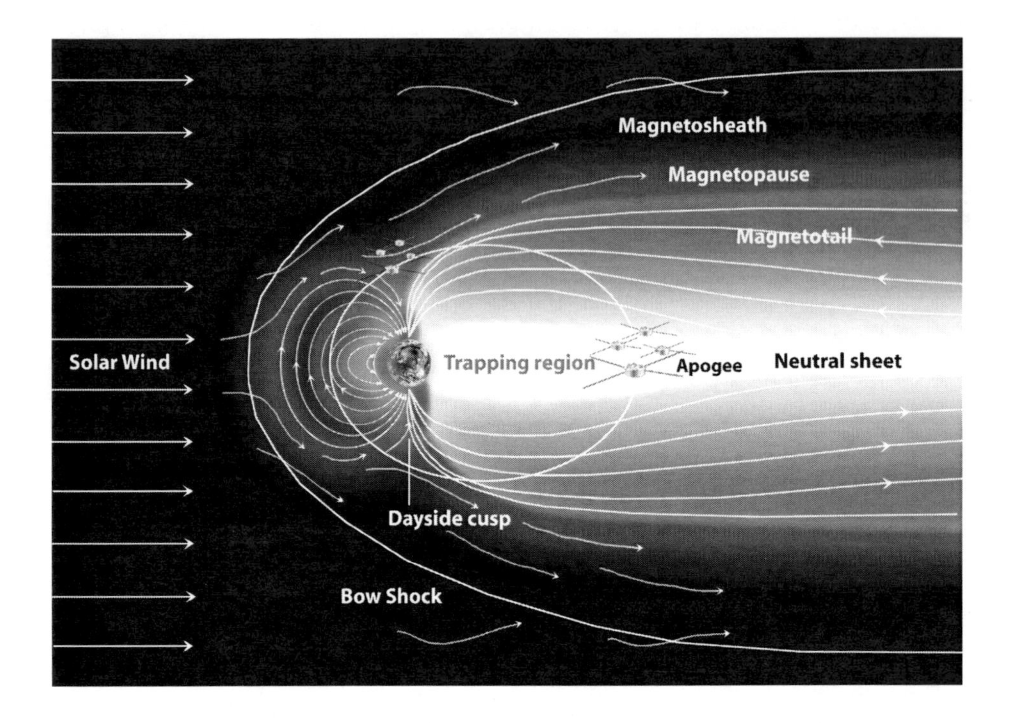

Figure 5.14. This diagram shows the solar wind compressing the Earth's magnetic field, and confining it within the magnetosphere. Here the arrows indicate the direction of the Earth's magnetic field. The outer boundary of the magnetosphere is the magnetopause (at a geocentric distance of ~ 6.5 RE on the dayside of the Earth, here). Solar wind plasma can enter the magnetosphere at the northern and southern cusps. The elliptical orbits of the four closely spaced spacecraft of the ESA Cluster mission are shown as the elliptical orbit. For clarity the apogee is shown at a geocentric distance of 11 RE whereas, in reality, it is at 19 RE (from Goldstein 2003).

(Courtesy of Apogee Books).

The interplanetary plasma beyond the magnetosphere, in the magnetosheath, is rather turbulent until the bow shock is encountered a few R_E further upstream. Here the magnetic field suddenly changes its magnitude and/or direction, and so do the solar wind velocity and temperature. The bow shock is rather like the shock wave formed ahead of a model aircraft in a supersonic wind tunnel. It is a region where charged particles are accelerated. An excellent introduction to this topic and to the entire physics of the magnetosphere is given in the 1995 book "Introduction to Space Physics" by M.G. Kivelson and C. T. Russell.

ESA's Cluster mission has four closely spaced satellites orbiting in formation in a different highly elliptical orbit (also shown in Figure

5.14). Properties of the plasma and energetic charged particles, and also of magnetic and electric field fluctuations and waves, are observed with a comprehensive suite of sophisticated instruments (see, for example, Gustafsson, et al.[6]). These are designed to study in detail the plasma physics of solar wind entry into the magnetosphere, especially in the vicinity of the cusp on the magnetopause.[7] On the night side they sometimes pass through the source region of strong natural electromagnetic waves,[8] at an audio frequency of a few kHz, between 0.2 and 0.4 times the local electron cyclotron frequency, f_{ce}. These waves are termed chorus; they travel through the magnetosphere in the whistler mode.[9] A hot research topic nowadays[10] is the acceleration of electrons to very high energies (~ MeV) by chorus waves. These so-called "killer" electrons can penetrate satellites and cause damage to electronic circuits inside. The damage may be either temporary (called a "latch up") or, even worse, permanent.

The northern lights, called the aurora borealis, and the southern lights, called the aurora australis, have been observed from above by astronauts on the Space Shuttle, as depicted in Figure 5.15. TV cameras sensitive to ultraviolet light on several satellites have also recorded Auroras, starting with Dynamics Explorer and, more recently, Polar. Images acquired by the instrumentation on these platforms clearly show the evolution, from minute to minute, of the location of the auroral oval, a ring of light around the geomagnetic pole at magnetic latitudes of ~ 67 degrees.[11] The aurora is produced when electrons with energies of ~ 10 keV moving down geomagnetic field lines strike the upper atmosphere.[12] They excite atmospheric atoms and molecules into higher energy states; these then fall back to their ground states, emitting visible light of a certain wavelength (or color). At times when the solar wind is disturbed by excessive solar activity, the auroral oval expands to lower latitudes. The initiation of a specific feature, e.g., a brightening, of the auroral oval, and its subsequent temporal and spatial development, are well investigated by studying sequences of such images.

Figure 5.15. Observed from the Space Shuttle are the southern lights (the aurora australis), in a band stretching from East to West, against a background of stars. These are caused by energetic electrons coming down the Earth's magnetic field lines and hitting atoms and molecules in the upper atmosphere. (Courtesy of NASA)

The plasmapause[13] at ~ 4 R_E from the center of the Earth is the geomagnetic field aligned outer boundary of the plasmasphere, which has recently been imaged by the IMAGE satellite (Imager for Magnetopause-to-Aurora Exploration).[14] The plasmasphere is filled by plasma flowing out of the ionosphere. Discovered experimentally in 1924 by radio experiments on the ground, the ionosphere at heights between ~ 80 and ~ 300 km reflects HF (High Frequency, 3 to 30 MHz) radio signals used for communications around the world. Radio signals at ~ 1.23 and 1.58 GHz from the GPS satellites,[15] which nowadays are essential for positioning and navigation on - and above - the Earth's surface, pass through the ionosphere. The ionosphere adversely affects them if it is disturbed following solar activity.[16]

During strong auroral displays, huge electric currents (~ millions of Amperes) flow around the auroral oval in the ionosphere at heights of ~ 110 km. These are fed by currents flowing down geomagnetic flux tubes in certain Local Time regions, the circuit being completed by upward currents elsewhere. Power generation for this gigantic electric circuit relies upon the dynamo action of the solar wind speeding past the Earth's magnetic field. The response of the near-Earth space environment to disturbances on the Sun is a most important topic, which is now termed "space weather" and which is discussed in the next section.[17]

Space Weather and its Effects

What caused the huge power outage in March 1989 in North America? Everything started with a major solar flare, which happened to be observed from the Kitt Peak Solar Observatory on 9 March. Several eruptions followed, sending X-radiation and clouds of ionized gas in the direction of the Earth, because of the flare's central location on the solar disk, as seen from Earth. A positive effect of this eruption was a splendid aurora seen at as low a latitude as the Florida Keys! A negative effect was the breakdown of the electricity supply that paralyzed the Canadian province of Quebec and caused major damage to the power grid in the USA.

The first evidence of a link between solar activity and terrestrial disturbances goes back to 1859 when Richard Carrington observed a sudden and intense bright light on the Sun. Some 24 hours later there was a strong magnetic storm at the Earth. Before that, Von Humboldt, who observed strong and irregular variations of the Earth's magnetic field correlated with bright displays of the northern lights, reported the first evidence of a magnetic storm in 1808. The main causes of magnetic storms are violent and sudden releases from the Sun of bubbles of plasma and strong magnetic fields that propagate far into interplanetary space. Such Coronal Mass Ejection (CME) events, and especially High Speed CMEs or multiple CMEs, are the major cause, but solar flares, filament eruptions, active regions and also enhanced solar wind speeds also contribute to powerful geomagnetic storms. As a CME cannot be observed directly from Earth, we need instruments aboard spacecraft to observe this phenomenon which was seen, for the first time, in December 1971 with a white light coronagraph on NASA's OSO-7 spacecraft (the seventh Orbiting Solar Observatory).

Later, the US National Space Weather Council (1995) clearly defined this phenomenon: "The term Space Weather generally refers to conditions on the Sun and in the solar wind, magnetosphere, ionosphere, and thermosphere that can influence the performance and reliability of space-borne and ground-based technological systems and can endanger human life or health". The National Oceanic and Atmospheric Administration (NOAA) classified storms as follows:

- Geomagnetic storms, disturbances in the geomagnetic field caused by gusts of solar wind.
- Solar Radiation storms: elevated levels of radiation that occur when the fluxes of energetic charged particles increase.
- Radio blackouts: disturbances of the ionosphere caused by X-rays emitted by the Sun.

Explosive solar events generate geomagnetic storms and cause ionospheric disturbances as well as magnetically induced currents, and also disrupt telecommunications and radio positioning as illustrated in Figure 5.16.

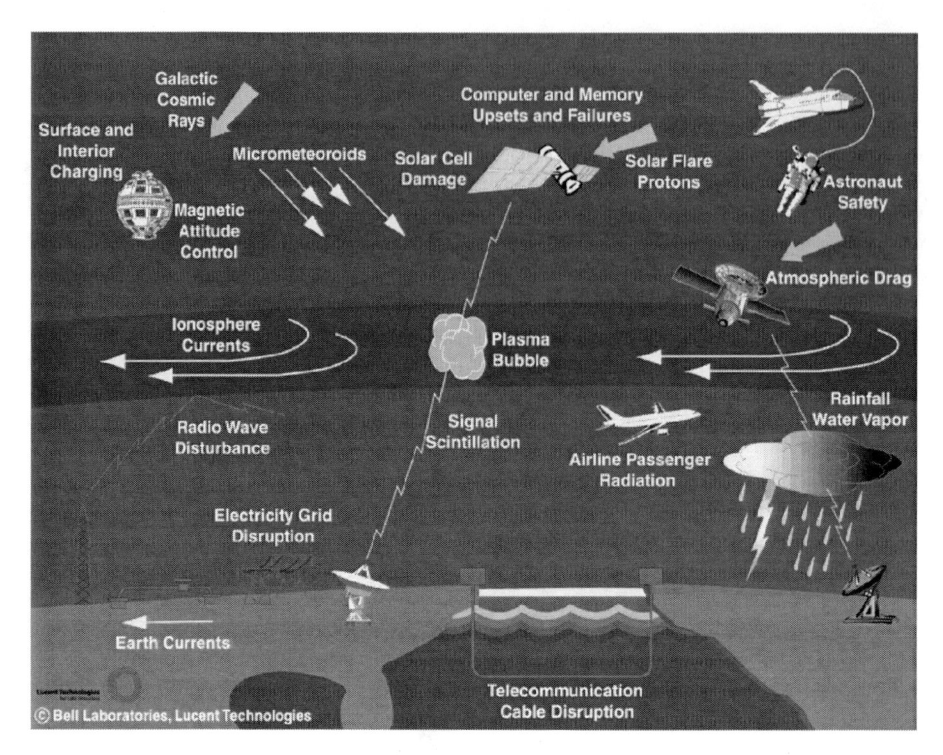

Figure 5.16. Space Weather Effects and Impacts on the Earth's Environment

(Based on original figure by L. J. Lanzerotti, Bell Laboratories, Lucent Technologies).

Explosive solar events have various effects on human beings as has been discovered from space exploration programs. Long-term exposure to the space environment can entail either short-term or delayed biological effects for astronauts living in space (e.g., on the ISS or the Space Shuttle) or carrying out extra vehicular activities (EVA) in space or on the surface of the Moon or Mars. Exposure to solar radiation (UV or X-rays) can lead to cancers or genetic changes. But current technologies cannot effectively protect astronauts against enhanced fluxes of energetic charged particles, especially heavy ions, during strong magnetic storms. Therefore, carrying out an EVA at the same time as a CME collides with the magnetosphere could lead to deadly events.

From the launch phase to their final orbit, spacecraft are not very well protected from radiation. Nowadays electronics components are becoming smaller and smaller, and therefore more vulnerable to energetic charged particles. Damage due to such radiation can be either transient or cumulative. Single effect upsets (SEU) are transient events that happen when highly energetic charged particles (> 50 MeV) penetrate spacecraft shielding, and then change the bit states of computer chips. This situation can lead to changes in onboard software and memory content, and later provoke a "latch-up" (temporary outage requiring reset action) or a permanent failure of an onboard computer. Noise in detectors (especially CCD detectors) can also be a temporary disturbance. It can lead to bad data acquired by an instrument, or even the loss of a spacecraft that uses a star tracker to acquire its position, since a bright point generated by an excess of photons could be misinterpreted as a referenced star. Bombardment by charged particles also has a cumulative effect on spacecraft components, causing degradation and accelerating the aging process of spacecraft parts. Solar panels are one example of sensitive materials whose lifetime is dramatically decreased each time a proton event occurs.

When the shock ahead of the CME front arrives at the Earth's magnetosphere, the magnetosphere is so compressed that a geostationary spacecraft is now in interplanetary space. The enhanced solar wind of the CME drags the spacecraft, moving it by ~ 1 meter typically, but by up to 50 meters sometimes. At the same time the spacecraft's magnetic sensors become confused and their data cannot be used for orienting the spacecraft.

LEO space debris tracking is also a critical issue, since the tracking process becomes inaccurate and must be reinitialized after each magnetic storm. For example, during the 1989 storm, the increased density of the atmosphere caused increased drag on an orbiting object so that it lost 30 km in altitude and there was a significant decrease in its lifetime too.

During a storm some terrestrial radio services become unreliable due to absorption by the ionosphere. Some of the many such applications here include aviation, coastal marine traffic, and search and rescue services. Mineral resource exploration and geophysical prospecting, using airborne magnetic surveys at high latitudes, is also upset during a magnetic storm. GPS signals are affected when solar activity causes sudden variations in the density of the ionosphere, which reduces the accuracy of positioning and navigation systems on the ground, in particular at high latitudes.

The Sun emits light at all wavelengths. It is the most variable at short and long wavelengths. The variability in solar emissions at radio frequencies can be so intense that the radio signal from a spacecraft can be overwhelmed. This could be critical when tracking a spacecraft, or when downloading telemetry from or sending commands to the spacecraft. Indeed if there is sufficient warning preventive action is taken. Communications and other in-orbit satellites are often powered down during such events since these powerful solar bursts can completely and permanently disable a satellite.

Commercial aviation today strongly relies on GPS and radio communications for several phases of flight. The risk of disturbances to radio propagation increases at high altitudes and high latitudes. Air routes crossing the Arctic have been developed for economic reasons, mostly from the US to China. However, when there is a magnetic storm, radio equipment on aircraft using these polar routes is very sensitive to radio disturbances, and communications can be interrupted. On these routes there is also an increased risk of exposure to radiation for crew and passengers alike, as there is too for avionic equipment.

When there is a storm alert, flights need to be re-routed to lower latitudes for obvious safety reasons. This automatically leads to delayed flights, higher fuel consumption and, of course, various financial losses. For example, in January 2005, four air routes were closed for the first time due to a strong storm. Several planes completely lost contact with the ground since no radio signals could get through, and no GPS techniques were operating; one airline announced a loss of more than $250M in four days.

There is growing evidence indicating that changes of the geomagnetic field can affect biological systems. Several research groups have begun to study relations between geomagnetic activity and psychiatric disorders. Animal life also seems to be affected by these storms, in particular the navigational abilities of pigeons and dolphins.

The electrical power grid infrastructure, particularly at high latitudes, is affected by storms. Because transformers can be destroyed by induced currents from solar storms and thus lead to serious power outages. There can be a domino effect on the mesh of transformers directly connected to the one that fails. The corrosion of oil and natural gas pipelines is another concern following a geomagnetic storm.

Because many Earth applications rely so heavily on the space environment nowadays, there is a need to predict space weather events that may endanger humans and their services. There are many different users of space weather alerts and information, with the requirements mainly being for:

- Space exploration, aviation, and satellite operators, and grid operators, etc., who need to know, in real-time, about upcoming events and also about clear periods,
- Space launch operators who tend to delay rocket launches when there is a high probability of a solar storm,
- Spacecraft re-insurance companies who need to know if the risk was predictable or not,
- Spacecraft designers who need to assess risks, analyze them and mitigate their effects by studying the database of past failures,
- Space tourism passengers who would not risk flying if a solar event is forecast,
- Military activities around the world and in space, and
- Airlines for which radiation monitoring is critical.

All these customers' needs require a better understanding of the space environment, the Earth's magnetic environment and geomagnetic activity, and on the capability to develop better models.

Space Observations beyond the solar system

Telescopes looking outward from orbit around the Earth observe the Universe across the electromagnetic spectrum, from radio waves to the infrared (IR), through the visible and ultraviolet (UV), to X-rays and gamma-rays; much new astrophysical knowledge has been found. Instruments on satellites make in situ observations of the properties of the plasma environment around the satellite. Instruments looking downward from Low Earth Orbit (LEO) or from a similar orbit around a planet observe the planetary surface and its atmosphere.

Background

Astronomers explore the Universe by observing the electromagnetic radiation emitted by the planets, stars, galaxies, the interstellar medium, and the intergalactic medium. However, most of this radiation is absorbed by the Earth's atmosphere (see Figure 5.17) and never reaches the surface of the Earth. As a result, for most of the history of mankind, our view of the Universe has been limited to the very narrow optical (visible) band of the electromagnetic spectrum. Another atmospheric window exists at radio wavelengths and, although radio waves from our galaxy were first detected in 1933, it was not until after World War II that techniques were available to explore the radio spectrum fully. A third partial window exists at infrared wavelengths. The first infrared observations were of the Sun in the early 1800's, but, due to the difficulty of developing sensitive detectors, it was not until the 1960's that ground-based infrared astronomy made a significant impact.

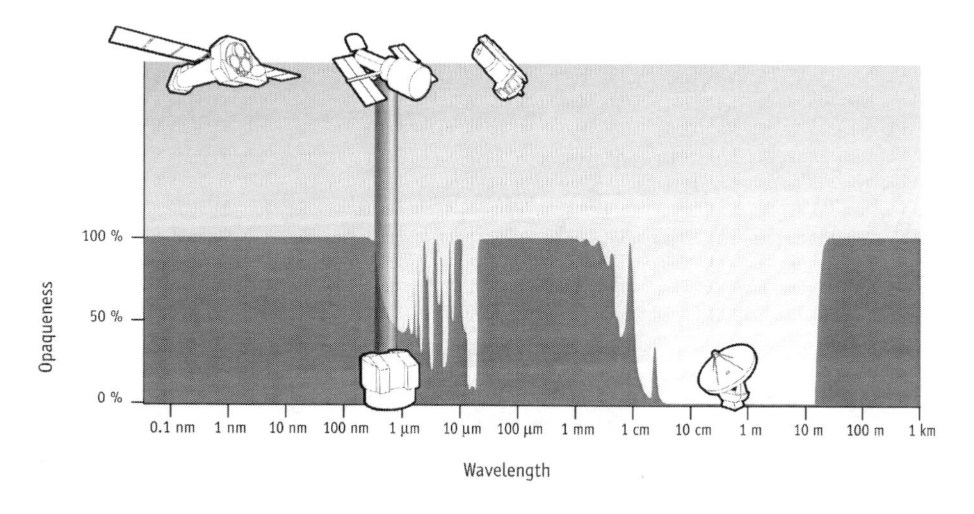

Figure 5.17. This diagram depicts the opaqueness of the Earth's atmosphere to transmission of electromagnetic radiation as a function of wavelength. The atmosphere is transparent to visible light, most radio waves, and part of the infrared. The three satellites images shown, from left to right, are the Chandra X-ray Observatory, the Hubble Space Telescope, and the Spitzer Space Telescope. (Courtesy of NASA/JPL-Caltech)

Since the 1960's, spacecraft carrying telescopes above the atmosphere have opened up the entire electromagnetic spectrum for astronomers to observe, and have permitted observations free from the distorting effects of the Earth's atmosphere. Over the past half-century space telescopes have revolutionized our knowledge of our Universe.

Compared to ground-based observations, observing with space telescopes is not easy and is very expensive. However, for observing most of the electromagnetic spectrum there is no alternative. Instruments in space have to work for long periods and unattended, in many cases in difficult environments, due primarily to the extremes of temperature and damaging radiation experienced. The mass, power and volume available for instruments are very limited on spacecraft. It takes many years to design, build, test, and launch a telescope into space. Instrumentation placed at the focus of space telescopes consists of spectrometers (both narrow and/or broadband), interferometers, and imagers. To find extra solar planets, other telescope techniques used in space include accumulating intensity data over specific wavelength regions, and producing light curves as functions of time to detect a planet occulting (or slightly blocking) light emitted by the observed star.

Space Telescopes

Telescopes are a means of collecting electromagnetic radiation from distant objects and focusing it on a detector. The detector absorbs the radiation and converts it into an electrical signal from which an image of the source can be produced or from which the intensity of the source as a function of wavelength can be measured. Telescopes can be classified as either "reflectors" (using mirrors) or "refractors" (using lenses). Today, most large telescopes in space are reflectors, and the primary collecting surface is usually parabolic in shape to convert a parallel beam of radiation to a single, sharply focused spot of light. The size of a telescope is described by the diameter, D, of the primary mirror. Most large space telescopes are of the Cassegrain design, in which light from the primary mirror is reflected back by a secondary mirror (hyperbolic), through a hole in the primary mirror, and focused behind the primary mirror. The advantage of this type of design is that the focal plane detectors can be mounted on the mirror support structure. To obtain good image quality over a large field of view, a modification called the Ritchey-Chretian (RC) design is used in which both the primary and secondary mirrors are hyperbolic. The Hubble Space Telescope (HST) launched in 1990 is an example of this type of telescope (see Figure 5.18). The angular resolution of a telescope is proportional to the observed wavelength divided by the diameter of the telescope mirror (λ/D).

Figure 5.18. An artist's impression of the NASA/ESA Hubble Space Telescope in orbit above the Earth (Courtesy of NASA).

Radio Astronomy

Although the atmosphere is transparent to most of the radio spectrum, one advantage that space offers is the ability to perform very long baseline interferometry at these wavelengths. Using two telescopes, one on the surface of the Earth and another in a highly elliptical orbit (apogee of 21,000 km), and measuring both the amplitude and the relative phase of the same radio signal at both telescopes, an angular resolution can be achieved which is proportional to the wavelength divided by the separation distance of the telescopes. This technique provides a very significant increase in angular resolution at radio wavelengths, even better than can be achieved by ground-based interferometers. An example of this type of mission is the Japanese satellite Halca, which was used in conjunction with a ground-based interferometer.

Infrared and Sub-millimeter Astronomy

Space offers tremendous advantages for infrared astronomy. Because the atmosphere absorbs most of the infrared spectrum, an infrared telescope in space can observe the entire infrared spectrum. Also the telescope mirror can be cooled (usually by liquid helium), to reduce the unwanted infrared background radiation by a factor of a million. Instruments at the focal plane include cryogenically cooled infrared array detectors, both for imaging and spectroscopy. Our knowledge of the birth and evolution of galaxies in the very early Universe, the nature of exoplanets, and the birth and evolution of stars in our galaxy has increased significantly because of observations with space infrared telescopes with even modest mirror sizes (< 1 m in diameter). (The earliest infrared space telescopes were flown aboard rockets and high-altitude balloons and aircraft.) Examples of satellite missions include NASA's Infrared Astronomy Satellite (IRAS), ESA's Infrared Satellite Observatory (ISO), NASA's Spitzer Space Telescope, and the Japanese mission AKARI. Spitzer was place in a solar orbit, gradually drifting away from the Earth, to reduce the heat load on the telescope from the Earth and the Moon and to permit more efficient observations. NASA's WISE, a mid-infrared all-sky survey, will be launched in 2010 into a polar orbit. SOFIA is a Boeing 747 aircraft with a 2.7-meter telescope that will initiate its science program in 2009. ESA's Planck and Herschel missions were launched in 2009 into L2 Lagrangian point orbits. Herschel observes the sub-millimeter part of the spectrum and Planck will measure the cosmic microwave background radiation produced by the Big Bang. Already these satellites are producing amazingly detailed results for that part of the sky that they will observe for many years with the hope of identifying Earth-sized planets. Planned for launch in 2013 is the James Webb Space Telescope (JWST), which will have a 6.5-meter diameter mirror. This very high resolution telescope will also orbit around the L2 Lagrangian point and thus will not have to contend with the filtering effects of the Earth's atmosphere.

Optical and Ultraviolet Astronomy

Even though optical astronomy can be done from the Earth's surface, an orbiting optical telescope has the advantage that it is not affected by the blurring effects of the Earth's atmosphere, and hence it achieves a much higher resolution and sensitivity. The Hubble Space Telescope (HST), launched in 1990, has produced a series of very important astronomical discoveries at both optical and ultraviolet wavelengths, and is a prime example of what can be achieved from space. Examples of these discoveries include accurate distance measurements to nearby galaxies, refined estimates of the Hubble constant, which specifies the rate of expansion of the Universe, new insights into galaxy evolution, the deepest image of our Universe ever made, and evidence for massive black holes in the centers of galaxies. The HST Cassegrain telescope has a 2.4 m diameter mirror. At its focal plane is a series of four instrument bays for astronomical cameras, photometers, spectrometers, and polarimeters. Because the telescope can be serviced by the Space Shuttle, the focal plane instruments have been replaced and/or repaired on several occasions.

Ultraviolet radiation at wavelengths less than approximately 320 nm is absorbed by the Earth's atmosphere. Therefore ultraviolet observations must be carried out from rockets or satellites and many such telescopes have been flown. These include astronomical missions with photometers, cameras, low and high-resolution spectrographs, and polarimeters. HST remains the primary ultraviolet observatory, but numerous smaller missions of moderate size have been flown including IUE, EUVE, FUSE and GALEX. Ultraviolet space astronomy has been especially important in improving our understanding of the nature of the interstellar medium, the nature of stellar atmospheres, properties of the interstellar dust, the composition of comets, and the properties of active galactic nuclei.

X-Ray Astronomy

X-rays are completely absorbed by the atmosphere, so telescopes and instruments to observe celestial X-rays were first flown on balloons and rockets, but are now flown primarily on spacecraft. X-rays originate from an extremely hot gas, are associated with very energetic phenomena, and arise from such objects as neutron stars and black holes. Normal Cassegrain telescopes cannot be used to focus X-rays because X-rays impinging perpendicular to any material are absorbed rather than reflected. X-rays can be focused by selecting a mirror material that reflects the radiation if it is incident on the mirror at small, or grazing, incident angles. The materials usually used are gold or nickel. The critical reflection angles are of the order of 1 degree. The highest resolution telescopes use a reflection from a parabolic mirror followed by a reflection from a hyperbolic mirror. The Chandra X-ray Observatory (CXO) is the prime X-ray mission today, consisting of nested sets of circular grazing incidence parabolic and hyperbolic mirrors. Other mission examples include the XMM-Newton observatory, Rossi X-ray Timing Explorer (RXTE), ROSAT, BeppoSAX, and Swift. X-ray detectors are required to determine the location of an X-ray photon in two dimensions and have reasonable detection efficiency. X-ray detectors at the focal plane of these telescopes include imaging proportional counters, charge-coupled device (CCD) detectors, similar to those in visible-light cameras, micro-calorimeters that can detect one X-ray photon at a time and measure its energy, and transition edge sensors, that are a more advanced type of micro-calorimeter. Both these latter sensors need to be cooled to liquid helium temperatures.

Gamma-ray Astronomy

Gamma rays are the most energetic and shortest wavelength photons in the electromagnetic spectrum. Because the atmosphere absorbs most all of them, cosmic gamma-ray observations must be implemented using high-altitude balloons and satellites. Gamma rays are so penetrating that they simply pass through most materials and cannot be reflected by a mirror like optical or X-ray photons. Some gamma-ray telescopes act as "light buckets" and detect photons incident on the sensitive area of the detector. These specialized telescopes use scintillators or solid-state detectors to detect a gamma ray and convert it into an electronic signal. Another class of detectors is based on the nature of the gamma-ray interaction process, using either imaging pair-production or Compton scattering to determine the gamma-ray arrival direction or a coded-mask in front of the detector to allow an image to be reconstructed. Gamma-ray detectors usually have poor angular and spectral resolution. The current satellite missions include INTEGRAL and GLAST. Gamma rays are generated in the most violent parts of the Universe, being the result of explosions or high-energy collisions. One of the most spectacular observations has been the detection of gamma-ray bursts, lasting from fractions of a second to some minutes. They appear to come from the distant Universe; therefore they are some of the most energetic events ever observed. NASA's Swift satellite is devoted to detecting these bursts and identifying their source.

Space Physics

In the field of space physics, space borne instruments primarily make measurements *in situ* but imaging systems at specific wavelengths are used to observe aurora and the Sun. The main instruments in this area consist of static magnetic and electric fields (DC), electromagnetic radiation with frequencies from Hz to MHz, and a variety of charged particle measurements. Charged particles (ions and electrons) are measured in specific directions with respect to the magnetic field (called pitch angles) over specific energy ranges extending from eV to GeV. Ion measurements, typically from mass spectrometers, can be made as total ions or delineated species such as hydrogen, helium, oxygen, etc., at various states of ionization. Recently, a number of space physics instruments have started to observe "hot neutral" ions which have their origin in charge exchange processes between a cold neutral atmospheric ion giving up an electron to a fast moving ion (typically accelerated within a magnetosphere) in a collision. These neutral atom detection instruments can create pseudo images of large spatial regions. Magnetospheric radio sounding has also been used to find the density distributions of remote regions of plasma to very low densities (~ 10^5 /m^3) and to very great distances from the Earth (up to 30,000 km).

Earth Observation

Earth science uses both active and passive remote sensing techniques to make observations of the integrated Earth system (i.e. the land, oceans, ice sheets, and atmosphere), especially from Low Earth Orbit. Passive sensors detect scattered, reflected and/or absorbed solar radiation and emitted thermal radiation from the Earth to acquire spectral, spatial and temporal measurements of the solid Earth, its ice caps, the oceans, vegetation and the atmosphere. Typical instrument pointing is nadir (straight down) or side-scan to cover large land areas, while some instruments use solar and lunar occultations in a limb-scanning mode to probe the atmosphere with high vertical resolution. Limb-scanning is typically used to make measurements of the vertical structure of aerosols, ozone, water vapor, and other important trace gases in the upper troposphere and stratosphere. Active remote sensing employs pulses of electromagnetic energy generated by the orbiting instrument (typically lasers and microwaves) transmitted to the Earth and scattered, reflected or absorbed and re-radiated by the atmosphere, oceans, or land surfaces. A tiny fraction of the scattered or reflected energy is collected by a telescope and detected. Typical lidar (i.e. laser) and radar measurements include sounding and profiling of the atmosphere to deduce the atmospheric composition, cloud and aerosol distributions, major/minor trace gas content and wind speeds and directions. Lidar and radar are also used to map the topography of the solid Earth, glaciers and polar ice caps, profile vegetation, measure sea surface heights, and probe the subsurface structure with ground penetrating radar.

Planetary Observation

For planetary science many of the same instruments used for Earth science and space physics are applicable. However, in addition to orbital remote sensing techniques, planetary landers and rovers carry out *in situ* measurements. These include geophysical instruments such as seismometers and heat flow probes, mass spectrometers and gas chromatographs for compositional and organics analysis, neutron and gamma ray spectrometers as well as X-ray diffraction and X-ray fluorescence instruments to discover the elemental composition of a sample. Compositional analysis can also be done with laser ablation spectroscopy and laser induced breakdown spectroscopy from several meters away from the rover on the planetary surface. Further, for planetary bodies with atmospheres, acoustic anemometers and lidars can measure vector winds and profile clouds and aerosols from the surface up to > 10 km altitude. Radio science occultation techniques use an orbiting spacecraft's Radio Frequency (RF) transmitter to make measurements from which the ionospheric and atmospheric properties of a planetary body may be derived. In such cases the spacecraft moves behind the planet while using its RF signal to communicate back to Earth. The planet's ionosphere and atmosphere cause dispersion and refraction of the RF signal that can be measured and used to better understand that intervening environment.

Astrobiology[18]

Astrobiology crosses the boundaries of all disciplines in order to answer some of the most fundamental questions: "Is there life outside of our planet?", "Does it look (biologically) like us?" and, "Are there other intelligent life-forms in the Universe?" Here we will start with an introduction to Astrobiology and set the conditions for life as we know it and discuss why this is an important endeavor. Finally, we will present an overview of some of the most exciting opportunities available in the Astrobiology research community.

What is Astrobiology?

Carl Sagan (Cosmos, 1985) wrote: "*The nature of life on Earth and the search for life elsewhere are two sides of the same question: the search for who we are.*" Thus astrobiology is the study of the origin, evolution, distribution, and future of life in the Universe. It addresses three basic questions: how does life begin and evolve, does life exist elsewhere in the Universe, and what is the future of life on Earth and beyond? By its very nature, astrobiology research is international, interdisciplinary, and intercultural and has many societal implications beyond the purely scientific. Its success depends critically upon the close coordination of diverse scientific disciplines such as molecular biology, ecology, planetary science, astronomy, information science, space exploration technologies, and other related disciplines. At the same time, astrobiology encourages planetary stewardship, the recognition of ethical issues associated with exploration and societal implications of discovering other examples of life, and envisioning the future of life on Earth and in space. Astrobiology has a strong emphasis on education and public outreach; it offers a crucial opportunity to educate and inspire the next generation of scientists, technologists, and informed citizens.

Seven science goals have been developed in order to try to begin to answer these questions that have been asked in various ways for generations: understanding the nature and distribution of habitable environments in the Universe, exploring for habitable environments and life in our solar system, understanding the emergence of life, determining how early life on Earth interacted with its changing environment, understanding the evolutionary mechanisms and environmental limits of life, determining the principles that will shape life in the future, and recognizing signatures of life on other worlds and on early Earth.

Conditions for Life

The search for life begins with an understanding of what are the basic necessities for life. These lead to the determination of the requisite conditions to support these basic necessities and ultimately define specific parameters that serve as our guide for locating and detecting life or the signatures of life. The most basic necessity to sustain life as we know it is access to liquid water, but the existence of liquid water alone does not guarantee that life will develop. Additional requirements include elements such as carbon, nitrogen and phosphorus and an energy source. The energy source may either use inorganic compounds (for chemoautotrophic life forms) or organic compounds (heterotrophic life forms), or light (phototrophic life forms).

Habitability, or the factors which when combined create the necessary conditions to support life, depends upon the characteristics of both the star system and the planetary body. The size and type of the central star most directly affect this, with the intensity of radiation output defining a minimum and maximum orbital radius from the star within which liquid water exists on a planetary surface. The Habitable Zone (HZ) of our solar system is from about 0.9 to ~ 1.5 AU. Only the Earth lies clearly within this boundary, with Mars being on its extreme outer edge. Distances less than the minimum result in super-heated surfaces incapable of supporting liquid water, such as found on Mercury, or conditions which cause a runaway greenhouse gas situation such as found on Venus. At distances greater than the maximum conditions are too cold to support liquid water (at least on open surfaces) as, for example, on the moons around the gas-giants.

A planets' capability to support liquid water can be modified to a certain extent by the type and characteristics of the planet (i.e. factors which include its size, geological activity, mineralogical composition, magnetosphere, and the type and thickness of its atmosphere). Furthermore, liquid water can exist well outside this classical habitable zone. The tidal heating of a moon around a planet outside the habitable zone can sustain liquid water, such as is exemplified by the Jovian moon, Europa.

While the minimum and maximum distances vary from star-system to star-system, generally, the larger the star, the hotter and quicker it burns.

This could mean that life will neither have the appropriate conditions nor the necessary time to develop. Planetologists interested in finding potential habitable planets in other star systems therefore focus their searches on stars that are known to have the correct characteristics. The gas-giants so far discovered around large stars may have habitable moons and, thus, these stars cannot be automatically eliminated from the search for far-away habitable planetary bodies.

Why is Astrobiology an Active Science Area Now?

In his book A Pale Blue Dot, Carl Sagan noted: *"The Earth is the only world known so far to harbor life."* The search for life in our solar system is a continuous process. There have been many ideas over the years, such as hidden cities under the thick layers of the clouds of Venus and the infamous canals of Mars first drawn by Giovanni Schiaparelli in 1877 and supposed by Percival Lowell in 1895 to be of artificial origin. In both cases, theories and myths were over-ridden as technology improved direct observations and after spacecraft had landed on these planets in the 1960's. The Venus missions, principally led by the former USSR, determined that the surface conditions were too harsh for life to survive due to high pressures, temperatures, and acidity. The US Mars Viking missions included one experiment designed to detect CO_2 respiration of active microbial life, and a second to detect organic carbon by super-heating a soil sample and measuring vaporized compounds via Gas Chromatograph-Mass Spectrometry (GC-MS). The results of these, while still controversial, have determined that neither life nor life signatures were detected. In more than forty years of searching for life within our solar system, the results continue to be frustratingly negative; however, recent results across many areas give plenty of optimism that our search for extra-terrestrial life, both inside and outside our solar system, is advancing, thus fueling our desire to continue the search.

The origin and rise of life on Earth is the subject of much research from scientific, philosophical and religious perspectives. From the scientific point of view, the conditions on early Earth that led to the increase of life has to be understood in order to judge the capacity for the rise of life in similar situations on other planetary bodies, both in and beyond our solar system. Currently it is estimated that life originated on Earth approximately 3.9 billion years ago. This is significant because it is a mere 700 million years after the formation of the solar system and barely 100 million years after the Earth's surface cooled enough to allow the formation of stable permanent oceans and the accumulation of organic compounds. Also 3.9 billion years ago was at the very end of the period of heavy bombardment during which large meteorites were colliding with the Earth and continuously re-vaporizing the oceans. That would possibly have destroyed any build-up of organic compounds (either those which were arriving to the Earth's surface from the very same meteorites or those which were being naturally created by Earth's reducing atmosphere of the time). This is a very surprising revelation, indicating that the genesis of life occurred very quickly once there were stable liquid oceans and access to organic compounds and an energy source (in this case, light from the Sun).

This is very important because it is believed that there was also liquid water on Mars, Europa, and Enceladus, all of which had access to the same organic compounds arriving from meteorites, and all at the same time as when life arose on Earth. Even if life does not exist on any of these planetary bodies today, it would appear that conditions and time were sufficient for the rise of life there. The discovery of fossils or other ancient life signatures would be of great significance to our understanding of the conditions leading to the genesis of life and the ease of its genesis. This begs the question: is the genesis of life common or is the genesis of life rare and, perhaps, unique to Earth?

The great number of thriving ecosystems in extreme environments on Earth demonstrates the durability and adaptability of life. The limits of life on Earth are broad; life is found, or has been found to be able to survive, under many of the most extreme environments imaginable: high and low temperatures, high and low pressures, high acidity (e.g. pH ~ 0), high salt, and high radiation. Hyperthermophilic and anaerobic chemoautotrophic bacteria have been found at the highest temperature and pressure environments at the deep-sea vents at the bottom of the Marianas Trench in the Pacific Ocean, at a pressure of more than 1200 atmosphere and at 113?C, and 3 km underground in the East Driefontein Gold Mine of South Africa. Endoliths (microorganisms that inhabit the interior of rocks) and hypoliths (that inhabit the underside of rocks) have been found under quartz rocks in the hyper-arid Taylor dry valleys of Antarctica and the hyper-arid Atacama Desert of Chile and Peru. The ammonia-oxidizing bacterium, *Nitrosomonas cryotolerans*, has been shown to be active in -18°C ice cores and thus to explain high N_2O levels found in ice cores from Lake Vostok, Antarctica.

The *Deinococcus radiodurans* bacterium is an example of a polyextremophile, which is one of the most radioresistant organisms known; it can survive cold, dehydration, vacuum, and acid. But *D. radiodurans* is not the only organism that is able to survive in a vacuum. Recent experiments on NASA's Long Duration Exposure Facility and the European Space Agency's BioPan space experiments have shown that *Bacillus subtilis* spores and halophiles in the active (vegetative) state can survive direct exposure to the raw conditions of space. The use of extreme environments as "analog" environments for other planets gives astrobiology a strong field component. No single environment equates exactly in all physical and chemical extremes to other planets, but some environments can be used to study the response of microorganisms to certain types of extremes. For example, the Atacama Desert (see above) is used as an environment to study life at extreme desiccation. Asteroid and comet impact craters are used to study the habitats created in these environments for microorganisms; volcanic environments yield insights into how microorganisms can access minerals as a source of nutrients and energy. The information from all of these environments provides astrobiologists with a means to focus the search for life elsewhere.

The expansion of the limits of life on Earth has had a dramatic effect on the expanding search for life in the solar system. While the many missions led by the former USSR found the surface of Venus to be uninhabitable by life as we know it, the conditions on Venus for its first 1-2 billion years may have sustained liquid water on its surface. Life could have had the chance to develop and, over time, communities could have grown in extreme niches as conditions changed to the current state. Some scientists believe that there are regions of the atmosphere where microbial communities could float at pressures, temperatures, and highly acidic conditions which are not all that dissimilar to conditions found in highly acidic thermal baths such as the Norris Geyser Basin in Yellowstone National Park, USA.

Mars, at the outer limit of the HZ, is a barren desert with no visible life, much like the Dry Valleys of Antarctica and the hyper-arid Atacama Desert in Chile and Peru. While Mars is unlikely to have life directly on the surface today, there is evidence that liquid water once existed on the surface of Mars for many millions of years at the same time as life first developed on the Earth. There is also evidence of liquid water from time to time in more recent years, which, if there are sufficient nutrient and energy supplies, could support a thriving subsurface microbiological community.

Jupiter's moon, Europa, is located far beyond the traditional HZ, but there is significant evidence that tidal forces cause enough internal heating to maintain a liquid ocean below the icy surface. The environment could be similar to that expected to be found in the liquid Lake Vostok, Antarctica, 4 km under the surface of the central Antarctic ice sheet.

Another active field of research is the search for, and study of, planets orbiting distant stars – extra-solar planets. As of early 2008, the count for the discovery of extra-solar planets stood at 277. There are a few Earth-like planets such as OGLE-2005-BLG-390LB, which is 5.5 times the mass of

Earth, located more than 20,000 light years away towards the center of the galaxy. Earth-like planets are expected to be in at least 1% of the star-systems that contain planetary systems. Further, it has recently been shown that organic material is found widely in the outer solar system, in comets and meteorites, and within the interstellar medium. This would indicate that the three basic conditions for life - stable planets, access to organic matter, and an energy source (light) - might be very common throughout our galaxy.

Opportunities in Astrobiology Research

Astrobiology research covers many diverse areas, but the major focus of astrobiology can be summarized in five searches:

1. The search for the origin of life on Earth. Understanding the origin of life on Earth, the major scientific question of all time, would provide us with a sound basis for considering how life might have begun on other worlds. Evidence about the origin of life comes from several fields. Laboratory simulations such as the Miller-Urey synthesis of organics under possible early Earth conditions test direct chemical models for the origin of life. The geological record provides evidence on both the conditions existing on the Earth and the timing when life began. Finally, the biochemical record of phylogenetics and metabolic pathways provides information about the nature of early life.

2. The search for the limits of life. Understanding the biochemical and ecological limits to life on Earth provides a basis for assessing the habitability of environments on other worlds. There are two complementary aspects of this search. The first is to understand the biochemical limits of organisms. These include temperature, UV and ionizing radiation, acid-base properties, and salinity. The second aspect, understanding under what range of conditions communities can survive, has seen remarkable discoveries in recent decades. Ecosystems have been found in unexpected places, such as deep below the surface in basaltic rocks, at the bottom of glaciers, and in deep-sea vents.

3. The search for life on other worlds in our solar system. A key goal for astrobiology is the detection of a second genesis of life. Having a second example of life would allow us to compare the biochemistry of our life against this new alien life. Such a comparison may provide deep insights into what aspects of our biochemistry are essential and what aspects are accidental. It may also allow us to address questions about the nature of life at the level of life itself. Indeed, there may be a number of questions about life that we will not be able to answer until we have another example for comparison with ours. In our solar system the most likely targets for

a search for life are, as mentioned earlier, Mars, Europa, and Enceladus. On Mars we might find living organisms in water reservoirs, if they still exist. Another location where we might find intact and preserved organisms, although probably dead, is in the ancient ice-rich ground in the polar regions. In both of these cases we would be able to study the biochemistry of the life that we find and compare it with the biochemistry of Earth life to determine if it indeed represents a second genesis. We cannot assume that life on Mars is a second genesis without proof, because we know that, as meteorites, rocks can exchange between Mars and Earth, and presumably in the other direction as well. These rocks could have exchanged microbial life between these worlds, making Earth life and Mars life identical.

There is some evidence to believe that Europa and Enceladus have liquid water below their icy surfaces. On Europa ocean water may be carried to the surface through the ridges seen on the surface ice. On Enceladus, the geyser of water jetting out from the South Pole may come from a deep, pressurized aquifer. Only further space probes to these locations can give us more definitive information. In both cases, any biological material in the subsurface water might be carried to the surface, providing a target for collection on a space mission. As on Mars, the direct biochemical analysis of any organic material collected may allow us to determine if it represents biologically produced organics and if the biology that produced them has a different biochemistry.

4. The search for evidence of life on other Earth-like planets around other stars by looking for oxygen or ozone. Current technology cannot directly determine the presence of life on such a planet; however, the detection of O_2 or O_3 in the atmosphere by spectroscopic techniques could be a strong indication of life. That could provide additional indications of life that has advanced considerably in terms of environmental evolution.

5. The Search for Extraterrestrial Intelligence (SETI). The long-standing search for radio signals from extraterrestrial civilizations continues and at an ever increasing rate due to advances in computer technology. However, no signal has been detected after forty years of effort. Experts in the SETI methods suggest that, if the current exponential growth rate in search capabilities continues, then the effective search will be completed in about another forty years. It will never be possible completely to remove the possibility of an undetected signal, and so the main SETI search may continue indefinitely.

1 Additional Readings, Planetary Science:
M. Marov (1986). Planets of the solar system. Nauka, Moscow. (Translated into German, Verlag –Nauka, 1987, and Spanish, MIR-Editorial, 1986).
D. Morrison and T. Owen (1988). The Planetary System. Addison-Wesley Publ. Co., Menlo Park/NewYork/Wokingam/Amsterdam/Bonn/Sydney/Singapore/Tokyo/Madrid/Bogota/Sntiago/San Juan.
M. Marov and D. Grinspoon (1998). The planet Venus. Yale University Press.
Encyclopedia of the solar system (1999). Eds. P. Weisman, L.-A. McFadden, and T. Johnson, Academic Press, San Diego/London/Boston/New York Sydney/Tokyo /Toronto.
J. Kelly Beatty, C.C. Peterson, and A. Chaikin (1999). The New solar system. Cambridge University Press.
C.J. Lada, and N. D. Kylafis (1999). The Origin of Stars and Planetary Systems. Kluwer Academic Publishers, Dordrecht/Boston/London.
W. Benz, R. Kallenbach, and G. Lugmair, Eds. (2000). From Dust to Terrestrial Planets. Kluwer Academic Publishers, Dordrecht/Boston/London.
M. Marov (2002) Collisions in the solar system: Implications for Planetary Atmospheres Origin. Space Times, AAS, issue 5, v. 41, pp. 10-15.
I.P. Williams, and N. Thomas (2001). Solar and Extra-Solar Planetary Systems. Springer, city.
M. Marov, H. Rickman, eds (2002) Collisions in the solar system. Kluwer Academic Publishers, Dordrecht/Boston/London.
Lewis, J. S. (1997) Physics and Chemistry of the solar system (revised ed.), New York: Academic Press.
Wilhelms, D. E. (1993) To a Rocky Moon: a Geologist's History of Lunar Exploration, Tucson, AZ: University of Arizona Press.
Spudis, P. D. (1996) The Once and Future Moon. Washington: Smithsonian Institution Press.
King, L.C. (1967). Morphology of the Earth, 2nd Ed. Oliver and Boyd Ltd., Edinburgh.

2 The orbits of planets and satellites around the Sun indicate key elements of celestial mechanics. The fundamental Newtonian law of gravity governs the motions of the planets around the Sun and of the satellites around their planets, whereas Kepler's three laws define the shapes and regularities of their orbits. Newton's law states that the gravitational force F is $F = G\ Mm/r2$, where M is the mass of the Sun, m is the mass of a planet, r is the distance between a planet and the Sun, and G = 6.67 x 10-11 m3 kg-1 s-2 is the universal constant of gravitation. Kepler's laws of planetary orbits are as follows (for details, see also Chapter 8):
A planet's orbit is an ellipse, with the Sun (our own star) at one focus of the ellipse,
The radius vector drawn from the Sun to a planet sweeps out equal areas in equal times, and
The squares of the periodic times of revolution P are proportional to the cubes of the semi-major axes a of their orbits: $(P1/P2)2 = (a1/a2)3$.

3 Van Allen, J. A. and L. A. Frank, 1959, Radiation around the Earth to a radial distance of 107,400 km, Nature, 183, 430- 434.

4 Ness, N. F., C. S. Scearce and J. B. Seek, 1964, Initial results of the IMP-1 magnetic field experiment, Journal of Geophysical Research, 69, 3531- 3569.

5 Goldstein, M.L., 2005, Magnetospheric physics: Turbulence on a small scale, Nature, 436, 782- 783.

6 Gustafsson, G., R. Bostrom, G. Holmgren et al., 1997, The electric field and wave experiment for the Cluster mission, Space Science Reviews, 79, 137- 156.

7 Fritz, T.A. and S. F. Fung (eds), 2005, The magnetospheric cusps: Structure and dynamics, Surveys in Geophysics, 26, 1- 414.

8 Santolik, O. and D.A. Gurnett, 2003, Transverse dimensions of chorus in the source region, Geophysical research Letters, 30(2), 1031, doi:10.1029/2002GL016178.

9 Green, J. L. and U. S. Inan, 2007, Lightning effects on space plasmas and applications, Chapter 4 in Plasma Physics Applied, Research Signposts, C. Grabbe (ed.), 12 pp.

10 Rodger, C. J. and M. A. Clilverd, 2008, Magnetospheric physics: Hiss from the chorus, Nature, 452, 41- 42.

11 Akasofu, S.-I., 1968, Polar and magnetospheric substorms, D. Reidel, Dordrecht, Holland, 280 pp.

12 Rees, M. H., 1989, Physics and chemistry of the upper atmosphere, Cambridge University Press, 289 pp.

13 .Lemaire, J. F. and K. I. Gringauz, 1998, The Earth's plasmasphere, Cambridge University Press, 350 pp.

14 Burch, J. L., 2003, The first two years of IMAGE, Space Science Reviews, 109, 1- 24.

15 Kaplan, E. D. (ed.), 1996, Understanding GPS: Principles and applications, Artech House Publishers, Boston, 554 pp.

16 Hunsucker, R. D. and J. K. Hargreaves, 2003, The high-latitude ionosphere and its effects on radio propagation, Cambridge University Press, 617 pp.

17 Lilenstein, J. (ed.), 2007, Space weather; research towards applications in Europe, Springer, Dordrecht, The Netherlands, 330 pp.

18 Additional Readings, Astrobiology:
Des Marais, David J., Joseph A. Nuth, Louis J. Allamandola, Alan P. Boss, Jack D. Farmer, Tori M. Hoehler, Bruce M. Jakosky, Victoria S. Meadows, Andrew Pohorille, Bruce Runnegar, Alfred M. Spormann.(2008). The NASA Astrobiology Roadmap. *Astrobiology*. 8(4): 715-730. doi:10.1089/ast.2008.0819.
Drake, Frank and Dava Sobel (1993). *Is Anyone Out There?* Souvenir Press Ltd.
Gilmour, Iain, and Mark A. Sephton (2004) *An Introduction to Astrobiology*. Cambridge University Press.
Lunine, Jonathan (2004). *Astrobiology: A Multi-disciplinary Approach*. Benjamin Cummings.
Sagan, Carl (1985). *Cosmos*. Ballantine Books.
Sagan, Carl (1994). *Pale Blue Dot: A Vision of the Human Future in Space*. Random House.
Woodruff, T. Sullivan III, and John Baross (2007). *Planets and Life: The Emerging Science of Astrobiology*. Cambridge University Press.
Online Astrobiology Resources:
NASA: http://astrobiology.nasa.gov/
European Astrobiology Network Association (EANA): http://www.astrobiologia.pl/eana/
Astrobiology Magazine: http://www.astrobio.net/news/
International Journal of Astrobiology: http://journals.cambridge.org/action/displayJournal?jid=IJA

Chapter Six
Satellites Serving Humankind

"Thanks to a few tons of electronic gear 20,000 miles above the equator, ours will be the last century of the savage, and for all mankind, the Stone Age will be over." —Arthur C. Clarke, Father of Satellite Communications

How Our Orbiting Satellites Serve Us

For a half-century now various "flying smart machines", namely satellites, have been able to access space to serve humankind. Hundreds of satellites circle the Earth each day in a variety of different orbits, providing services in a surprisingly large and growing number of applications.

The word *satellite* has an interesting origin. It was Galileo who first used the ancient Latin word *satelles* to describe the moons of Jupiter. This Latin word meant a servant, guard, or attendant of a powerful master or lord in ancient Rome. Thus a satelles scurried around the city to do his master's bidding and to provide protection to the household he served. The plural *satellites* meant a retinue of attendants and guards. This seemed an appropriate name for these flying objects, racing around the mighty planets. It is also a most fortunate and accurate term for our 'artificial satellites', as they were originally called in the early days after Sputnik was first launched on October 4, 1957.[1] Space satellites truly are the servants of humanity, providing communications, broadcasting, weather data, navigation, timing, mapping, search and rescue, and a variety of other services to humanity.

Today the practical applications of Earth orbiting satellites are almost endless. Satellite applications are a vital aspect of our use of space. Some satellites provide vital governmental services like weather monitoring and prediction, or search and rescue. Other satellites provide major commercial services like telecommunications and broadcasting that regenerate over US$120 billion in revenues. Figure 6.1 documents the huge variety of space applications today.

The satellite industry involves not only the satellites but also the launchers that put them into orbit, the ground tracking antennas, telemetry and control systems that keep the satellites operating, and the insurance companies that protect against losses and liability claims, etc. These various parts of the industry must all work together for successful satellite services to thrive and grow.

Launch services exist primarily to support the satellite applications markets or the launch of defense-related communications and surveillance systems. Satellite launches for scientific missions and space exploration are actually fewer in number. Launches with humans aboard, although they receive lots of coverage in the press, are only a small percentage of the total number of missions into space. The total commercial market for Earth-orbiting applications satellites, including satellites services, launch services, command and control, is currently well over US$200 billion.[2] (It has been estimated that the commercial and governmental combined space markets and related applications and services will grow to a staggering US$2 trillion by 2050.[3])

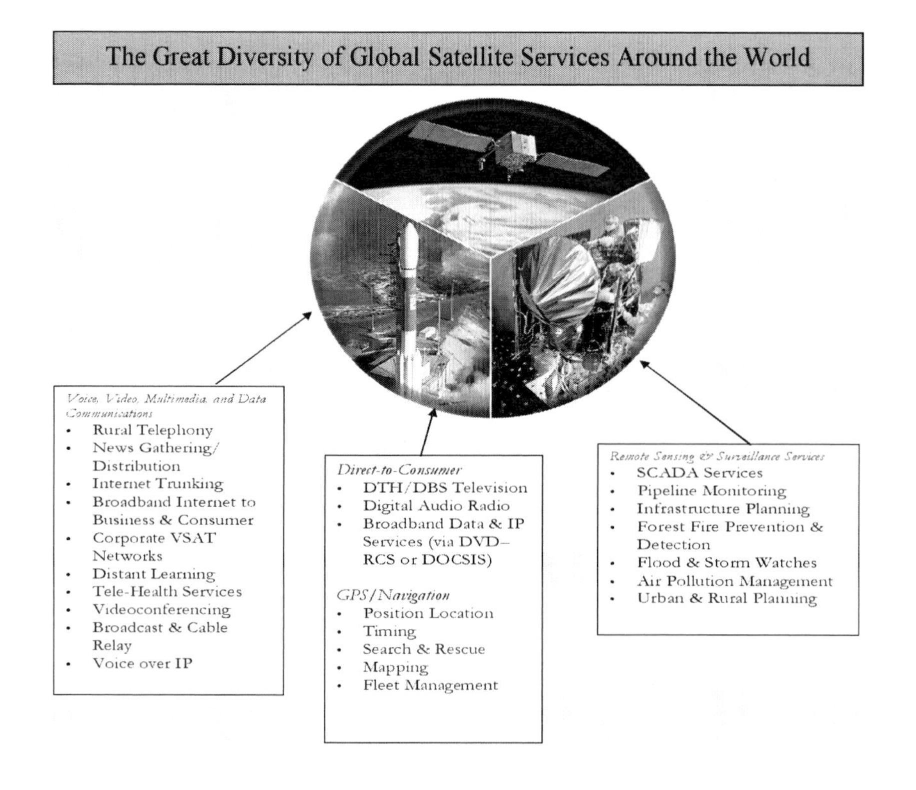

Figure 6.1. The many uses of our servants in the sky

(Courtesy of the Satellite Industry Association).[4]

Application satellites also served as the original model for international cooperation in space. Indeed the first satellites Sputnik 1 and 2 (launched by the USSR) and Explorer 1 (launched by the US) were launched in late 1957 and early 1958 as part of the International Geophysical Year. Many space satellite systems such as those deployed on behalf of the World Meteorological Organization of the United Nations (WMO), Intelsat, COSPAS-SARSAT (search and rescue systems), and the Argos environmental telemetry system were early examples of the importance of international cooperation in space. These satellites for telecommunications, meteorology, search and rescue provided practical services to all peoples of the Earth, demonstrating that our access to space is relevant and important to all nations and regions of our planet.

Satellite applications, a truly interdisciplinary arena, are often used together on a cooperative basis. All satellite systems share a common physical science and engineering basis; common elements include launch and deployment, power systems, data processing and transmission, use of assigned frequencies, and systems design, plus policy, legal, and commercial issues. All of these elements and areas of expertise are involved in the successful development of application satellites. Some of these satellites are used both for commercial services and for governmental and defense purposes - "dual use". Many commercial applications were developed originally for military purposes and then evolved to become commercially viable. In other cases, commercial satellites were adapted to support military communications or other services.

For instance, the GPS (Global Positioning by satellite System), originally developed for military services, now supports a host of commercial applications, and the military are now a small minority of total users. The satellites developed for defense surveillance purposes pioneered technologies that are now used for commercial remote sensing.

Satellite communications and broadcasting is by far the largest commercial space activity, making up over 90% of the commercial space arena in terms of revenues.[5] Sir Arthur Clarke, the famous writer of science fiction and science fact, is most widely known for writing the screenplay for the film *2001: A Space Odyssey* and other science fiction classics. But he is also often called the "father of satellite communications". In 1945, twelve years before the launch of Sputnik, Clarke published an article in *Wireless World* that outlined how to use the special geosynchronous orbit. This is a unique 24 hours a day orbit[†] where satellites appear to "hover" in the same location over the equator as the Earth turns on its axis. Clarke explained how three satellites in this orbit could cover the world for global broadcasting and communications.[6] This orbit where most communications satellites have been deployed over 40 years is sometimes called the Clarke orbit.

Arthur C. Clarke's original vision of the geostationary orbit has made space, especially the comparatively easy-to-achieve geosynchronous orbit, very profitable "real estate". The space industry employs many tens of thousands of people around the world, and provides routine communications and broadcasting services around the planet. The vision of Arthur C. Clarke has played a tremendous role in creating the global village we live in today.

Views From Above - The Synoptic Overview

Satellites give us the best means of viewing, monitoring, and understanding our spaceship Earth. Remote sensing satellites, which provide us with constant weather, climate, planning, and military reconnaissance data all around the globe, have an interesting history, intertwined with dual use military systems. Today, weather satellites constantly monitor the Earth, providing real-time information that we rely upon for trade, commerce, disaster warning and daily life.

Remote sensing satellites of various types image our planet in a variety of portions of the electromagnetic spectrum and at various resolutions (i.e. degree of precision, or clarity) on a constant basis.

It was Plato (ca 450 B.C.) who quoted Socrates as saying "... we are unable to attain to the upper surface of the air; for if anyone should come to the top of the air or should get wings and fly up, ...he would see things in that upper world; and, if his nature were strong enough to bear the sight, he would recognize that is the real heaven".[7]

Space remote sensing has changed the way we think about our planet home. It has also provided a stabilizing influence in geopolitics through the use of "national technical means", or spy satellites. Satellites supply a daily service, with applications as diverse as cartography, fire monitoring, ocean current mapping, iceberg tracking, urban planning, renewable and non-renewable natural resources, archaeology, ecology, urban planning, and agricultural crop management.

Precise navigation and timing systems, such as the U.S. Global Positioning System (GPS), the Russian Glonass system, the Chinese Beidou and the future European Galileo system, are another major category of satellite applications that has quickly become part of our modern lives. The ability to determine, accurately and routinely, our position and time using a small receiver has revolutionized our ability to navigate on land, sea, air and in space. Satellite navigation and timing services are used routinely for car and truck navigation, for the tracking of wildlife or hazardous materials, for regulatory enforcement, and to time-stamp every "packet" which crosses the internet. They are even used on Hollywood film cameras for ease of editing. The current satellite navigation market exceeds US$20 billion per year market, with 20% growth per year and over 70,000 GPS receivers sold per month.[8]

Space telemetry systems, such as the ARGOS environmental positioning and telemetry system, transmit routine environmental data from marine buoys around the world's oceans, provide remote environmental monitoring, and track endangered species and hazardous materials. Satellite search and rescue systems like the international COSPAS-SARSAT system have saved thousands of lives each year, and have made the worldwide search and rescue process both more efficient and safer.

Each of these capabilities on its own provides practical benefits and important commercial markets, but it is the functional integration of these tools that is creating many of the new applications and emerging commercial markets. Geomatics, or Geoinformatics, are terms used to describe the integration of these related but previously separate capabilities. Integrated mobile communications, positioning, data, and imaging are having a major impact on human societies around the world. Consider the integrated use of space-related tools in Google Earth, now accessed by over 300 million people around the world.[9]

For precision agriculture, weather satellite data are combined with GIS information of natural resources and agricultural productivity, GPS-controlled irrigation and fertilizer application, and crop monitoring from satellites into an integrated information system. The synergy provided far exceeds the benefit of each individual component. However, there are very important legal, policy, economic, dual use, and social implications in the continuing development of these tools.

Critical Services from Today's Communications Satellites

Satellite communications is actually a broad term that covers a wide variety of markets and technologies. Broadcast television and entertainment, broadband Internet services, mobile service to airplanes, ships and hand-held and portable units, are just some of the services. Figure 6.1 notes the true diversity of satellite services now available.

Thus the field of satellite communications is actually divided into a number of different categories of service that operate at different radio frequencies and use different types of antennas on the Earth or the oceans. These services are defined by the United Nations agency known as the International Telecommunication Union that is responsible for assigning bands of radio frequencies for the various applications. The three most important communications satellite services are: (i) fixed satellite services (FSS) where user antennas are at fixed locations and support all types of telecommunications services; (ii) broadcast satellite services (BSS) where entertainment and news services are broadcast to end users (typically to very small and low cost terminals) in their own homes and offices; and (iii) Mobile Satellite Services (MSS) where satellite services are provided to end users with mobile transceivers (like specially equipped cell phones with antennas large enough to relay messages to and from orbiting satellites.

Broadcast Satellite Services—

News and Entertainment to and for Everyone

The largest single sector in the entire satellite applications industry, in terms of revenues, is that known as broadcast satellite services (BSS) or, as it is sometimes called, the Direct Broadcast Satellite (DBS) Service. These broadcast satellite systems, with estimated total revenues well over US$100 billion a year (estimated for 2008), provide news and entertainment and even health and educational information to a growing percentage of the world's population. They provide video (television) services to every continent on Earth - even Antarctica. There are, however, also direct broadcast systems for audio (radio) services.[10]

Some of these direct broadcast systems such as Worldspace, XM and Sirius are designed to provide radio broadcasts (even though they can deliver some compressed video if broadcast at low rates and stored on a computer). Also the BSS/DBS satellites can deliver high definition television, conventional television in a digital or analog format and/or audio and radio programming. There are hundreds of millions of subscribers for direct broadcast services around the world. Europe, with over 100 million subscribers, is where this type of satellite service is most concentrated, but it continues to grow on all continents. Typical terminals ("dishes") to receive satellite television broadcasts are in the range of 35 cm to 1 meter. Radio broadcasting systems require units that are only ~10 cm in size; they are now being built into new cars in the US.[11]

Mobile Satellite Services (MSS)

The fact that people can now buy a hand-held "satellite transceiver" that can communicate to and from a satellite, and which is very much like an extension of cell phone service, is truly remarkable. Mobile satellite services first started with communications to ships at sea via the so-called Inmarsat (once the International Mobile Satellite Organization system and now Inmarsat Ltd.) some three decades ago. Then this service extended to commercial airliners and defense-related aircraft. Beginning in the 1980s was a land-based mobile satellite service. To close the link between the satellite and very small user terminals, significant power is needed. Some of the first of these systems, like Iridium and Globalstar, operate in low Earth orbit constellations (with a combined network of some 50 to 70 satellites orbiting the Earth). Other satellite systems have deployed very large antennas in space from the Geo orbit. Figure 6.2 shows the Thuraya satellite, built by the Boeing Corporation, that beams sufficient power to close the link to the very small transceivers on the ground.

Figure 6.2. The Thuraya mobile satellite with giant 12-meter antenna

(Courtesy of the Boeing Corporation).[12]

The powerful beams from the Thuraya satellite allow communications to the small hand-held transceiver shown in Figure 6.3. Its small antenna can communicate to the satellite, and is extremely small compared with the first giant antennas used at the start of the satellite age in the 1960s. The first satellite "Earth stations" that communicated with the first, very small satellites weighed hundreds of tons and required large round-the-clock staffs to maintain them.

Figure 6.3. Mobile Satellite Service Transceiver Thuraya2520 characterized as world's smallest satellite phone (Courtesy of Thuraya).[13]

This transformation is thus even more dramatic in scope and scale than the computer revolution characterized by "Moore's Law" (named after the co-founder of the Intel Corporation). This law states that the performance of computer chips doubles every 18 months. This amazing standard of exponential improvement has been achieved in the computer industry over the last two decades. But satellite Earth terminals have achieved improvements in performance equivalent to a doubling of overall performance every year. The contrast between the hand-held satellite phone in Figure 6.3 and the 30-meter diameter and multi-ton Standard A Earth stations that were used by the earliest Intelsat satellites in the 1960s is indeed one of the most remarkable technological transformations imaginable. This change, sometimes called "technology inversion", has seen the satellites grow from small units to very large multi-ton satellites while the antennas on the ground have shrunk by a factor of more than a million times. The size of these initial Earth stations is shown in Figure 6.4.

Figure 6.4. The Jamesburg 30-meter multi-ton Earth station as used in the1970s

(Graphic courtesy of www.iots.com/comsatlegacyproject).

Fixed Satellite Services (FSS) and Applications

The initial satellite services were in the fixed satellite services (FSS) band. These services started out with trans-Atlantic telephone services and simple black and white television relays, but today these services, in the age of the Internet and advanced IT, are much wider. Broadband satellite services today have exploded into a wide range of activities. The following are a few examples of the vital services that satellites now provide across the world.

Electronic Funds Transfer (EFT) and Business Services

The speed and volume of electronic funds transfer today is tremendous; the World Bank reports that transfers on the order of US$200 trillion took place in 2007. This is about 2.5 times the entire collective Gross National Product (GNP) of all the countries in the world.[14] Global and national commerce depends on rapid and secure electronic transfers taking place, and rapid authentication of credit card purchases at service stations, grocery and drug stores and shops. Fiber and satellite networks all over the world make this possible. It is not widely known that satellites are used routinely to verify credit cards and support Very Small Aperture Terminal (VSAT) networks among automobile dealerships, retailers, banks, and distributed businesses everywhere. Satellites connect Wi-Fi/Wi-Max hotspots, and support corporate training networks, decentralized corporate enterprise networks, teleworking virtual private networks (VPNs) and more. Countries without extensive fiber networks are particularly dependent on satellites for such business services, but such satellite networks are found all over the world.

Cable Television

One of the surprising facts about FSS networks is that cable television systems would not exist without satellite distribution to provide the programming at the "head end" for cable television networks. Premium movie channels like HBO, Showtime, Starz, news channels like Viznews, Reuters, CNN and specialized channels on cooking, travel, animal life are all distributed to local cable television channels via satellite. So called Earth station "antenna farms" that receive national and international television programming can be found at the head end of cable television stations. Large cable television companies, known as multi-service operators, are closely tied to the satellite industry to bring them their 24-hour a day program from around the world. Large television networks are equally dependent on satellites to bring them news and sports from all over the world. The coverage of the Olympic games, with up to two billion viewers, is dependent on hundreds of on-site satellite uplinks to provide the programming in local languages to over 100 countries around the world. Communications satellites are at their best when they are used to broadcast over large areas, or connect a very large network, or provide a link to a remote or unexpected new location. Satellites represent the flexible technology needed to perform these types of tasks. Much is written about the competition between satellites and fiber technology, but in many cases it is a hybrid network that combines satellites and fiber to create the optimum solution. Satellites have remarkable capabilities, whether they are beaming the Olympics to the world, providing tele-education or tele-health to a remote village, or supporting telework for a multi-national corporation linking together tens of thousands of employees.

International Development

Increasingly it is fiber networks that carry the heaviest traffic across the Atlantic or Pacific or provide the critical telecommunications networks in large cities. Clearly optical links provide the highest capacity and most cost-effective pipes connecting large cities. Yet, for many locations where there is limited development or steep mountain ranges, or tropic jungles, or deserts, or many islands to interconnect, satellite networks rise to the forefront. Satellites, in short, represent the key technology that transcends barriers and allows flexible and instantaneous interconnection. Island countries in the Pacific have achieved surges of economic progress by being able to use satellite connections to negotiate better prices for their imports and to sell their exports at international market rates rather than accepting the price offered by local steamer companies. Communications to and from over half of the world's 200 countries and territories today depend heavily on satellite communications. This is true not only for international connections, but for communications to rural and remote parts of their own lands.

Algeria established a domestic satellite network by leasing capacity from the Intelsat international satellite network in the early 1970s.Thus Algeria became the first of the developing countries to find a cost-effective way to establish a national television and telephone network across the Sahara Desert. Today scores of countries in Africa, South America, Asia, Oceania and the Middle East have leased capacity from international satellite networks to create local networks for television, radio and telecommunications. Other larger, developing and industrializing countries, such as Argentina, Brazil, China, India, Indonesia, Malaysia, Mexico, Nigeria, the Philippines, Taiwan, Thailand, and Turkey have launched their own domestic satellite systems.

Tele-Education and Tele-Health Services

Most people think of communications satellites as promoting broadcast entertainment, news, sports, business services, and government and defense-related services, but satellites have proved to be a powerful tool to promote and deliver education and health care services. At first these satellite systems were developed in the 1970s and 1980s to provide services in the remote parts of Canada, of the USA (Alaska and Appalachia), and the USSR. As international satellite systems and domestic satellite systems spread around the world, the use of satellites to deliver education and health care has spread widely. There are satellite networks that are critical infrastructure for the University of the West Indies, the University of the South Pacific and other regional groupings. The Chinese National TV University that started in the mid 1980s as part of the Intelsat Project Share (Satellites for Health and Rural Education) is now the largest of such tele-education networks in the world, with over 10 million students and over 90,000 terminals all over rural China.[15] Indonesia's domestic satellite system known as Palapa supported business and telecommunications, but it also continues to support a national satellite-based educational program. The Republic of South Africa has a major distance learning program via satellite called Mindset. Mexico's government has now deployed one of the most modern and extensive satellite education programs in partnership with the Via Sat Corporation. Argentina, Brazil, Colombia, Ecuador, and Venezuela also have satellite education programs, using either dedicated national satellite systems or leased satellite capacity. Brazil's program for the Amazon covers the vast rainforests to provide educational services.

India has launched its own satellite dedicated to health and education. This program known as Edusat serves millions of students and thousands of villages. Satellites are used in developed countries for education and training, with most of the countries of the Organization for Economic Cooperation and Development (OECD) having a variety of programs. Overall, many tens of millions of students receive important educational programs via satellite.

The delivery of health care via satellite is much more difficult and demanding than tele-education. The technical requirements for health care services via satellite, in terms of interactive communications and the resolution of video images, are severe. Also questions of liability and misdiagnosis may arise. The use of satellites to distribute health care information, to provide basic information on nutrition, treatment for diarrhea, prevention of AIDS and sexually transmitted diseases (STDs) have already helped save thousands of lives.[16]

Eyes in the Sky—Remote Sensing

Remote sensing is the art, science, and technology of obtaining reliable information about an object, a feature, or other phenomenon by recording, measuring (without being in physical contact) and interpreting data from a distance. For instance, our eyes and ears are remote sensing devices. The data used in remote sensing can be visible light that falls on the human eye or a photographic film that is developed and interpreted by the brain. Space remote sensing systems have been developed that use sophisticated devices to measure electromagnetic radiation outside the normal visual range, in the shortwave, thermal infrared, RADAR (Radio Detection and Ranging), microwave, LIDAR (Light Detection and Ranging), and ultraviolet, as well as multi-spectral sensors that record data from a wide range of the electromagnetic spectrum at once. Systems can be of various different designs. Active systems send their own burst of energy out to be recorded when received at a sensor, whereas passive systems simply record the energy at a sensor that comes from the Sun or that is emitted from the Earth.

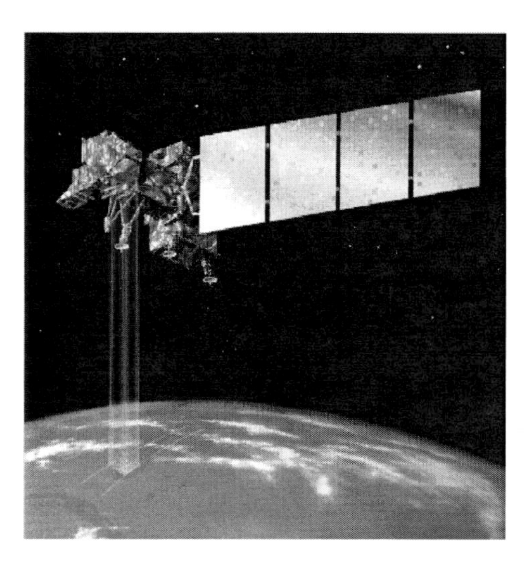

Figure 6.5. The Landsat 7 Earth resources satellite (Courtesy of US. Department of Commerce, National Oceanic and Atmospheric Administration (NOAA)).[17]

The data that are acquired in space must be stored, transmitted to Earth, and then processed into useful information using a variety of digital image processing techniques. That is then used for a wide variety of scientific, engineering, commercial, national security or military tasks. There are hundreds of applications, including:

- Military reconnaissance and intelligence gathering, target recognition, target tracking,
- Earth resources, including agriculture, forestry, geology, oil and minerals prospecting,
- Environmental monitoring and Earth Systems Science, including ecology, global studies, ecosystems analysis and disaster studies,
- Cultural, urban and demographic studies, including urban growth and planning, human dimensions of environmental change and archaeology; for instance, the McDonald's Corporation is one of the biggest users of remote sensing data, not so much to monitor potato crops as to decide where to locate future franchise restaurants,
- Ocean and marine studies, including water quality, fishing, pollution monitoring and oceanographic science,
- Atmospheric studies, including weather and meteorological analysis, weather forecasting, severe storm tracking, pollution monitoring, hazards such as volcanic eruptions, and upper atmospheric science, and
- Geodesy, geodetics, surveying and mapping the Earth.

The same technologies are used in planetary remote sensing and for astronomical studies, and there are many commonalities with these systems that image other worlds.

History and Development

The very first humans engaged in remote sensing when they surveyed their environment from a hilltop to locate food, threats, or to gather resources. The first aerial remote sensing was accomplished from balloons in the late 1700's.[18]

The first aerial photograph was taken by 'Nadar' of Paris in 1858 from a height of 80 meters. Balloons, kites, and even pigeons were used to carry early film cameras aloft to obtain the first aerial views of our world. Alfred Nobel designed a photo rocket in 1897 that took aerial pictures in Sweden and Dr. Julius Neubronner patented a miniature pigeon camera in 1903. The aftermath of the April 16, 1906 earthquake and fire in San Francisco was photographed from a large kite, with pictures such as Figure 6.6 being published in newspapers around the country.

Figure 6.6. Aerial Kite photo of the 1906 San Francisco earthquake

(Courtesy of Centennial of Flight).[19]

World War 1 provided a major impetus in the development of both military and civil remote sensing, with the development of practical aerial photography, aerial cameras, new films, and the techniques to interpret these photos and produce accurate maps. The oldest surviving aerial photographs are of Boston, Massachusetts, taken by S. A. King and J. W. Black in 1860. The first aerial photographs taken from an airplane were by the Wright Brothers, and the first "black and white" infrared film was produced in the 1930s. The Eastman Kodak Company developed the first commercial infrared film in 1938, which expanded our ability to see beyond the visible part of the spectrum for the first time.

In the Second World War new technologies for aerial photography and remote sensing were developed, including color infrared film used to locate camouflage. As well as the many military reconnaissance photographs taken and maps made, thousands of pilots and photo interpreters were trained; surplus airfields, aircraft and cameras led to the rapid development of aerial photographic remote sensing as the primary means of mapping and interpreting our world. The cold war and the dawn of the space age provided the impetus for the development of remote sensing from space.

Eyes from Space, Remote Sensing in the Space Age

Space is the high frontier from which to conduct global meteorological, military, and environmental analysis. In the 1950s both the Soviet Union and the United States developed space photographic systems for military reconnaissance and national security purposes, and all modern space remote sensing systems are the direct descendants of these. The Soviet Zenit spy system, using the same Vostok capsule design as launched Yuri Gagarin, was approved in November 1958 and first launched in December 1961. Ten of the first twenty Cosmos launches were secret photoreconnaissance satellites. The now declassified CORONA system of the USA, also approved in 1958, first recovered film in August 1960 after thirteen failures in a row. Over the life of the program, which ended in May 1972, CORONA had over 100 successful missions and photographed over 14.2 million square km of the Earth. Later, the US Skylab and Soviet Salyut, Almaz, and Mir orbiting labs carried extensive remote sensing systems having diverse objectives. The great majority of all remote sensing satellites ever launched have been military reconnaissance systems, and modern national security imaging systems are currently operated by many nations.

The potential civil benefits were also evident. President Eisenhower's Science Advisory Committee reported in 1958 that satellites will, in time, turn their attention downward. The Committee thus projected that satellites held promise for meteorology and the eventual improvement of weather forecasting. In fact, the most beneficial aspect of space remote sensing is probably the development of weather "eyes in the skies" to warn us of dangerous weather conditions and to use new observations to improve weather forecasting. TIROS (Television and Infra-Red Observation Satellite), the first dedicated weather satellite, was launched in 1960. This began as a Defense department program but was transferred to the new civilian NASA agency in April 1959. TIROS acquired its first image on April 1, 1960; this is shown in Figure 6.7.

Figure 6.7. The first image from the Tiros Satellite in 1960

(Courtesy of the SSPI Space Timeline).[20]

The first dedicated civil Earth remote sensing satellite was the NASA Earth Resources Technology Satellite (ERTS-1), launched on July 23, 1973. The system was later named Landsat, and the Landsat program continues with Landsat 7, which is now in operation, although in a degraded state. The French entered the civil remote sensing market in 1986 with SPOT 1 (Satellite Pour l'Observation de la Terre) launched on February 22, 1986. Since the advent of the SPOT civil and commercially based system, many other civil remote sensing systems have been launched by a number of countries, plus several private companies around the world. Seventeen countries have mid-to-high spatial resolution remote sensing satellites in orbit today, with a total of 31 optical systems; another 27 satellites should be launched by 2011. There are also 4 active radar systems in orbit, with Canadian and ESA programs being quite prominent; another 9 missions are planned by 2011. These numbers are for commercial and civil satellites only, and do not include military systems.[21]

The optical systems can be divided into 2 high and 3 medium spatial resolution groups:

- 13 very high spatial resolution (0.41 to 1m) and 9 high (1.8 to 2.5m)
- 14 high-medium (4 to 8m), 10 medium (10 to 20m) and 7 low-medi um (30 to 56 m) spatial resolution.

This combination of systems gives a wide range of potential applications. A *Forecast International* study in 2006 projected the delivery of 139 imaging satellites worth US$16.3 billion over the next ten years.[22] A major question that may arise from these projections is: why so many similar satellites are being developed and launched? What is driving this? Major drivers include national pride, military dual use and high technology, economic development and competitiveness.

Orbits of Remote Sensing Satellites

There are two basic orbits that are used by Earth remote sensing satellites - geosynchronous and Sun-synchronous polar orbiting. As noted earlier, the geosynchronous (GEO, or Clarke) orbit provides a satellite whose position is fixed in relation to the Earth's surface with respect to the equatorial plane. Although the geosynchronous orbit is primarily the domain of telecommunications satellites, meteorological satellites also use this orbit. This is a splendid orbit from which to look down on the Earth to have a constant and synoptic "weather watch" of over a third of the world's surface. Geosynchronous weather satellites are currently operated by the

USA, Europe, India, China, Japan and Russia. These satellites ring the Earth and provide weather information at all longitudes on a constant "24/7" basis.

The Sun-synchronous polar orbit is a much lower orbit, usually from about 500 to 800 km altitude. In this Low Earth Orbit (LEO), the satellite passes over the Earth's surface from pole to pole on a repeating basis, as the Earth rotates under the satellite's polar orbit; one orbit of the satellite takes 90 minutes or just over. Satellites in polar orbit give a more detailed view of the Earth and its atmosphere, so that this is the domain of most remote sensing satellites.

Weather Satellites

Figure 6.8 Eumetsat weather satellite image from GEO (Courtesy of Eutmetsat).[23]

TIROS III made the first observations of a hurricane from space in July 1961; the global hurricane watch program that continues in operation today was started using TIROS IV in September 1962. Daily global weather coverage started with TIROS IX in January 1965, and continues today.

There are currently seven geosynchronous weather satellites: the US GOES East and West over the Eastern Pacific and Western Atlantic over the Americas, Meteosat over Africa and Europe, Indian Insat, Chinese Fen Yung, and Russian GOMS over the Indian Ocean, and the Japanese Himawara over the western Pacific. Having weather and environmental data over the entire Earth for 24 hours a day is vital for forecasting severe weather events, as well as for gathering data showing changes in local and global environmental patterns. The individual observing instruments differ in design, but generally record and transmit data in the visible, thermal and water vapor parts of the spectrum; they observe cloud top temperatures, sea surface temperature, and winds. The coverage can be global or the instruments can zoom in to view smaller areas of special interest.

The satellites and their data are only one part of a complex system to study the Earth's weather on an operational basis that includes ground receiving and processing stations, as well as facilities to analyze and distribute the data. Weather Fax (WEFAX) is a direct communications transponder service provided through the satellites. It involves the retransmission of low-resolution imagery and other weather data from the satellites to relatively low cost receiving units located on ships at sea or on land in remote locations. WEFAX enables the greatest possible number of individual users to have access to weather imagery and related data around the world. More detailed data are also sent to ground receiving stations, e.g., for use by weather forecasters, with the data being distributed by television, radio, and other news media.

Polar Orbiting Systems (POS)

The polar orbiting satellite systems have much higher spatial resolution than is possible from GEO, and so giving more detail; the spatial resolution of one "pixel" (picture element) ranges from 1 km down to 1 m or even less. POS satellites view the same part of the Earth at least twice a day from their low polar orbits. There are many national, military, and commercial systems that are used for a wide variety of applications, some general and others with specific purposes. There has been an explosion in the number and types of remote sensing satellites, including small satellites, with many from developing nations such as Liberia, South Africa, and South Korea - remote sensing is no longer the domain of the super powers. Recent developments in technology have enabled these systems to have a much broader community of users; this trend will continue. The website associated with this book gives a detailed list of current and planned future space remote sensing systems.

Remote Sensing Applications

According to a recent survey, 41% of remote sensing data are used for national security and defense, 17% for mapping and cartography, 15% for civil governmental use, 5% for transportation, and 4% for environmental analysis. Agriculture, wildlife, resource exploration, forestry and all other uses are only 1-2% each.[24] There are very many interesting applications; the integration of satellite imagery with GPS, GIS, *in situ* data, and other sources greatly increases the utility of the remote sensing data. Some of the primary remote sensing applications are discussed in the next section.

Global Issues

Our world faces a wide variety of global issues today. Satellite data are, in many cases, the best means to study and monitor the health of "spaceship Earth" in an integrated and comprehensive manner. Satellite data are currently acquiring daily and global information on a wide range of atmospheric, land, water, and ice phenomena. These include information on global vegetation cover, ocean productivity, global biosphere change, ocean temperature, ocean wave height and direction, ozone levels, atmospheric pollutants, snow and ice extent. Major global issues, e.g., stratospheric ozone depletion, sea level rise, ocean pollution, deforestation and the human dimensions of climate change can be studied using a combination of satellite data, *in situ* data, and GIS information.

There are many national and international issues relating to the acquisition, storage, and use of remote sensing data. Private companies operating remote sensing satellites maintain copyright on the data; there are a variety of legal and policy issues relating to these data and their applications discussed in chapter 9 of the book. There are also complex issues relating to data structures, data standards, metadata (descriptive information about the data), and descriptions of data processing that are all part of the remote sensing data analysis process.

Resource Exploration

Remote sensing data are used for a wide range of geological, minerals, energy and mining applications. Surface features can provide important clues to subsurface deposits of gas, oil, minerals and water. Geological analysis and prospecting has been done using aerial photography for many years, but space imagery provides important advantages including much more synoptic coverage, a wider range of spectral bands, and the integration of multiple data sources in the GIS environment. Bare rock and soil can be distinguished in the Near-Infrared (NIR band) at wavelengths of 2.0 to 2.4 μm. Individual minerals can be identified using digital image analysis. Extremely large areas can be surveyed and analyzed using these techniques, but ground verification is always required.

Anthropological Research

One of the more interesting applications of space technology is the study of human origins - anthropology and archaeology. Satellite images provide anthropologists and archaeologists with a powerful tool to map changing settlement patterns, modern vegetation and infrastructure in remote areas. Whilst archaeologists have used aerial photographs to locate archaeological sites since the 1920s and Charles Lindbergh did aerial prospecting for sites in the southwestern US and Central America in the 1930's, modern, ultra high-resolution satellite images empower archaeologists to search for buried structures, old roadways and field boundaries from their offices. Remote sourcing tools have been used to locate a wide variety of archaeological features using Google Earth.[25] These satellite-detected sites were later verified, by making "ground truth" visits, as valid archaeological locales.

Another interesting example is the research done in Africa on mountain gorillas. Radar-based remote sensing from space provided vital base mapping of the Virunga mountain gorilla reserve in Rwanda, Uganda, and the Democratic Republic of Congo. These locations are often shrouded in clouds and mist, which radars aboard satellites can penetrate. Vegetation maps were created, and GPS units were used to track gorilla movements as well as evidence of illegal poachers. These data were entered into a Geographic Information System (GIS), which is used to protect the gorillas, track poachers, manage the park and support scientific analysis of the gorilla population.[26]

Flora and Fauna Monitoring

Vegetation monitoring is one of the oldest types of remote sensing data analysis. Remote sensing data are analyzed using digital image processing techniques to determine land use and land cover classifications accurately. Their uses include forest management, where timber resources can be monitored for disease and health. Agricultural areas can be routinely inspected from space to determine the health of crops, disease or pest infestations, when crops are ready for harvest, and even to forecast overall crop productivity. Several international organizations, national governments and private corporations use space-based remote sensing, combined with GIS, to monitor global commodity crops such as wheat, oil seeds, cocoa, and rice. They produce monthly crop forecasts for global commodity markets, companies and individual countries.

Precision Agriculture

Remote sensing, GIS, and GPS data can be combined into an integrated system to provide significant benefits to agricultural production at a local scale. Precision agriculture is a new technique where farmers integrate their equipment with precise GPS receivers and detailed GIS information about the soils and fields to improve the management of land and to obtain a more efficient use of machinery, personnel, water and pesticides. The state of crops monitored over the growing season is combined with detailed soil information and previous yield data to indicate specific actions for the farmer. Tractors carrying computers and GPS equipment then apply seeds, fertilizer, pesticides and water for each specific square meter of the ground, rather than simply applying large amounts over entire fields, to achieve greater yields at lower cost. Reduced pesticide and fertilizer use provides net economic benefits ranging from US$5 to US$14 per acre. Precision agriculture significantly lowers the environmental impact of pesticides and fertilizer, and reduces the water required for irrigation, while simultaneously improving the overall yield and reducing both direct costs and the environmental impact.

These are only a few of the many applications of remote sensing data. Military and defense, urban planning, and forest inventory activities, along with meteorological, oceanographic, and atmospheric studies are amongst a wide variety of commercial and scientific applications conducted around the world on a routine basis. The application can range from mundane (e.g., where to put a new shopping mall) to humanitarian (e.g., aiding disaster relief after the recent earthquake in China or the monsoon in Miramar.)

GEO, the Group on Earth Observations

Remote sensing and geomatics are now becoming increasingly international in scope. With so many nations launching remote sensing systems there is a danger of having many expensive and redundant capabilities. A fairly recent development has been the establishment of GEO, the Group on Earth Observations. GEO is a partnership of 72 national governments and 52 international and regional organizations, formed in 2002. The goal of GEO is to provide a framework for improved cooperation and coordination of programs and data, and to address nine areas of Earth observation "societal benefits" including disasters, health, energy, climate, water, weather, ecosystems, agriculture and biodiversity. GEOSS is the GEO system of systems, which addresses common user requirements, data, and processing data into useful products from a global and coordinated viewpoint.

The GEONETCast system is designed to disseminate space, airborne, and *in situ* data and products around the world. This will use existing commercial satellite telecommunications capabilities to broadcast Earth observation data around the world to low cost reception stations for a wide variety of users.

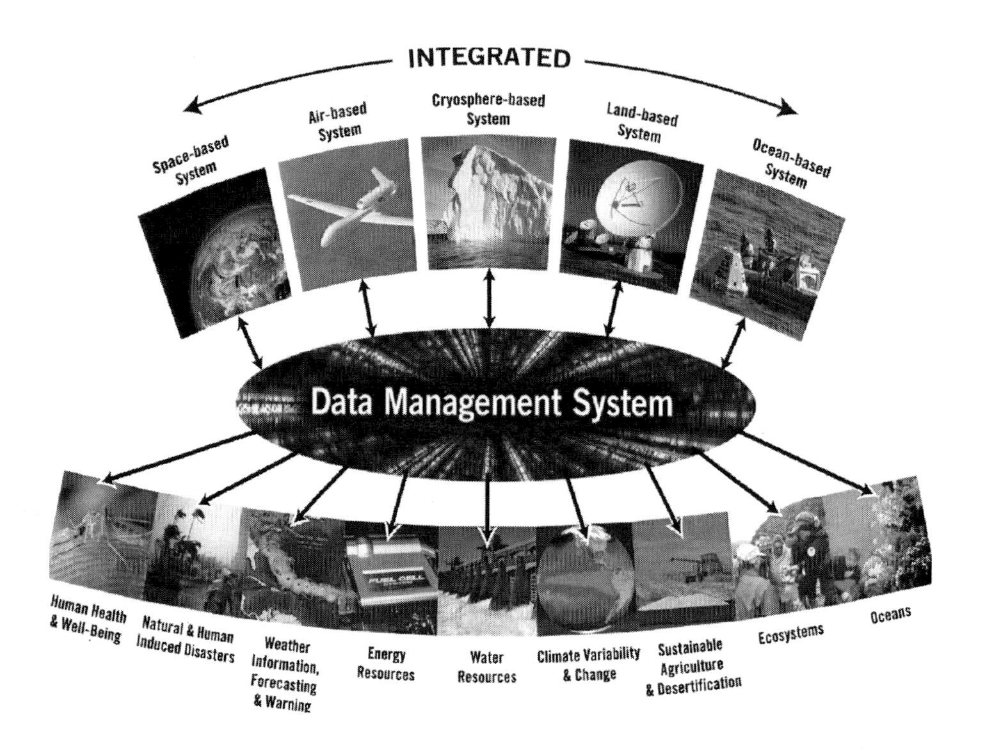

Figure 6.9. Integrated GEO systems of systems approach [27]

(Courtesy of Group of Earth Observations).

Our Amazing Tools in Space

The Fundamentals of Remote Sensing Data Handling

Space remote sensing data are transmitted via a telemetry system to the ground for analysis and distribution. The general sequence of procedures includes:

1. Acquiring the raw data collected by a sensor on the satellite platform
2. Storing the data onboard and then sending them to the ground using telemetry
3. Preprocessing at the ground facility to remove systematic errors and

to prepare the data for analysis
4. Entering the data into a computer analysis system
5. Georeferencing the data to a given map coordinate system, or projection
6. Enhancing and digitally manipulating the data, such as classifying land use or vegetation cover, water temperature, or atmospheric pollution content
7. Conducting ground verification field work or other external data analysis
8. Exporting the processed data into a Geographic Information System (GIS) for integrated analysis with ancillary data, maps, other imagery or related data, and
9. Conducting the analysis or scientific research, presenting and publishing the results, etc.

Sensors

There are many types of remote sensing sensors. Dental X-rays, airport metal detectors, ship and weather radars, cell phone cameras, and police traffic radars are all examples of common remote sensing systems used around the world on a daily basis. For space remote sensing, our sensors have common features to record the electromagnetic waves that have interacted with a target. All types of physical matter, such as vegetation, the bare soil, or the surface of the ocean interact with electromagnetic energy in different ways. They can absorb, reflect, scatter and emit energy at many different wavelengths so that we can detect, identify, classify, and describe the target material without being in direct contact with it.

The total energy received by a remote sensing instrument is recorded in various wavelength bands, which are referred to as spectral bands. Measurements over several bands can be combined to make up a spectral signature, a response pattern unique to each material. The source of the energy that is detected divides remote sensing sensors into two categories: passive and active. Passive sensing works by collecting energy from the Sun or thermal heat from the remote energy source, e.g., the atmosphere. If the satellite illuminates the target using its own source of radiation, such as radar, then the system is termed an "active system".

In modern electro-optical systems, the sensor produces a signal, an electric current that is related to the number of photons that are striking a photoelectric detector. Albert Einstein was awarded the Nobel Prize in 1921, not for his much greater discovery of relativity but for his work on the photoelectric effect. Simply stated, this is the emission of electrons from a negatively charged plate of some appropriate light-sensitive material when a beam of photons strikes it. The signal is converted into a digital number, stored onboard and later transmitted to the ground for analysis by a computer and an analyst.

The electromagnetic energy of various wavelengths emitted by the Sun or an active transmitter on a satellite interacts with the atmosphere, land, ocean, or vegetation in different ways. Some energy is reflected back to the satellite, some is scattered, and the target absorbs some. On its way back through the atmosphere part of the energy is absorbed and scattered, but some does get back to the satellite and falls on our detector. The detector acting as a photon counter generates an electrical current that depends upon the amount of energy that reaches it. That is recorded on the satellite and converted into a number that is sent to the ground for analysis.

Resolution

Each type of remote sensing sensor has unique characteristics and is useful for a particular application. There are four types of resolution, which characterize our sensor, and which we now consider.

Spatial Resolution

This defines how small an object can be distinguished on the ground, and how large an area is covered in each orbital pass. In digital systems this is defined by the picture element, or pixel size, of the "instantaneous field of view" (IFOV), which is equal to the smallest area on the ground that the detector can pick out. The swath width is the full width of the ground (typically >100 km) observed as the satellite moves along its orbit.

Spectral Resolution

This denotes the particular portion of the electromagnetic spectrum that is sampled by the detector. For example, near infrared is very useful for vegetation analysis, while mid-infrared is more useful for geological prospecting.

Radiometric Resolution

This defines how accurately the digital sensor records the data, usually measured in terms of the number of bits of data. Older sensors were 6 bit, meaning that they could measure the amount of energy falling on the detector in the range from 0-63, or 2 to the 6th power. Later systems were

8 bit (256 categories), and now we have 10 bit sensors with a data range from 0-1023. More refined measurements allow us to characterize the target more accurately and hence determine the composition of the target better.

Temporal Resolution

This is the time interval - usually some days - between successive observations of the same place on the Earth as the sensor travels on our Sun-synchronous polar orbit; its inverse is the revisit rate for the on-board sensor. The time interval is related to the altitude of the orbit, the swath width on the ground and the latitude (how far North or South) is the region where we require data. By way of contrast, geosynchronous satellites have a temporal resolution of several minutes, as the sensor constantly scans the Earth's surface. If we are using optical or multi-spectral scanners, rather than radar, and the area we are trying to monitor is covered by clouds, it could be a long time before we can obtain good data, even if the revisit rate is quite high.

Each sensor has a different combination of resolution parameters that define the spectral and other capabilities of that detector. We have to decide which sensor, which orbital altitude and inclination, and which data type are best for our particular application, and then choose these four resolutions of the sensor in an optimal way.

Active Remote Sensing Systems

There are important differences between active and passive sensor systems. Passive systems generally record the number of reflected solar photons falling on a detector in a given time. But what happens at night? They cannot work. Active systems generate their own scene illumination, by sending a burst of energy that is reflected off the target and recorded by the sensor. These systems, often radar systems operating in the microwave region of the spectrum, at wavelengths between 2 cm and 0.25 m, can operate at any time of the day or night, and can even see through clouds and rain, and penetrate a short distance under water or below the ground. Active systems provide a day/night rain/shine remote sensing capability that is very complementary to passive systems.

Radar systems produce very different information from that deduced from passive systems like Landsat or SPOT. Radar sensing satellites send a burst of coherent energy to the target and measure the time and intensity of the returned energy. The radar systems operate at very high power. For example, the ESA ERS-1 system could only operate the radar for some 8-12 minutes each 90 minute orbit, due to power and data storage limitations. The returned data are very complex, and depend upon the roughness of the surface and its moisture content. They require much processing in a dedicated ground-based processor to create visible images. Synthetic aperture radar uses the motion of the satellite to "synthesize" an antenna that is over a kilometer long, to obtain a good spatial resolution.

LIDAR (LIght Detection And Ranging) is another active system, which transmits pulses of coherent visible light to the target. It uses reflected or scattered coherent light from a laser on the satellite to determine the distance to and from a target, and/or to provide information about the target. While mainly being used in aerial remote sensing at this time, LIDAR has a wide range of applications, including investigating aerosol characteristics or winds in the atmosphere, locating geological faults, mapping historical or archaeological sites, measuring ocean waves, and producing detailed topographic maps. The joint US NASA/French CNES CALIPSO satellite (Cloud-Aerosol Lidar and Infrared Pathfinder Satellite Observation) was launched in 2006, and provides the global observation of clouds and aerosols to understand their role in the Earth's weather and climate. The Mars Orbiting Laser Altimeter (MOLA) was an instrument on the NASA Mars Global Surveyor that produced detailed elevation maps of Mars.

Future Trends in Remote Sensing

Environmental analysis at many spatial scales using remote sensing data is one of the fastest growing segments of the space industry, and makes up an important segment of many national space programs. There are many new satellite systems in operation today, and many more are in development around the world. Many nations are interested in developing remote sensing capabilities due to the dual use nature of the systems and their own military and national security uses. Commercial ultra-high resolution systems are now available with spatial resolution of 0.6 meters. They also bring with them a variety of complex political and diplomatic issues.

In the future there will be many more and many varied remote sensing systems, including micro-satellites built for specific purposes. With the reduced cost and size of computers and their increased capabilities, we will see many more scientific, commercial and dual use applications, especially in the developing world. There is a trend towards integrating remote sensing data with GIS, the Internet and GPS. Major issues of data archiving, storage, access to data, and modeling will have to be faced.

How far can remote sensing go? The tremendous popularity of Google Earth (with over 200 million downloads of the software) certainly shows us that there are many uses for space remote sensing and GIS data. How we process the massive amounts of costly satellite data that will be available in the future, who will have access to these data, and how we balance the commercial, military, and societal needs will be difficult decisions to make in the future. New international policy and legal issues will continue to be addressed as new technical capabilities are developed. In the end, remote sensing systems and data are tools in the hands of human beings, with all of our weaknesses and frailties. What benefits we derive from these tools, or what damage they cause, is entirely up to us.

Global Navigation Satellite Systems

One of the most important space applications systems of the last twenty years is the Global Positioning System (GPS). One of the satellites in this medium Earth orbit (going out to 20,000 km altitude) constellation is shown in Figure 6.10.

Figure 6.10. An artist's representation of a US GPS satellite in orbit [28]

(Courtesy of Boeing).

The Early Days

Global Navigation Satellite Systems (GNSS), also referred to as PNT (Precise Navigation and Timing) systems, are a growing area of space technology that is providing enormous benefits to humankind as well as having huge scientific and commercial potential. At present there are three operational GNSS systems, the original US Global Positioning System (GPS), the Russian Glonass system, and the Chinese Beidou regional system. These systems, and the several others in development (European Galileo and Indian IRNSS), are revolutionizing how we navigate around our world, and provide precise timing information. They are being swiftly integrated with remote sensing and GIS technologies for an amazing variety of direct benefits to society.

The need to determine our location and to know the time accurately has been a constant problem for our species from the earliest days of human societies. The stars and the Sun traditionally provided us with the ability to determine our position and time, and were the first "space assets" which we used on a daily basis. But they do not function when it is cloudy. Satellites provide us with a superb vantage point for positioning, navigation, timing, and telemetry for a huge variety of practical applications.

The first published concept for satellite navigation was a fanciful short story called *The Brick Moon,* which was published by Edward Everett Hale in 1869. The story was about a plan to launch a brick satellite into a polar orbit over the Greenwich meridian that would allow navigators in the Atlantic to measure their position East and West of Greenwich in the same way that they used the North star to measure position North and South. The satellite was to be 90 meters in diameter and would be launched into a 6,000 km orbit. If successful, a second was to be orbited over New Orleans. These two satellites would provide for safe navigation across the Atlantic Ocean and would be paid for by insurers, shipping companies, and by public subscription.

The practical origins of satellite navigation go back to Sputnik in 1957. Scientists at the Johns Hopkins University found that they could determine the orbit of the first artificial satellite by Doppler analysis of the radio signal that it transmitted. They also realized that, if the orbit of a satellite were precisely known, they could then determine their position on the Earth by reversing that same technique. This led to the first satellite navigation system, the US Navy TRANSIT system. Using a constellation of 6 polar satellites in 1,000 km orbits, it was first launched in 1959, and declared operational in 1964; it was only retired on December 31, 1996. The system was designed for military and submarine navigation, and it was soon joined by the Soviet Kosmos-Tsikada system, operational in 1974. They both used the Doppler technique, measuring the change in the frequency of a signal transmitted from a satellite. While they represented major advances, neither system was considered to be sufficiently accurate. Both required 10-15 minutes of tracking the satellites, used large and expensive ground equipment and only provided a navigation fix several times each day.

The US proposed a new system in 1973 to address these limitations, called the NAVSTAR (NAVigation System using Timing And Ranging) Global Positioning System. The first launch was in 1978. This mammoth project, usually referred to simply as GPS, which now totals over US$20 billion in cost (currently US$750 million per year) to the American taxpayers, provides constant three dimensional position, navigation, time and speed, anywhere on the Earth, aboard aircraft and up to low Earth orbit, with high precision and using a very small and inexpensive receiver. The system can accommodate an unlimited number of users; there is published information on the signal structure, and the civilian code is broadcast in the clear to be used by anyone around the world. Over 40 million units have been delivered to date, with some 70,000 units being sold per month. GPS receivers made up a commercial market of over US$22 billion per year in 2008, equaling the total investment in the system by the US.

The GPS system is truly useful to all in our modern digital world, with the US military now making up a very small minority of users. There are no export controls on GPS goods and no licensing fees are collected by the US. GPS hardware is inexpensive to develop and manufacture, and there are complete GPS on a chip systems measuring less than 50 x 10 mm that are commercially available. Originally developed for US governmental use only, President Ronald Reagan opened the system to free, worldwide use in 1983 after the Korean Air Lines KAL 007 disaster.

Three Segments of the GPS Network

There are three segments that make up the GPS system: the space segment, the control segment, and the user segment.

The GPS Space Segment

The space segment consists of a minimum of 24 satellites with four in each of six orbital planes inclined at 55 degrees. These are in a 12-hour circular orbit at 20,200 km altitude. There are usually more than 24 satellites, often up to 31, as new satellites are launched to replace older designs. They are in these high orbits for several reasons, including more accurate orbits, more survivable locations, and a larger area of coverage. Each satellite has multiple atomic clocks, which are accurate to a few nanoseconds. The satellites constantly transmit three types of data via radio signals on two carrier frequencies. The Coarse Acquisition (CA) code is a string of bits that repeats roughly every 1/1000 of a second and is used for civilian navigation. The Precision code (P code), is a much longer string of code that repeats every 7 days and is encrypted for military use only. The system also transmits system data including the satellite's clock time and ephemeris information about the status and location of the satellite. The frequencies used range from 1176.45 MHz to 1575.42 MHz. These streams of data are used to determine both time and place to a very high degree of accuracy for both.

The Control Segment

The satellites are controlled and monitored by the US Air Force from Schriever Air Force base in Colorado Springs, Colorado. There are four control and monitoring stations evenly spaced around the world: in Hawaii in the Pacific Ocean, the Diego Garcia atoll in the Indian Ocean, Ascension Island in the South Atlantic, and Kwajalein Atoll in the Pacific. These sites uplink information to the satellites and monitor their status. They are all located in areas that are under strict security and control. The system is operated by the US Air Force but is under joint control of the Interagency GPS Executive Board, which is co-chaired by the US Department of Defense and the U.S. Department of Transportation. Since the GPS systems are used for the take-off, landing and navigation of aircraft the US Department of Transportation clearly has an important role to play here.

The GPS User Segment

The system is designed so that all the complex and expensive technology is located in the space and control segments, so that the user's receiver can be portable and inexpensive. The original requirement for the entire US Army was thought to be a total of 700 units (the total commercial GPS sales in 2006 was over 720,000 units worldwide). The original GPS receivers weighed 8 kilograms, had a one channel receiver, and cost US$45,000, but today we have a variety of small, inexpensive 12 channel receivers that can be purchased for US$100 or less. More capable receivers can include hand-held or tablet PCs, with GIS, satellite imagery data and wireless connectivity, or car navigation systems. Many GPS receivers are located in unexpected places, such as on the top of bridges to measure deformation, buried in the ground to measure continental drift, attached to homing beacons to track migrating birds, or in cell phones. The precise timing capability is an inexpensive and valuable way to manage telecom systems, pagers, financial transactions, electrical power grids, and the Internet.

How the GPS System Works

GNSS systems work using a technique called satellite ranging to measure our distance from the satellites. Using the atomic clocks on the satellites and the code transmitted, we can determine when the code left the satellite and the time that it takes the message to arrive at our location. Knowing the speed of light, we can determine our distance from that satellite. Knowing our distance from four satellites gives us our precise location, and time, which can then be used to navigate to a certain position.

Sources of Error

There are a variety of sources of error in this system, the largest of which is interference caused by the ionosphere and atmosphere. Whilst we accurately know the speed of light in a vacuum, in the constantly changing ionosphere its speed is slightly, yet continuously, changed and this alters our calculated position. Clock errors are another source of error, as is the configuration of the satellite constellation when the readings were taken. All these and other errors cause an uncertainty in our actual position. We obtain sub-centimeter accuracy with the best surveying systems, fifteen meters with standard commercial receivers, and three meters with systems that use ancillary information called Differential GPS (DGPS) calculations.

The Upgraded GPS System Capability

The most significant improvement of the system since the 1970s started in 2005 with the launch of a new generation of satellites with many improvements. This will guarantee uninterrupted service through 2016, with improved positioning and resistance to jamming.

The Russian GLONASS System

The Russian GLONASS (GLObal NAvigation Satellite System) was developed by the Soviet military and is comparable to the US GPS system. The system was begun in 1982 and achieved 24-hour global capability in

1996. It operates using a similar system to GPS, with 24 satellites in three 19,100 km altitude orbital planes that are 120 degrees apart. The constellation began to lose satellites after the collapse of the Soviet Union, but Russia is now rebuilding the system and envisages a fully populated satellite system by 2010. Russia established a cooperative relationship with India's ISRO (Indian Space Research Organization), starting in 2004, to include Indian involvement in development costs and launches.

Europe and the Galileo System

Some European leaders are very uncomfortable with the primacy of the US GPS system and reliance on this system for so many military, commercial, and public activities. They see a major new technology and commercial market for satellite positioning and navigation, and are keen to play a role without relying on the US. They proposed the Galileo system to compete with (or perhaps, more precisely, to complement) the US GPS system. Galileo consisting of 30 satellites in three 20,000 km orbits was originally proposed in 1999. The original deployment date was to have been 2008, but this has been delayed a few years. The system has suffered from a variety of political and economic setbacks, but is now funded and moving ahead. When launched and operational, the system will provide several levels of service, including a free 6-meter capability, a subscription service with guaranteed access and higher precision, and a government-only security capability similar to the GPS P code. It will also have an enhanced Search and Rescue system called SARGAL to complement the existing COSPAS-SARSAT system.

China: The Beidou and Compass Systems

China was originally a partner in the Galileo system, but it has developed its own system called Beidou (the Northern Dipper). This is a different architecture, with three geosynchronous satellites first launched in 2000. It is a regional system covering all of China and the surrounding areas and requires the user to both receive and send information to the satellites. This limits the number of concurrent users. China is also developing a GPS-like system called Compass, with 35 satellites (5 geosynchronous and 30 in inclined orbits similar to GPS, Glonass, and Galileo).

India

India has approved a project to develop their own independent GNSS system, the "Indian Regional Navigation Satellite System" (IRNSS), with the first launch planned for 2009 and a complete constellation operating by 2011. This will be a hybrid system, including 4 geosynchronous satellites and 3 in a high inclination orbit, and will have a 20-meter precision within India. India is also cooperating with Russia in the development and launch of the next generation Glonass system, and is a partner in the Galileo program as well.

Japan

Japan has proposed an innovative Quasi-zenith Satellite System (QZSS) that will address the problems of using GNSS in the deep "urban canyons" of the major Japanese cities. Satellite signals are often blocked or reflected by tall buildings, making car navigation unreliable. The satellite constellation will consist of three or more satellites in inclined geosynchronous orbits. These satellites would trace a "figure of 8" on the ground around a point on the equator. Three satellites would always place one over Japan so that a GNSS signal is always available high in the sky. This system would be interoperable with both the US GPS and European Galileo systems.

The Future of Precision Navigation and Timing Systems

Applications for this technology are amazingly broad and continue to grow. We not only navigate on land and sea, in the air and in space, but we control unmanned aircraft and ships using GPS as well. We conduct high precision measurements of the Earth and monitor deformation of fault zones and volcanoes. Other scientific applications include direct measurements of the ionosphere, the gravity field, and the atmosphere. We track everything from golf carts to hazardous waste shipments, and are remapping our world more accurately than we ever dreamed would be possible. Emergency response vehicles are routed through traffic in the fastest way, farmers are obtaining better crop yields with less fertilizer and pesticide use, and wildlife are tracked on their migrations. We can even buy watches that allow us to track our children, or patients with dementia. As computers become even smaller, faster, and cheaper, GNSS systems will become integrated into more aspects of our lives. There are very complicated national security, dual use, and policy issues with GNSS, as is illustrated by the large number of redundant systems in development.

Satellites provide vital timing, telemetry, navigation, and scientific data for a wide variety of purposes. These data can be easily integrated with GIS and remote sensing data, and popular Internet offerings like Google Earth. Many new applications and uses will be developed, and the commercial market may confidently be expected to grow.

Space and Security –

How Satellites Keep Us Alive and Our Planet Well

Satellites and our access to space provide a variety of direct and practical benefits to humankind. Many of these systems have been designed specifically to assist us in problem situations including disasters and emergencies. Weather satellites provide constant monitoring of severe weather and hurricanes around the globe. Military reconnaissance systems can stabilize a dangerous world by providing accurate information on military situations. However, more sophisticated defense related space-based IT, communications, surveillance and weapons systems may make the world more bellicose and willing to use arms in international disputes.

Weather Satellites

Satellites provide an excellent vantage point for constant weather surveillance. Current GEO weather satellites are continuously imaging the weather systems around the globe for forecasting and severe weather monitoring. They can provide warnings to the public about hurricanes and cyclones, tornadoes, floods, fires, and high winds that may soon occur - see Figure 6.11.

Figure 6.11. Eumetsat image of a typhoon in the Indian Ocean region

(Courtesy of the Eumetsat). [29]

ARGOS Environmental Telemetry System

The ARGOS positioning and telemetry system is an excellent example of a practical and low cost benefit of our access to space. It is a satellite-based two-way location, time, and data collection system worldwide. GPS seems to be a passive system, and GPS receivers only calculate position; they cannot communicate or transmit data. ARGOS, jointly developed by the US and France, uses French hardware on US polar orbiting environmental satellites. The system is dedicated to environmental studies and provides very cost effective, low data rate global telemetry and positioning data. ARGOS hardware is on two NOAA satellites and can receive transmissions from ARGOS transponders on the ground over a moving area 5,000 km in diameter. This provides an average of 8 readings per day at the equator and 28 per day at the poles. ARGOS uses a Doppler location technique like the Transit and COSPAS-SARSAT systems. Each transmitter can have up to 32 sensors and can transmit 15 kb up to the satellite each time the satellite passes over the device. The positioning accuracy of the system is, on average, 350 meters, but modern systems also include a GPS receiver with better accuracy.

There are many applications of ARGOS data. These include tracking containers and dangerous cargo and self-contained remote telemetry stations that can transmit housekeeping data from dams, power plants, pipelines and other remote or high-risk locations. Remote river monitoring stations in Africa send data daily. Hundreds of drifting or tethered marine buoys around the world carrying instruments to measure atmospheric pressure, sea surface temperature, and air temperature return their data via ARGOS. More specialized instrumented buoys can measure wind speed and direction, wave amplitude and spectrum, and ambient acoustic noise. Several nations use ARGOS to monitor individual vessels of fishing fleets in their territorial waters and exclusive economic zones, in order to provide near real-time trends in fish stocks.

ARGOS buoys are also used in the International Ice Patrol program in the North Atlantic Ocean. Special ARGOS devices are dropped from US Coast Guard aircraft onto icebergs in the open ocean; these are then tracked to protect shipping. Perhaps the most important use is in wildlife tracking. Tiny ARGOS transmitters can be placed on birds, mammals, and even sea turtles, whales or sharks. These measure the depth and duration of dives for fish, or altitude and airspeed for migrating birds. Biologists have revolutionized their understanding of migrating animal species around the world by tracking long duration migrations of arctic caribou, storks, sea turtles and other species. In one example, biologists tracked black storks (Ciconia nigra) on their 4,000 km migration from Europe to Africa each year. One individual stork flew 5,230 km from Prague in the Czech Republic to wintering grounds in Central Africa, with migration distances ranging between 150 and 350 km per day.

COSPAS-SARSAT

This is a dedicated satellite search and rescue system that is designed to take the 'search' out of search and rescue. It is also an excellent example of international cooperation in space, even at the height of the cold war. It was originally a collaboration of the US, USSR, France, and Canada, and now includes many other nations. The idea for the system was the 1970 aircraft accident in Alaska; two members of the US Congress were never found, even after a massive search and rescue program. Studies show that aircraft accident survivors have less than a 10% survival rate after 2 days, but a better than 50% survival rate if rescued in less than 8 hours. Search and rescue is often done in very bad weather conditions and so is a very dangerous activity. A technology automatically reporting the position of a shipwreck or an aircraft accident means the difference between life and death.

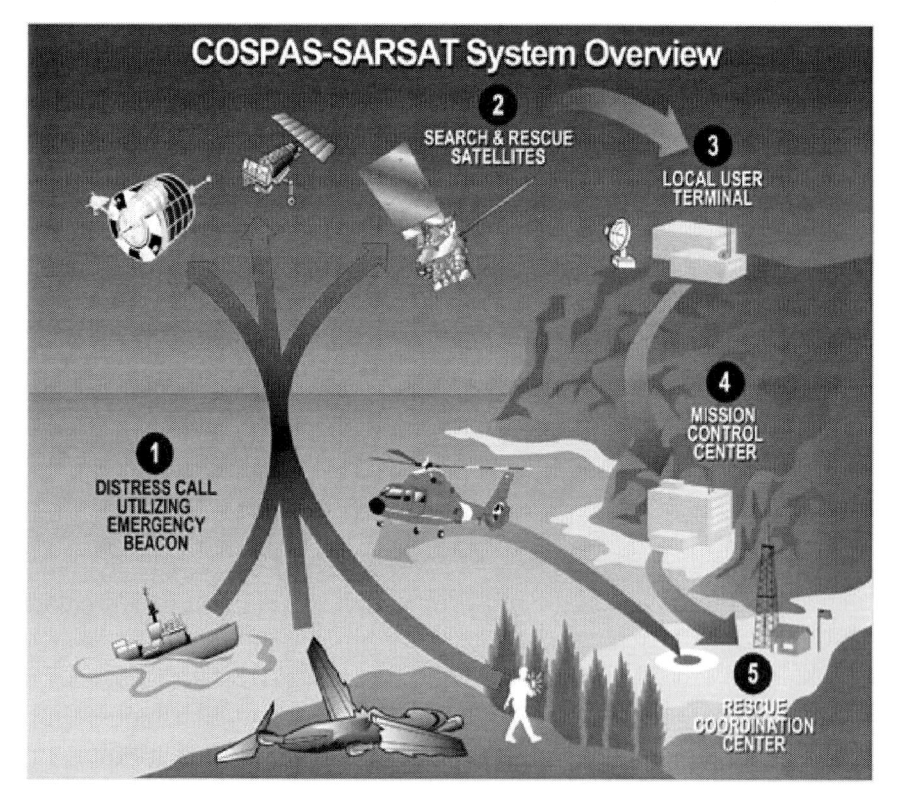

Figure 6.12. How the COSPAS-SARSAT system works [30] (Courtesy of SARSAT).

The system uses the same Doppler technique as the Transit, Tskada, and ARGOS positioning systems. Small transponders on a ship or aircraft can be automatically activated in an emergency to broadcast an alert message containing the time, position, name of the vessel and type of emergency. When a US NOAA polar satellite or Russian Nadejda satellite passes overhead carrying a COSPAS-SARSAT receiver it receives this message and then relays it on to the nearest unmanned Local User Terminal (LUT). There are over thirty of these located around the world. Once the alert signal is received on the ground a message is automatically transmitted to the nearest national Mission Control Center, which then initiates the search program. The transponder signal can also be used as a homing device for the search teams.

The system was first launched in 1982 and was operational in 1984. Over 19,000 people have been rescued in over 5,300 individual operations. In 2004 alone the system saved 1,748 people in 466 operations. Recent innovations include adding GPS receivers to the equipment to obtain better positions and a GEOSAR system that receives nearly instant alert messages using geosynchronous satellites.

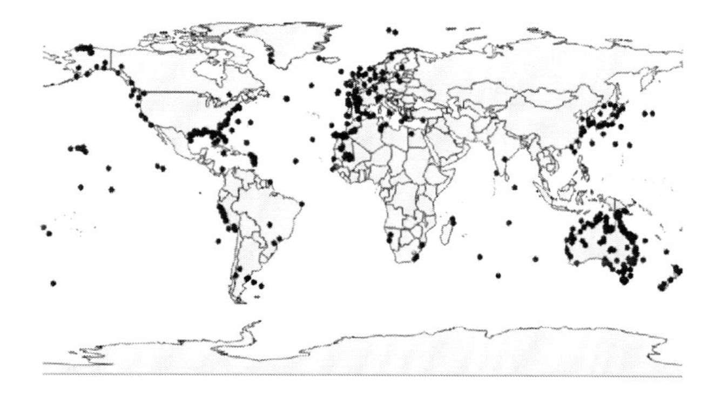

Figure 6.13. COSPAS-SARSAT rescues in 2004 [31] (Graphic Courtesy of SARSAT).

Disaster Applications

Space technologies play a vital role in preparing for, and responding to, disasters. This is a growing area of satellite applications; many national space agencies are becoming directly involved in providing data to support disaster operations worldwide. In a disaster, spatial awareness is vital. Immediate questions are: what has happened, what is the extent of the damage, where can rescue facilities be opened, when are equipment and supplies arriving, who is responding, what airports are open, where should search and rescue begin, where are safe shelter sites, and what evacuation routes are open? Having answers to all of these questions is vital when responding to disasters. Space technologies are important sources of current information, particularly as the answers can change rapidly as the disaster unfolds. Weather satellite data and forecasts, mobile satellite telecommunications, GIS data, recent satellite imagery of infrastructure, and GPS all play important roles in the preparedness and mitigation phases of disaster planning. In response and recovery activities after a disaster, high-resolution imagery, GIS databases, mobile satellite telecommunications, and GPS mapping all assist in damage assessment, rescue and recovery, logistics, and providing shelter and medical treatment.

The Tampere Convention on the Provision of Telecommunication Resources for Disaster Mitigation and Relief Operations of 1998, with its 60 signatories, is the international community's realization that modern telecommunications are vital in humanitarian assistance. The International Charter on Space and Major Disasters was created to provide a unified system for acquiring space satellite imagery, and delivering this imagery and derived products, such as updated maps, in a timely fashion. The system was declared operational on November 1, 2000. National representatives can call a single telephone number to acquire various satellite images and products on an emergency basis.[32]

Our ability to create new practical applications from satellites is an extraordinary story of innovation. In the fifty years since the launch of Sputnik thousands of satellites have been developed, launched and operated for a wide variety of purposes. Hundreds of satellites providing telecommunications, positioning, timing, navigation, and imaging, along with their associated ground infrastructure and operators, are truly servants of humanity, providing vital services and information for our changing world.

Solar Power Satellites (SPS)

One of the greatest challenges of our time is to find a way to supply humanity with plentiful, clean and "green" energy as the global population continues to grow. Wind, geothermal, tidal, Ocean Thermal Energy Conversion (OTEC), and solar (photovoltaic) cell systems are clearly part of the answer. Another contributor is a solar power satellite. The Sun's great power is available in space 24 hours a day and can be beamed, as microwave energy, to locations back on Earth. A further new technology is to create a new light-weight "lens" that can capture the Sun's rays and concentrate it efficiently so that solar cells deployed in space can capture the energy equivalent of 1000 Suns. This way of capturing "green and renewable" energy in space has been discussed for decades. However, several key factors have changed to make SPSs economically viable, including (a) spiraling energy costs for conventional gas and petroleum sources; (b) new space technologies and new materials for huge, not too expensive, and light-weight lenses; and (c) improved solar cell technologies that achieve higher efficiency and better durability at reasonable cost.

These new systems could deliver Gigawatts of power to the Earth. If designed properly they could function for many decades, beaming energy to various locations around the world. Today, telecommunications and broadcast satellites plus remote sensing and other application satellites provide crucial services and represent huge commercial enterprises. Tomorrow, solar power satellites could become the most important space application of all. They could lower energy costs, increase energy supplies and help fight global warming, because Solar Power Satellites (SPS) would not release greenhouse gases or disturb the ozone layer. The challenge of the future is not only to generate cheap and plentiful energy, but also to have energy sources that avoid climate change and protect humans and other animals from genetic mutation.

Satellites in the Service of Humanity

This chapter has sought to demonstrate the wide range of ways that application satellites serve humankind. From communications, to remote sensing, to space navigation, to search and rescue, and environmental and weather services, these satellites serve humanity. Satellite systems now support finance, agriculture, airlines, ocean liners, international trade, military and defense, coast guard and dozens of other interests essential to human enterprise, scientific investigation and national security.

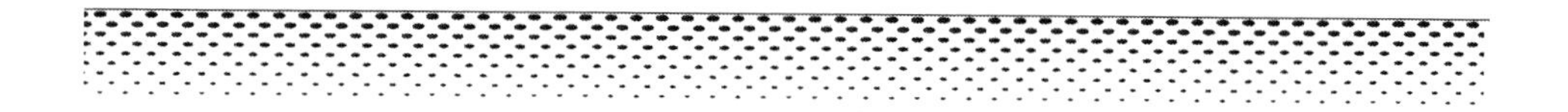

† The orbital period is actually 23 hrs and 56 minutes in duration because the Earth is moving around the Sun. There are two terms that are often used almost interchangeably to describe the "Geo", or Clarke, orbit - geosynchronous and geostationary. The geosynchronous orbit, where the satellite moves in synchronism with the Earth's rotation in the East-West direction, is relatively easy to maintain. Excursions off the equatorial plane in the North-South direction (i.e. "inclined orbit") occur on a daily basis so that the satellite moves in an "S-shaped" curve above and below the equatorial plane. Because this is difficult to control, the "geostationary" orbit (E-W and N-S) is hard to maintain.

1 Society of Satellite Professionals International (SSPI) International Space Events Timeline (See www.sspi.org)
2 Futron Corporation, Space Foundation, Satellite Industry Association 2007 composite figures See www.futron.com and www.sia.org.
3 Futron Corporation www.futron.com
4 Satellite Industry Association, Satellite 101 Presentation, 2007, www.sia.org.
5 Joseph N. Pelton, Basics of Satellite Communications (2006) International Engineering Consortium, Chicago, Illinois, also see www.sia.org and www.futron.com.
6 Arthur C. Clarke, "The World of Communications Satellites", *Aeronautics and Astronautics*, February, 1964.
7 Fowler, Harold North. 1966 Plato in Twelve Volumes, Vol. 1 [109e" translated by Harold North Fowler; Introduction by W.R.M. Lamb. Cambridge, MA, Harvard University Press; London, William Heinemann Ltd.
8 ABI Consultants "GPS World Markets: Opportunities for Equipment and IC Suppliers", http://www.abiresearch.com/products/market_research/Satellite_Positioning_Systems_and_Devices
9 Hardwick, B. "Google Earth, Satellite Maps Boost Armchair Archaeology" Digital Places Special News Service National Geographic website. http://news.nationalgeographic.com/news/2006/11/061107-archaeology.html.
10 The Futron Corporation-Satellite Industry Association (SIA) Annual Report on Satellite Communications, www.futron.com and www.sia.org.
11 Joseph N. Pelton, *Basics of Satellite Communications*, *op cit*.
12 Graphic provided by the Boeing Corporation, see www.boeing.com.
13 Graphic provided by Thuraya Corporation, see www.thuraya.com
14 International Bank for Reconstruction and Development (IBRD) Annual Report on Electronic Funds Transfer, www.worldbank.com)
15 Report on Project Share, "The Chinese National Television University", *Intelsat Technical Report*, Washington, D.C.
16 *ibid*
17 See www.noaa.gov
18 Centennial of Flight website. See http://www.centennialofflight.gov/essay/Lighter_than_air/Early_Balloon_Flight_in_Europe/LTA1.htm and http://www.centennialofflight.gov/essay/Lighter_than_air/Napoleon's_wars/LTA3.htm
19 *ibid*
20 *op cit*, SSPI Space Events Timeline.
21 "The Market for Civil & Commercial Remote Sensing Satellites Forecast International, Inc. May 24, 2006 Forecast International 22 Commerce Rd. Newtown, CT 06470 USA)
22 *Ibid*
23 Image courtesy of Eumetsat. See www.Eumetsat.int
24 ASPRS remote sensing survey 2006. http://www.asprs.org/news/forecast/
25 *op cit*, B. Hartwick,
26 Scott Madry, "Mountain Gorilla Protection: A Geomatics Approach "Gorillas in the Database". http://www.informatics.org/gorilla
27 Graphic courtesy of the Group on Earth Observations http://earthobservations.org/
28 Graphic courtesy of Boeing www.boeing.com
29 Graphic courtesy of Eutmetsat www.eumetsat.int
30 Graphic courtesy of SARSAT www.sarsat.org
31 *Ibid*
32 International Charter Space and Major Disasters http://www.disasterscharter.org/main_e.html

Chapter Seven
Systems and Spacecraft

"Men are weak now, and yet they transform the Earth's surface. In millions of years their might will increase to the extent that they will change the surface of the Earth, its oceans, the atmosphere, and themselves. They will control the climate and the Solar System just as they control the Earth. They will travel beyond the limits of our planetary system; they will reach other Suns, and use their fresh energy instead of the energy of their dying luminary."

— Konstantin Tsiolkovsky

Since the dawn of the space age, the exploration of space has been championed as the technological high ground – a perilous environment demanding the best of humankind's creative talents to explore. Indeed, at first assessment, engineering a spacecraft, or designing a space mission, or architecting an entire space program, appears to be as daunting a feat as any in the realm of human endeavor. Yet no matter how large the problem, skilled engineers have developed methods to break it down into more easily understood parts, and then re-integrate these parts back into a working whole.

Such is the world of complex space "systems". The seemingly endless number of components that allows us to explore space is just as well defined by their interactions with each other as they are with their standalone characteristics. The result is that the machines and missions of space provide the ultimate "systems" challenge. The engineering pieces of a "space mission" include system elements such as spacecraft subsystems, astrodynamics, launch vehicles and project operations. In this chapter, the most fundamental element of a space mission will be discussed, namely the spacecraft. The spacecraft, be it an Earth orbiting satellite, an interplanetary satellite, or a manned spacecraft, is the most critical aspect of a space mission. But first, let's discuss systems and the "big picture." [1, 2]

The Big Picture – What is a System and Why is That Important?

During the Renaissance and the so-called "Enlightenment," scientists, doctors, inventors, mathematicians and even philosophers began systematic sharing of their studies and experiments. Royal Academies of Science aided a process of intellectual sharing of research from country to country. This process philosopher Teilhard de Chardin has characterized as the creation of a "noosphere"—a new type of environment where systematic sharing of knowledge and scientific curiosity can truly blossom. But true "systems thinking" is a recent development.

Prior to the mid-1900s, problems of natural science or engineering were still largely viewed as discrete events that could be addressed independently of one another. It was not until the 1940s when Hungarian biologist Ludwig von Bertalanffy and his contemporaries began to apply systems thinking to such diverse fields as physics, biology, psychology, engineering, computing, and management. This way of thinking has become a discipline for seeing the "big picture," and a framework for seeing broad interrelationships among the pieces of a system rather than a narrow view of the individual piece-parts. About the time that systems thinking was becoming codified as "systems theory" a new era of space exploration began to open up as Sputnik was launched in October 1957. Systems theory, a scientific and philosophical approach to studying a complex problem, and space exploration, a complex scientific and engineering problem, were the perfect marriage of problem and solution. Systems theory now permeates our existence and, as a result, the word "system" has become one of the most overused and perhaps abused words in the English language.

Space Systems – A Three-Level Model

There are many models that allow the "deconstruction" of a complex space system into sub-parts of ever-increasing detail. Most of these models are associated with product or work breakdown structures (WBS), and are used to organize the work or hardware products of a particular spacecraft or space mission. Such a breakdown structure might first divide a spacecraft into subsystems (e.g., propulsion, structures, guidance and control, power, etc.), and those subsystems into assemblies (e.g., main solar array assembly), and those assemblies further into components (e.g., a latch mechanism), and so on. These models are useful once a spacecraft or space mission is well defined, but less useful for the initial conceptualization of space systems in the abstract. To view space systems from a "big picture" point of view, we can use a simplified three-level model that addresses the three topmost elements of the space system hierarchy – the spacecraft, the space mission, and the space program architecture.

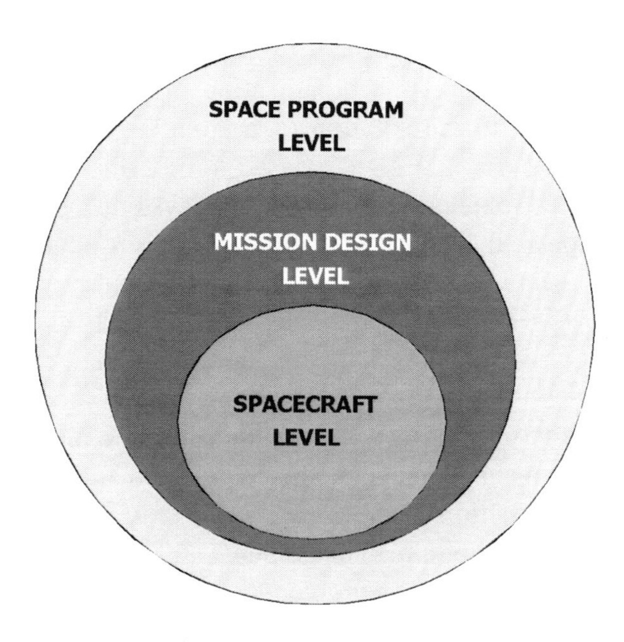

Figure 7.1. The three-level model of space systems.†

The model shown in Figure 7.1 is simplified on purpose so that we can use it to discuss the highest levels of space system integration. The core of this model is the spacecraft itself. Though a spacecraft is in itself a complex integration of subsystems, assemblies, and components, this model uses it as the smallest unit of integration in this discussion. A spacecraft, regardless of the precision of its design and craftsmanship, cannot accomplish a mission by itself. The second level of the model, the space mission, combines the spacecraft with the other necessary components, including the launch vehicles, ground infrastructure, mission operations, payload, and the application of astrodynamics that govern the physics of the mission. As with the previous "level", an expertly designed mission may not ever get beyond the drawing board if not placed within the context of broader space system architecture. The Architecture is the highest level of integration in the three-level model and represents the coordination of a mission with other missions, political pressures, budgets, project schedules, national priorities, and commercial forces. Each of these levels will be explained in greater depth.

Each of the levels of the model presented in Figure 7.1 is associated with a "systems" process that involves the breakdown of tasks into ever more precise subtasks and then optimization of each activity. The optimized integration of the spacecraft subsystems (propulsion, structure, communications, etc.) is often referred to as "spacecraft systems engineering". Designing and assembling a spacecraft is an excellent example of "systems engineering." This requires the organization of multiple disciplines in a systematic approach to the solution of a complex engineering problem. The term "Space Mission Design" is used to describe the coordination of spacecraft, launch vehicles, ground support systems, planetary physics, and other mission elements. This design process is used to achieve specific scientific or technical goals. Space mission design requires integration of these many diverse components. If done properly the result will be designs that represent the minimum cost and risk. The Program Architecture level of integration in the complex space system model is the coordination of a series of missions with external influences, and falls under such titles as "program management" or "space program architecting". Each of these integration functions, comprising spacecraft systems engineering, mission design, and program management, are processes used to create integrated solutions to complex problems.

Similarly, the individuals who perform these integration functions are given titles that are employed throughout the industry. At the spacecraft level, the individual who understands the detailed interrelationship among the many subsystems that make up a spacecraft often holds the position of spacecraft "Systems Engineer". The individual who understands the detailed interrelationship between and among the many vehicle and operational components that make up a mission is often known as the "Mission Designer" or "Mission Engineer". And at the uppermost level of the hierarchy, the individual who understands the program-level interrelationships, which include missions, politics, national policies, and other factors can occupy any number of important leadership positions, including Program Manager, Administrator, Minister of Space or Space Program Architect. Thus a team of engineers and program managers combine their experience and "systems skills" to perform the integration function within a space mission.

Spacecraft are the Core of the Three-Level Systems Model

The *spacecraft* is the core building block in the space systems hierarchy and stands alone as the central element in the three-level space system model in Figure 7.1. It is the fundamental element that ultimately achieves the mission, and it is upon this performance that the mission's success is judged. Individual spacecraft come in many sizes, ranging from kilogram-sized "micro" and "nano" -spacecraft, to refrigerator-size remote sensing satellites, to bus-sized communication satellites and up to aircraft-size space shuttles for transporting human crews and very large payloads. Regardless of size, all spacecraft share common characteristics – a multitude of subsystems. Systems engineers oversee the design and integration of these subsystems to produce a final assembly that performs as a single entity.

Spacecraft are inherently complicated. Their many interacting subsystems require the expertise of an experienced engineer to optimize and integrate the many spacecraft subsystems. This integrator, or Spacecraft Systems Engineer, must be expert in the details of many engineering systems, and must be able to envision how a change to one system will affect all other spacecraft systems. In the vignette below, let's not worry about the definitions of each of the subsystem components called out in the question. The point is that even though the spacecraft is the core of the three-level systems model, it comprises many smaller subsystems and components. The Systems Engineer has to know how all of these things work together and how they are all interrelated.

The Relative Importance of Spacecraft Subsystems

Here's a simple test to gauge your Spacecraft Systems Engineering aptitude: You are part of a team designing a mission to land on Mars. Which spacecraft subsystem is the most important for this mission: Power? Propulsion? Structure? Communications? Attitude control? Command and Data Handling? Guidance, Navigation, and Control? Thermal? Power? The Payload?

Discussion: The structural engineer is first to speak up and suggests that the structural system is the most important because it has a high mass. He further suggests that some mass can be saved on this mission by flying a lower "g" entry profile through the Martian atmosphere. The Spacecraft Systems Engineer, however, knows that the lower g entry will affect flight path angle, the guidance system, the integrated heating on the entry system's thermal protection system, the vehicle shape, the placement of communication antennas, and the packaging of the landing vehicle inside the new aero shape, …in short, she recognizes that a single change in one vehicle subsystem can likely affect every other spacecraft subsystem. And she also knows that there is NO single most important subsystem – all are equal and understanding the interaction among them is most important.

"Systems Engineering" is practiced as both a process and as an art. Both are necessary in the production of a spacecraft, but it is critical to know when each should be applied. The process of Systems Engineering is the subject of many textbooks and professional courses, and most often involves a number of sub-processes such as Functional Analysis, Functional Allocation, Interface Definition, Requirements Development, and Technical Performance Measurement.

Each of these sub-processes often results in data products that usually take the form of documents. While it is important to understand the Systems Engineering Process and to produce the paperwork and computer files that methodically document the design process, this process alone will not produce a spacecraft. While the process of Systems Engineering is systematic and concerned with documentation, the art of Systems Engineering is hands-on, interactive and dynamic. This "art" can be compared to a symphony. In this metaphor a multidisciplinary collection of subsystem and specialty engineers represents the orchestra, and the Spacecraft Systems Engineer can be thought of as the conductor that achieves coherence and balance. The Spacecraft Systems Engineer's job is to artfully lead the team through a creative design process that begins with a set of spacecraft or mission requirements and ends with an "optimized" design for a spacecraft capable of achieving those requirements in the best possible way, at the least cost and the minimum risk.

The creative systems engineering process is helped by the use of coordinative facilities. These may be known as a "Concurrent Design Center" or many other similar names. This type of facility allows spacecraft team members to collaborate interactively in real-time as the details of the spacecraft design unfold. As a change is made to one subsystem, it is instantly visible to all other members of the concurrent design team, who can update their design in near real-time. Two examples of such a facility is ESA's Concurrent Design Facility at ESTEC, in Noordwijk in the Netherlands, and the NASA Jet Propulsion Laboratory "Team-X" in Pasadena, California. Both of these facilities are laid out in a fashion similar to that of the diagram shown in Figure 7.2, and allow a team of co-located engineers and designers to use collaborative software tools to perform rapid design cycles of spacecraft and space missions. Regardless of the facility, the art of systems engineering always requires an experienced systems engineer to guide the team members through the creative process, to mediate disagreement, and to keep the process focused on producing a product that meets its intended requirements with the correct balance of performance, cost and risk.

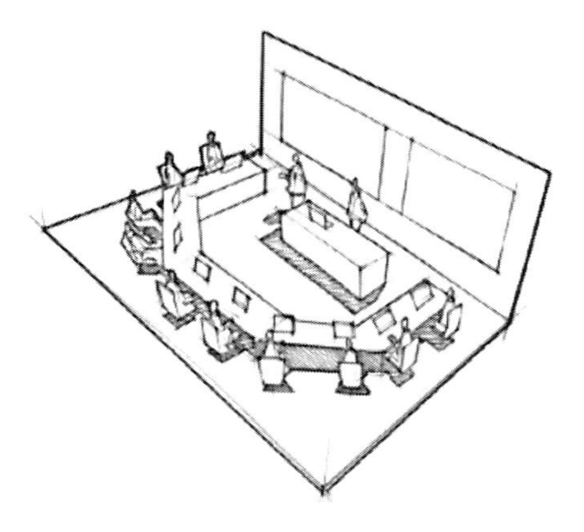

Figure 7.2. A concurrent design facility

(Courtesy JAQAR Concurrent Design Services, Rotterdam, The Netherlands).iii

Spacecraft Anatomy – What's In There and How Is It Put Together?

A spacecraft can be an Earth-orbiting satellite or a probe sent into deep space within the solar system or beyond. The spacecraft is conceptually divided into two main elements, namely the "payload" and the "bus." To distinguish between these two main elements is simple. Think of yourself using the most common method of inter-city public transportation, a bus. A bus takes you where you want to go, provides environmental control like heating and cooling and, when needed, a place for you to sit. The bus also has some crude level of communications to indicate where you are, what stop is coming up next, and so forth. In this simple example, you, as the passenger on the bus, are the payload. From the perspective of you as the payload, the bus exists to get you where you want to be and to cater for your needs while you are on board. The spacecraft bus exists to provide a "seat" to the payload traveler.

In this case the "bus" provides to the spacecraft payload many things: power, propulsion for movement, environmental control, and communications in the form of data transfer for the particular payload being carried to the place(s) it needs to be and do so in an efficient way.

We can think about many different kinds of payloads. These include all of the space applications services such as discussed in Chapter 6 that generate over US$100 billion in revenues and vital governmental services on planet Earth. These types of payloads include communication and broadcasting systems (think satellite phone or television transmissions), navigation, meteorological monitoring, or remote sensing units equipped with cameras, sensors or radar systems.

Then there are spy and reconnaissance payloads and a variety of different systems for strategic purposes. There are also astronomical observation platforms, like the Hubble Space Telescope or the James Webb Observatory. Also there are probes, manned or unmanned, that are designed to explore the solar system or beyond. Over ten thousand payloads have now been launched. These range from tiny nanosats to crewed vehicles that have gone to the Moon, as well as a series of space laboratories and international space facilities able to sustain astronauts in space for prolonged periods of time.

In all of these examples, the cameras, antennas, radar systems, telescopes, laser systems, transmitters, space laboratories, and life support systems all make up the payload. The rest of the satellite is the part called the bus and provides all the power, orbital maneuvering, temperature control, pointing functions, structural support, and anything else that might be required by the payload to perform successfully the tasks that make up a particular mission.

The elements of the spacecraft that make up the payload and bus are all interrelated and designed based on the mission requirements. The diagram in Figure 7.3 shows the various spacecraft subsystems and their relationship to the design of the space mission. Each of these subsystems will be discussed in the following sections. Don't worry; we'll also discuss the orbit, launcher, ground segment, and mission operations concepts as well, in the next chapter. The space environment was covered earlier, as were the various payload types.

A fundamental subsystem that is not explicitly shown in Figure 7.3, however, is the onboard computer system, which controls everything the satellite does. The onboard computer system can be thought of as the "brain" of the satellite because it monitors everything that's going on in the satellite. In this respect the on-board computer system is like your central nervous system and its functioning is critical to mission success. It receives a variety of signals and inputs on a continuous basis. As a result of the messages it receives from the various subsystems, it produces signals and commands that tell the various subsystems what to do. This is just what your brain does when it tells your hand to reach out and pick up a glass of water or stroke a soft fuzzy puppy.

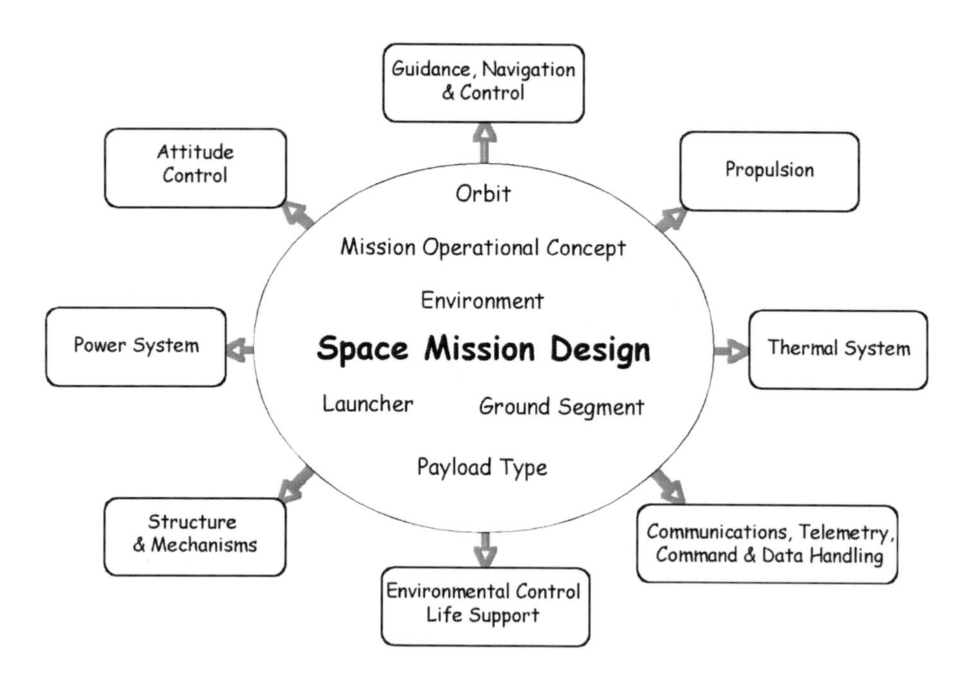

Figure 7.3. Bus subsystems and their relationship to the space mission and its design.

Table 7.1 is a summary of the spacecraft subsystems and their major functions. Except for subsystems that are associated with human space flight, particularly environmental control and life support systems, all of these subsystems will be discussed in this chapter.

SUBSYSTEM	PRINCIPAL FUNCTIONS
Structure and Mechanisms	Support structure Moving mechanisms Protection from space environment Adapters and covers
Guidance, Navigation and Control (GN&C)	Orbit determination Orbit control
Attitude Determination and Control (ADCS)	Attitude determination Attitude control Payload pointing
Tracking, Telemetry and Control (TT&C)	Systems health monitoring Communications Command processing Spacecraft tracking
Command and Data Handling (C&DH)	Data processing Command distribution
Power Supply (PSS)	Power generation, transformation and distribution
Thermal	Equipment thermal control
Environmental Control and Life Support (ECLSS)	Life support for human space missions
Propulsion	Orbital maneuvering Attitude control

Table 7.1. Spacecraft subsystems and their key functions.

The Satellite's Brain – The On Board Computer and Its Function

Just as you could not function without your brain, a spacecraft could not successfully function or execute its mission without the onboard computer system. The computer, or computers, onboard a spacecraft control everything it does. (Since the computer, or the brain, is so important there is often redundancy, or back up processors, in case there is a failure.) Examples of functions that the onboard computer executes include generating commands to tell the attitude control system how to point the satellite in the appropriate direction so that the payload collects data. It also sends commands to servomotors to point an antenna in the appropriate direction so that data collected by the payload can be transmitted to a ground station. Further the onboard computer sends commands to heaters to keep a payload or some other subsystems at the appropriate temperature; or a whole host of other functions that require monitoring and management. Basically, the onboard computer system thinks of everything needed to do the mission plus takes care of all the housekeeping functions, like power regulation and thermal management.

Computers and microprocessors are found throughout the spacecraft. These processors control very complicated interrelationships between subsystems and are also reprogrammable, which is handy when something goes wrong and has to be fixed, or "reprogrammed", after a satellite is launched. Virtually all of the spacecraft subsystems are linked to the onboard computer via data acquisition and control (DAC) systems and the spacecraft (S/C) data bus. It is via these means that the computer receives data and transmits commands to change things or to execute a specific function.

Next let's consider the telemetry, tracking and control (TT&C) and command and data handling (C&DH) subsystems also included in Table 7.1 These are two related subsystems that work with the onboard computer. Telemetry is simply data sent to the ground concerning the operation and health of the spacecraft. The data can come from any subsystem, but is typically spacecraft engineering data. The telemetry tells how well the spacecraft is working—or not working. Tracking is the determination of the actual location in space of the spacecraft. Control is simply the operation of the spacecraft through the execution of required functions. The TT&C system may also include a communications system. Its main functions are to: (i) control all the spacecraft subsystems; (ii) process commands; (iii) acquire data; (iv) format and store data; and (v) perform ranging. The "master control" computers are located in the TT&C unit.

A command is just an instruction to a subsystem to execute a particular function. A command can be sent from the ground or from an onboard computer. Data handling is the gathering and formatting of data from all the sensors and instruments onboard the spacecraft. The Command and Data Handling (C&DH) system is a relatively recent term added to the jargon of space systems engineering. The existence of such a subsystem implies a separate radio communications subsystem. All of the above functions work together in a seamless way to ensure the smooth execution of all satellite function to achieve the mission objectives.

In Control – How Does the Satellite Know Where It Is, Where It's Going, and Which Way to Look?

Probably the most complex subsystem on a satellite is the control system. There are many control systems on board a spacecraft. Let's focus on the attitude determination and orbital control system. That is the system that determines where the satellite is, where it needs to be, how to get it there, and which way it's pointing. A vast body of work in the area of automatic control theory has been developed over the years. We are now able to design and build systems, including very complicated satellites that are capable of autonomous operation involving extremely complex tasks. In fact, of all the onboard computers, the guidance and control computer is the "smartest" one! Control theory involves outrageously complicated mathematical relationships and techniques. But let's start by beginning with a simple system to convey the key concepts and then discuss how these are applied to spacecraft.

The first significant work in automatic control was James Watt's centrifugal governor for the speed control of a steam engine in the eighteenth century. The basic principle of the Watt's speed governor is illustrated in the schematic diagram of Figure 7.4. The amount of steam admitted to the engine is adjusted according to the engine speed through the mechanism that uses centrifugal force. The speed governor is adjusted such that, at the desired speed, the amount of steam is just enough to generate the desired speed. If the actual speed drops below the desired value due to disturbance, then the decrease in the centrifugal force of the governor causes the control valve to move downward, supplying more steam, and the speed of engine increases until the desired value is reached. On the other hand, if the actual speed increases above the desired value, then the increase in the centrifugal force of the governor causes the control valve to move upward, decreasing the supply of steam, and the speed of engine decreases until the desired value is reached.

In Watt's speed governor, feedback control was achieved by the mechanisms using the centrifugal force. Today, those mechanical means are replaced by electrical means, such as sensors (gyros, accelerometers, tachometers, and so on), computers, and motors. However, the principle of the feedback control has never been outdated. Although the sophistication of the sensors, the precision of the mathematics and the fine tuning of thrusters makes attitude control an exact capability, the concepts of using feedback to orient the spacecraft is built on century-old concepts.

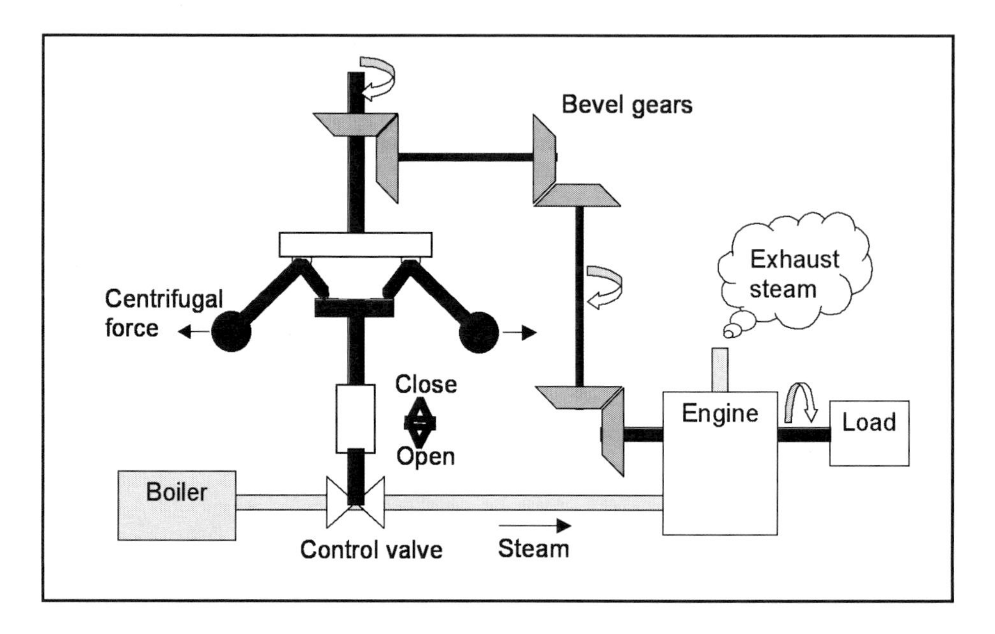

Figure 7.4. Schematic diagram of the James Watt's speed governor.

Pitch, Yaw and Roll: Angular Momentum

The key to maintaining a satellite in its proper orientation in an X, Y and Z-axis coordinate system is the proper application of angular momentum. Figure 7.5 illustrates the three-axis reference system that is critical to orienting the payload systems accurately. To work properly observation cameras, telescopes, broadcast or communications antennas, etc. need to be pointed precisely with regard to three-dimensional space. For spacecraft payloads to provide useful services in space, having the exactly precise pitch, yaw and roll angles is critical.

Finally, there is one other coordinate frame we need to consider, and that is the spacecraft coordinate frame. This frame is similar to the coordinate frame generally associated with any flying vehicle, like an airplane. If you are sitting in an airplane facing forward, the x-axis of that airplane goes straight out the tip of your nose and the nose of the aircraft in the general direction of forward travel. Now, if you raised your right arm and pointed to your right along the direction of the right wing, that direction defines the y-axis. Finally, the z-axis is perpendicular to the x- and y-axes and points straight down toward the ground. The origin can be located at the center of the aircraft, or any other convenient location. In similar fashion, we define the spacecraft coordinate frame with the x-axis pointing along the direction of travel in the satellite orbit, the y-axis points in the direction of the "orbit normal" (think of the right wing), where normal also means perpendicular. Finally, the z-axis points towards the center of the Earth. In general, the origin of the spacecraft coordinate frame is located at the center of mass of the spacecraft. The center of mass is the point where all of the mass of the object is concentrated. When an object is supported at its center of mass there is no net torque acting on the body and it will remain in static equilibrium: think about balancing a broomstick on your finger. We can now define the location of the spacecraft center of mass relative to the inertial frame and the Earth-centered frame. We can also define the location of any component or subsystem inside the spacecraft relative to the spacecraft-centered coordinate frame.

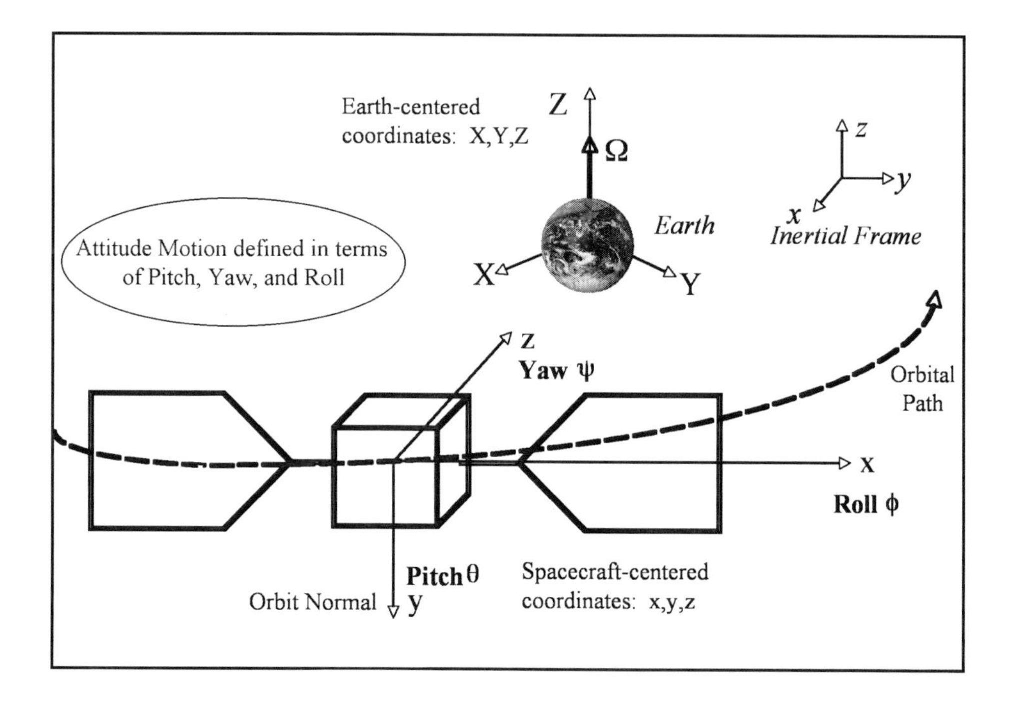

Figure 7.5. The various coordinate frames used in spacecraft attitude and orbital control are shown. The three primary coordinate frames include an inertial (fixed in space) frame designated as x, y, z and an Earth-centered frame that rotates in synch with the Earth.

Finally we need to address the concept of angular momentum because that is the principle on which the attitude control of a satellite is based. Now, the moment of inertia is the measure of the tendency of a solid body to resist rotational forces and it is analogous (in relation to linear forces) to mass. A rotational force is called a torque. When you screw or unscrew a light bulb into or out of a socket, or use a screwdriver to insert or remove a screw, you are applying a torque.

Just remember, if a torque is applied for a certain amount of time, then the *angular momentum* of the body will change in proportion to the applied torque. Another way to express this is to say that, if *no* torque is applied, the *angular momentum* of the body will be *unchanged*. Angular momentum is the primary tool used to accomplish attitude control. To explain how this works let's undertake a thought experiment. Remove a bicycle wheel from the bicycle and hold it in both hands by the axle. Have a friend spin the wheel hard so that it's spinning really fast. Now, try to rotate the wheel to the left or right. Do you feel how it resists? Using spinning masses (which we call inertial wheels or momentum wheels) or by spinning the entire spacecraft, or part of it, we can control the direction the spacecraft is pointing by taking advantage of this effect.

Control System Sensors, Algorithms, and Actuators

Controlling the vehicle attitude and orbit requires a number of specific capabilities. There is first a need for sensors to measure vehicle attitude. Secondly, there have to be actuators to apply the torques needed to re-orient the vehicle to a desired attitude. Finally, there need to be control computers equipped with algorithms to carry out the calculations needed to determine how to command the actuators, based on the sensor measurement inputs.

There are actually many different kinds of sensors used on spacecraft to make these measurements; here is a list of sensors required for attitude and orbital control.

> ***Gyroscopes*** – These devices sense rotation in three-dimensional space, without relying on the observation of external objects. Classically, a gyroscope consists of a spinning mass, but it also typically includes a "Laser Gyro". These laser gyros utilize coherent light reflected around a closed path to maintain stability. Gyros require initialization by some other means, as they can only measure changes in orientation from a pre-determined reference point. In addition, all gyro measurements are subject to drift, and can maintain orientation for limited times only, typically tens of hours or less.

Horizon Indicator **-** An optical instrument that detects light from the 'limb' of the Earth's atmosphere, i.e., at the horizon. It can be a scanning or a staring instrument. Infrared sensors are often used because they will work even on the dark side of the Earth. This type of sensor can provide orientation with respect to the Earth on two orthogonal axes. This type of sensor, however, tends to be less precise than sensors based on stellar observation.

Orbital Gyrocompasing **-** This type of sensor operates in a similar way to a terrestrial gyrocompass. The terrestrial gyrocompass uses a pendulum to sense local gravity and respond to the gravitational affect by alignment with the Earth's spin vector and thus to point North. The orbital gyrocompass uses a "horizon sensor" to sense the direction to Earth's center, and a gyro to sense rotation about an axis normal to the orbit plane. Thus, the horizon sensor provides pitch (x-axis) and roll (y-axis) measurements, while the gyro provides yaw (z-axis).

Sun Sensor **-** A device that senses the direction to the Sun can be as simple as some solar cells and shades, or as complex as a steerable telescope, depending on the mission requirements for accuracy of pointing and orientation.

Star Tracker **-** A star tracker is a sensitive optical device measuring the direction to one or more stars, using a photocell or solid-state camera to observe the star. There are 57 bright navigational stars in common use; one of the most used is Sirius (the brightest). However, for more complex missions entire star field databases may be used to identify orientation. One of the risks this entails is that star trackers may become confused by sunlight reflected from the exhaust gasses emitted by thrusters.

Control ***algorithms*** are the computer programs that receive input data from the vehicle sensors and derive the appropriate torque commands to the actuators to rotate the vehicle to the desired attitude. The algorithm can be very simple, e.g., proportional control, a complex nonlinear estimator or many in-between types, depending on mission requirements. Typically, the attitude control algorithms are part of the software running on the hardware that receives commands from the ground and formats vehicle data telemetry for transmission back down.

Just as there are a number of different kinds of sensors onboard the spacecraft, there are also many different kinds of actuators that can be used in attitude and orbit control. Here is a list of some of them.

Thrusters - Thrusters, which are often monopropellant rockets, must be organized as a reaction control system. An array of thrusters provides stabilization torques in all three directions so that the array can control all three axes of roll, pitch, and yaw. The obvious limitation of thrusters is in fuel usage. To keep the spacecraft pointing in the right direction, a tiny blip of thrust is applied in one direction and, a few tens of seconds later, an opposing blip of thrust is needed to keep orientation errors within limits. To minimize fuel limitation on mission duration, auxiliary attitude control systems are used to reduce vehicle rotation to lower levels. These can be smaller, lower thrusts, "Vernier" thrusters that accelerate ionized gases to extreme velocities electrically, using power from solar cells. The key here is that the solar cells provide power that is not exhausted like rocket fuel is.

Spin Stabilization - The entire space vehicle itself can be spun up to stabilize the orientation of a single vehicle axis. This method is widely used to stabilize the final stage of a launch vehicle. The entire spacecraft and an attached solid rocket motor are spun up on what is called a "spin table". This allows the spacecraft to be rotated along the "thrust axis" by the attitude control system on which the spin table is mounted. When final orbit is achieved, the satellite may be de-spun by various means, or left spinning. Spin stabilization of satellites is only applicable to those missions with a primary axis of orientation that need not change dramatically over the lifetime of the satellite and when there is no need for extremely high precision pointing. It is also useful for missions with instruments that must scan the star field or the Earth's surface or atmosphere.

Momentum Wheels - These are electric motor driven rotors made to spin in the direction opposite to that required to re-orient the vehicle. Since momentum wheels make up a small fraction of the spacecraft's mass and are computer controlled, they give precise control. Momentum wheels are generally suspended on magnetic bearings to avoid bearing friction and breakdown problems. To maintain orientation in three-dimensional space a minimum of two must be used, with additional units providing single failure protection.

Control Moment Gyros (CGMs) - These are rotors spun at constant speed, mounted on gimbals to provide attitude control. While a CMG provides control about the two axes orthogonal to the gyro spin axis, three-axis control still requires two units. This type of gyro is a bit more expensive in terms of cost and mass, since gimbals and their drive motors must be provided. The maximum torque (but not the maximum angular momentum change) exerted by such a gyro is greater than for a momentum wheel, making it better suited to large spacecraft. A major drawback is the additional complexity, which increases the number of possible failure modes.

Solar Sails – These thrust as a reaction force induced by reflecting incident light, and may be used to make small attitude control and velocity adjustments. This application can save large amounts of fuel on long-duration missions by producing control moments without fuel expenditure. Large communications satellites with large solar arrays can supplement their stabilization in this manner as well.

Gravity Gradient Stabilization - In orbit, a spacecraft with one axis much longer than the other two, or a long skinny spacecraft, will spontaneously orient itself so that its long axis points at the planet's center of mass. This system has the virtue of needing no active control system or fuel for orientation. The upper end of the vehicle feels less gravitational pull than the lower end. This provides a restoring torque whenever the long axis is not pointing in the direction of gravitational pull. Unless some means of damping is provided, the spacecraft will tend to oscillate about the axis of gravitation pull. Sometimes tethers are used to connect two parts of a satellite, to increase the stabilizing torque.

Magnetic Torquers – These devices are permanent magnets. They exert a moment against the local magnetic field, so this method works only where there is a magnetic field (such as the Earth's) to react against. Such a conductive tether can also generate electrical power, at the expense of orbital decay. Conversely, by inducing a counter-current, using solar cell power, the orbit may be raised.

Pure Passive Attitude Control - Gravity gradient and magnetic field pointing can be combined to form a completely passive attitude control system. Such a simple system has limited pointing accuracy, because the spacecraft will oscillate around the center axis of the attractive force or forces. This drawback can be at least partially overcome by adding a viscous damper. This can be a small can or tank of fluid mounted in the spacecraft, possibly with internal baffles to increase internal friction.

Reaction Control Systems

A reaction control system is another key spacecraft subsystem that is composed of an array of thrusters. As noted above its purpose is to provide attitude control and steering; it can be thought of as just another actuator.

Actually, it is a system of actuators that are turned on and off by the attitude or orbital control system. A reaction control system is capable of providing a small amount of thrust in any desired direction or combination of directions. It is also capable of providing torque to control the rotation (roll, pitch, and yaw). This is in contrast to the spacecraft main engine, which provides thrust in one direction, but is much more powerful, and is used for orbital control such as changing orbits or shifting a geosynchronous communications satellite from one region to another.

Typically these systems use combinations of large and smaller (vernier) thrusters, to allow different levels of response from the combination of different sized thrusters. Reaction control systems are used for

- Attitude control during re-entry
- Station-keeping in orbit
- Close maneuvering during docking procedures
- Control of orientation, or 'pointing the nose', of the spacecraft
- As a backup means of de-orbiting.

Such a control system can be incorporated into the payload stage and each of the stages of a multiple stage vehicle. In some missions and designs a reaction control system is built into only the uppermost stage. In this type of design it operates throughout the flight and provides the control torques and forces for all the stages. Liquid propellant rocket engines with multiple thrusters are most frequently used, not only on launch vehicles but also on the great majority of all spacecraft. Cold gas systems were used in early spacecraft designs. Solid fueled apogee kick motors were used in the early generations of communications satellites as well. In the last decade electrical propulsion systems have been more widely used, primarily on application spacecraft with extended lifetimes of 15 years or even more. The life of a reaction control system may be short (when used on an individual vehicle stage), or it may see use throughout the mission with a lifetime of many years when part of an orbiting spacecraft.

These control systems can be characterized in many ways. Thus payload control systems may be described by the magnitude of the total impulse, the number of thrusters, their thrust level, and the directionality of the thrusters, and by their duty cycles. The *duty cycle* refers to the number of thrust pulses, their operating times, the times between thrust applications, and the timing of these short operations during the mission. For a particular thruster, a 30% duty cycle means an average active cumulative thrust period of 30% during the propulsion system's flight duration. These propulsion parameters can be determined from the mission requirements. Thus the mission plan will spell out the guidance and control approach. This usually specifies such things as: (i) the desired accuracy, (ii) flight stability, (iii) the likely thrust misalignments of the main propulsion systems, (iv) three dimensional flight path variations, and (v) the perturbations to the trajectory. Some of these parameters are often difficult to determine so some margin must be provided against contingencies.

Attitude and Orbital Control

Attitude control is the control of the orientation of a spacecraft, or other flight vehicle, either relative to the celestial sphere or to a gravitational body influencing its flight path, like the Earth. The vehicle attitude must be controlled about the three mutually perpendicular axes in a coordinate frame, each with two degrees of freedom (clockwise and counterclockwise rotation), giving a total of six degrees of rotational freedom. These six degrees of rotational freedom are as follows. First there are pitch torques that can raise or lower the nose of the vehicle (i.e. the y-axis). Second there are yaw torques that can induce a motion to the right or the left side (i.e. the z-axis). Third, the roll torques rotate the vehicle (i.e. about its x-axis), either clockwise or counterclockwise.

To apply a true torque it is necessary to use two thrust chambers of exactly equal thrust and equal start and stop times, placed an equal distance from the center of mass firing in opposite directions. A minimum of 12 thrusters is needed in this system, but some spacecraft with geometrical or other limitations on the placement of these nozzles or with provisions for redundancy may actually have more than 12. The same system can, by operating a different set of nozzles, create different rotational effects. With clever design it is possible to use fewer thrusters.

Orbital control comprises two main functions, namely *navigation* and *guidance*. Navigation answers the question "where am I?" and guidance answers the question "where am I going?" The navigation function determines the actual position and velocity. The guidance function computes the commands to the thrusters required to send the vehicle to the desired location. Figure 7.6 helps to explain the basics of orbital control. By now you have figured out that engineers can't do anything without a picture!

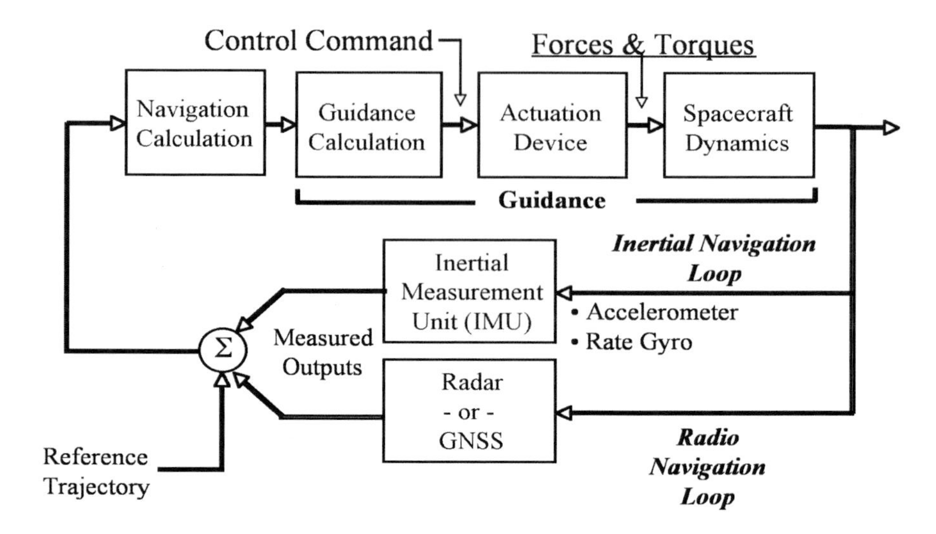

Figure 7.6. Block diagram of an orbital control system showing the guidance functions and the navigation loops.

In an attempt to control spacecraft we always begin with a reference trajectory. This reference model is either stored in the memory of the onboard computer or up-linked from controllers (people) on the ground. The reference signal is compared to the sum of the measurements of the vehicle position, velocity, and acceleration. This data when taken together tells you where the spacecraft actually is and a guidance calculation algorithm calculates what has to be done to get to where the spacecraft needs to go.

Once the calculations are complete the algorithm then generates commands that drive the actuation devices (thrusters for orbital control) so that the spacecraft can get to where it wants to be. After the appropriate thrust is applied further calculations are made to determine if fine-tuning is necessary to make small corrections.

It is important to note in Figure 7.6 that there are two navigation loops. One is based on measurements made onboard the spacecraft, called the internal navigation loop, and the other is typically based on radar measurements taken from the ground and transmitted to the spacecraft. More recently, global navigation satellite systems, like GPS, have been used to replace the "radio navigation loop," so that all measurements, as well as verification measurements, can possibly be made onboard the spacecraft. However, care must be taken when using onboard GPS systems because of the rapidly changing GPS satellite positions relative to the spacecraft itself.

Further, because the GPS is designed to determine locations at or relatively near the Earth's surface, some allowances for measurement inaccuracies must be made as well.

Power Up

Spacecraft Power Sources, Regulation, Distribution and Storage

There are no electrical outlets on a spacecraft - imagine the length of the extension cord required to provide power to an orbiting spacecraft! So, all power requirements onboard the spacecraft must be met by onboard power subsystems. The main functions of the power subsystem are listed below.

- Generate the required power to the spacecraft for the duration of the mission
- Regulate and distribute the power to all subsystems/users onboard the spacecraft
- Support power requirements for average and peak electrical loads
- Provide transformation from DC to AC, if required
- Protect against power supply faults or surges.

The mission requirements dictate the necessary performance of a reliable and safe power subsystem, including power levels, frequency, voltage, and capacity. The mission lifetime imposes another requirement on the power supply system in terms of acceptable power degradation over time. How much can the level of power that the power supply system provides degrade over the life of the mission and still do what's necessary? Mission constraints on the power subsystem include mass, size and volume, cost, schedule, and testability and repair possibilities, which are few on orbit!

During the course of a mission, the power levels required will vary. It takes much more power to move a pointing instrument, like a telescope on a gimbal, for example, than it does just to collect the data from the instrument without moving it. Typical power levels, N, that are required vary as a function of flight time. The peak power is the maximum power required, which the power supply must provide. Over the course of the mission, the average power is exactly that – the average of the power levels used over the course of the flight. Power is supplied by the spacecraft primary power source, such as a photovoltaic (solar array) subsystem. A secondary power source, such as a battery, provides extra power when needed and compensates during times of peak power consumption. The battery system is particularly critical if the spacecraft spends some of the time in eclipse, behind the Earth from the Sun where solar cells cannot generate any power. Table 7.2 provides typical average and peak power requirements for various spacecraft.

There are three energy sources used to generate power on board spacecraft. These are solar, nuclear, and chemical. Solar power is generated using photovoltaic solar cells (e.g., made of silicon) that convert sunlight into electricity. Solar energy in the form of heat can also be used in solar dynamic systems. Nuclear reactors have also been used on spacecraft, as have systems that convert the energy released by radioactive isotopes into electric power. Chemical power systems comprise the familiar battery system, as well as fuel cells.

The traditional spacecraft power system consists of a solar voltaic generator, incorporating solar panels to collect and convert the solar energy; chemical batteries; and the power regulation, control, and distribution system. The solar generator is sized for the average power output requirement, with a little extra added for safety, deterioration of performance over the spacecraft's flight time and for battery charging. The solar generator may comprise several solar arrays. Solar array performance is generally characterized by a quantity called efficiency, which is the ratio of the power in (from the Sun) divided by the power out (power produced). Typical efficiencies are 10-25%, but newer technologies using gallium arsenide materials and more junctions, especially in the ultraviolet, can exceed these levels. Solar arrays are either fixed to the spacecraft surface or to Sun-pointing flat panels, which are deployed once the spacecraft is on-orbit. It is also possible to add reflectors so that the solar cells "see" the equivalent of two or even three Suns.

Spacecraft Type	Average Power, kW	Peak Power, kW
Small Satellites	0.1 to 0.3	0.2 to 0.4
GEO ComSats	1.5 to 5.5	2.0 to 6.5
LEO ComSats	0.5 to 0.8	0.7 to 1.2
LEO Remote Sensing Satellites	2.0 to 6.5	2.8 to 8.7
Interplanetary Probes	0.3 to 0.5	0.8 to 1.0
Space Shuttle	10 to 15	13 to 17
Space Station	25 to 70	50 to 150
Micro - satellites	10^{-3} to 10^{-1}	10^{-1} to 1
Lunch vehicles	0.3 to 5	0.3 to 5

Table 7.2. Spacecraft power requirements

(Note: LEO = Low Earth Orbit; GEO = Geostationary Orbit).

The batteries, a chemical power source, provide a secondary power source that can be used during times of eclipse and to meet peak power requirements. Batteries convert chemical energy into electrical energy and are also characterized by efficiency. For batteries, the efficiency is the ratio of battery electrical capacity (in watts x hours) divided by the mass (in kilograms). Rechargeable batteries are typically nickel-cadmium, nickel-hydrogen, sodium-sulfur, silver-zinc, or lithium-ion. Another type of chemical power source, typically used for human spaceflight, is a fuel cell. Fuel cells work at low temperatures and generate power based on the reaction between hydrogen and oxygen. They have high energy densities, but are very expensive.

Finally, for a classical spacecraft power subsystem, the power regulator ensures that all the power subsystems are working correctly, controls battery charging and discharging, and the dumping of excess power. While the power generation devices are the main part of the power subsystem, other components responsible for power processing and power distribution (cables, fault protection, power switches, DC to AC converters, and voltage transformers) add up to 30 to 45% of the power subsystem mass!

As a final thought to our earlier comment about there being no unimportant subsystem, aerospace corporations often put their most experienced engineers on what is seen as the most difficult and challenging part of the mission, and their junior engineers on the "routine" parts of the program. In actual practice it turns out that the more "mundane" subsystems in the program can lead to spacecraft failure. In the Intelsat 5 program defects in the DC to AC power converters led to a satellite failure. Since one of the most common failures in spacecraft relate to power systems, particular care needs to be given to the design and manufacture of these systems.

Keeping Cool and Staying Warm – Spacecraft Thermal Control

Another key subsystem on the spacecraft is thermal control.[3, 4] The space environment is harsh and temperatures vary from very hot to very, very cold. Considering that almost all instruments and electronics are designed to work at "room temperature" (where humans are quite comfortable working!), it's no wonder that thermal control is very important to keep everything on the spacecraft working right.

Is Outer Space Hot?

Space can be extremely hot and harmful if you are facing toward the unfiltered radiation of the Sun, and extremely cold if you are facing away from the Sun. As a spacecraft moves around a planet, it will go through eclipses. Therefore, the temperature experienced by the spacecraft will vary from very hot to extreme cold. This further complicates the thermal control of the spacecraft because we would like to maintain the spacecraft and its components within a reasonable range of temperatures.

Many of the spacecraft components are very sensitive to the temperature. Some of these components will not operate, or their performance will be significantly reduced if the temperature operating range is not maintained. Even worse some components can literally be destroyed, or cease to function, when very hot or very cold. Although the main problem of the thermal engineer is to get rid of the excess heat, some components such as batteries may need to be kept warm to stay operational. Infrared cameras need to be kept cold to increase their sensitivity. Table 7.3 provides typical operating temperature ranges.

Components	Temperature range (°C)
Electronics	-10 to +50
Batteries	-5 to +20
Microprocessors	-10 to +50
Solar Arrays	-105 to +110
Propellant	+ 7 to +55
Infrared Detectors	-260 to -80
Bearing Mechanisms	-45 to +65
Structures	-45 to +65

Table 7.3. Typical operating temperature ranges for different spacecraft components.

Now let's discuss the thermal environment for a spacecraft in Earth orbit. Figure 7.9 shows that thermal energy from the Sun is transferred to the spacecraft by radiation. We know of three modes of heat transfer: conduction, convection and radiation. In conduction, heat flows from one point to another at a lower temperature through a medium (e.g., two metal blocks put in thermal contact). In convection, heat flows by means of a fluid medium flowing past the heat source (e.g., boiling water in a kettle). Finally, radiation is characterized by the transfer of energy by means of electromagnetic waves emitted by a hot body. This mode does not require a medium to propagate; solar energy arriving on Earth has traveled through the vacuum of space.

On Earth, we experience other modes of heat transfer in addition to radiative heat transfer. The spacecraft will also be heated by albedo radiation, the solar radiation reflected by the Earth, about 30% of the incoming solar radiation. The Earth is itself a source of radiation that contributes to the spacecraft heating. The heat produced by electronic components will further heat the spacecraft. You can now understand why thermal engineers are mostly concerned with rejecting excess heat with regard to LEO spacecraft.

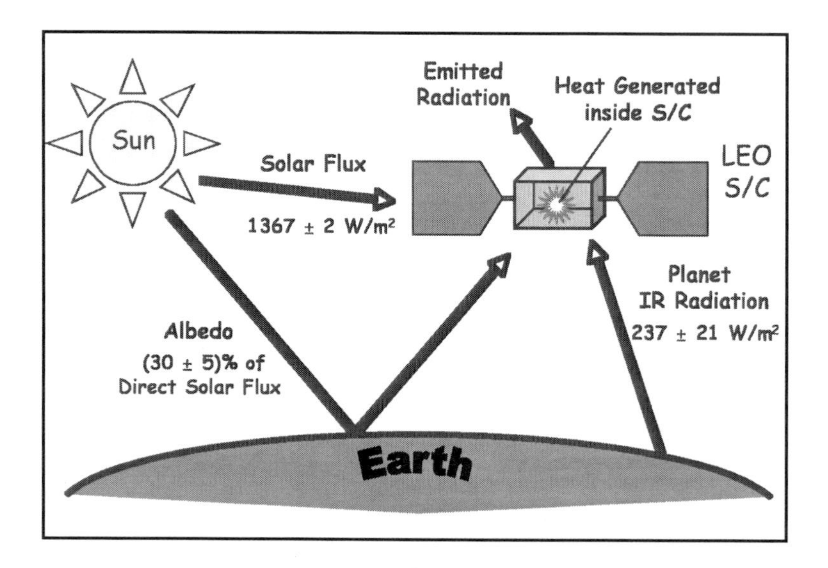

Fig 7.7. Typical low Earth orbiting (LEO) spacecraft with various heat sources.

Thermal Equilibrium

If two bodies at different temperatures are brought into thermal contact with each other, they will eventually reach the same temperature. The two bodies come into what we call "thermal equilibrium". An intuitive concept that we observe operating in daily life is the basis of one of the fundamental laws of thermodynamics. It is possible to determine the thermal equilibrium temperature of a spacecraft in orbit by using basic physical laws, and knowing that the incoming radiation to an object can be absorbed, reflected or transmitted.

Figure 7.8 shows the equilibrium temperature, T, of a spacecraft that is assumed to be a uniform sphere with a radius of one meter. For this example, the solar constant is taken to be 1390 W/m^2 (typical during northern hemisphere winter). The influence of the coating materials of the spacecraft is significant; the external materials, particularly their absorptivity (α) and their emissivity (ε) as determined by their color, are extremely important. In Figure 7.8, G_s is the heat generated inside the spacecraft, Q_w is the energy input to the satellite, and σ is the Stefan-Boltzmann constant; this Figure not only indicates how to create the correct thermal budget for a spacecraft but it also provides a useful hint for you to choose the right surface properties such as color for your clothing when the seasons change. White is the best for hot summer days, and gold will keep you very warm!

It seems that it is quite straightforward to manage the spacecraft temperatures by choosing the right external surface. In reality, the problem is much more complex due to the varying heat input and output. As the spacecraft goes around in its orbit, it goes in and out of eclipse and its orientation to the Sun will change. This will bring temperature variations with respect to time and major temperature gradients in the spacecraft itself. In addition, the spacecraft is unlikely to be a perfect sphere; its complex geometry needs to taken into account to make accurate predictions of the temperature distributions around the spacecraft. There are several software packages to tackle such a complex problem, the well-known ones include those named SINDA, TMG, and ESATAN.

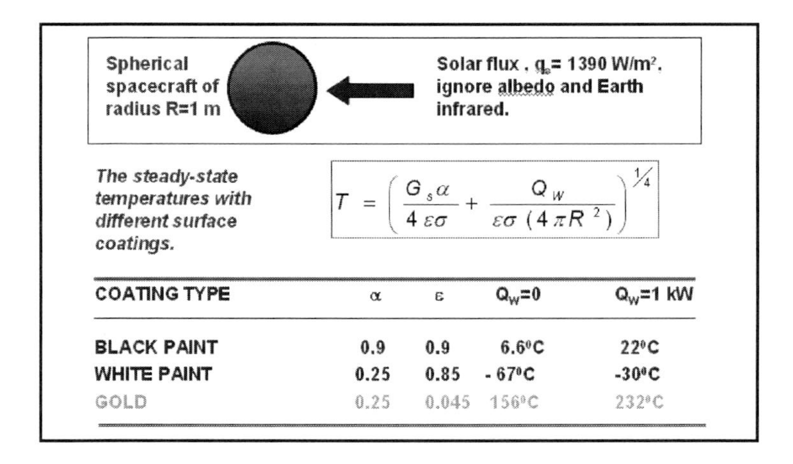

COATING TYPE	α	ε	Q_W=0	Q_W=1 kW
BLACK PAINT	0.9	0.9	6.6°C	22°C
WHITE PAINT	0.25	0.85	- 67°C	-30°C
GOLD	0.25	0.045	156°C	232°C

Figure 7.8. This figure shows the thermal equilibrium temperatures for a spherical spacecraft with various surface properties.

Thermal Control Hardware

At this point, it should be clear to you that the thermal engineer's simplest tool is choosing the appropriate coating or spacecraft surface finishes.[5] This is more or less what we do before leaving our home to go to work, depending on the weather conditions. Coatings and surface finishes are passive systems; they have no moving parts, nor do they require any power to operate. Engineers love passive systems since they do not require maintenance, and they are more reliable than the alternative active systems. Therefore, a thermal engineer first tries to solve the thermal design problems using passive systems. If that doesn't work, he has to try using active systems. It is common that passive and active systems are used together.

Passive Thermal Control Hardware

Multi-Layer Insulation. Multi-layer insulation (MLI) is frequently used in spacecraft to minimize the heat exchange of a component with other components or with space. A typical MLI has 20-25 layers per centimeter of blanket thickness. Mylar or kapton sheets with aluminized surfaces are widely used in this way. Several layers increase the thermal resistance to radiation, generating very low effective emissivities for the MLI blanket. The layers may be separated by a nylon net to decrease the thermal conductance. MLI insulation gives the spacecraft its shiny appearance. It is often used to insulate sensitive components, such as cameras, so as to minimize thermal gradients.

Coating and Paints. There are several coatings and paints available with varying reflectivity values. The thermal engineer chooses a combination of these materials to manage the temperature of a spacecraft. One important consideration is the degradation of these surfaces during the mission life, but this can be predicted. The thermal engineer often introduces some additional margin in the design to make sure that the surfaces continue to be effective until the end of the mission. Well-known coatings include Second Surface Mirrors (SSM) and Optical Solar Reflectors (OSR). The general principle is that the surface should reflect the incoming solar energy with low absorptivity. It should emit the absorbed energy as infrared radiation, with high emissivity.

Heat Pipes. Heat pipes have been effective tools for the spacecraft thermal control since the 1980s. The operational principle of a heat pipe is based on the capillary forces produced in a wick structure. As shown in Figure 7.9, the working fluid in a heat pipe, usually ammonia saturates the wick. When the heat is applied to one end, the working fluid evaporates and the vapor so formed is pushed to the other end of the heat pipe. This end is connected to a radiator, or a cold sink, where the working fluid condenses. The condensate is finally brought back to the evaporation section because of the capillary forces generated inside the wick structure. Heat pipe operation is passive and the thermal conductivity is very high. Therefore, heat pipes have been extensively used for more demanding thermal control challenges, and are still being improved. Specific applications of heat pipes include diode heat pipes, variable conductance heat pipes, and oscillating heat pipes. Newer versions of heat pipes, e.g., loop heat pipes, are being used more and more in space industry. They can handle higher heat loads, up to kilowatts. Although heat pipes are very sensitive to gravity and do not operate at adverse tilts (i.e. where the evaporator is above the condenser), these new design heat pipes can operate against gravity. This is a very important feature to note when testing the spacecraft on the ground.

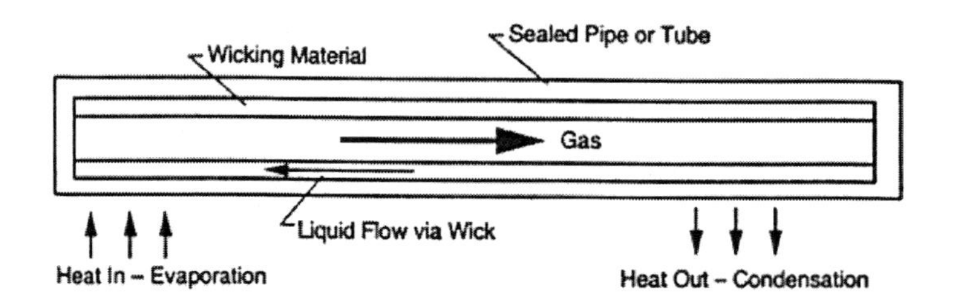

Figure 7.9. The operating principle of a typical heat pipe.

Active Thermal Control Hardware

Heaters. Heaters in the form of a film resistance are widely used in a spacecraft to maintain a component at a desired temperature. A general approach is to keep the component temperatures low by using passive techniques and to control the temperature within a narrower range using heaters controlled by thermostats. As these have moving parts and require power from the spacecraft, this type of thermal control system is clearly an active system that adds weight and drains energy. Although the thermal engineer, for these reasons, prefers passive systems, there are cases where it is impossible to accomplish the performance required without using some active thermal control hardware.

Thermoelectric Coolers. Thermoelectric coolers are based on the so-called Peltier Effect. Here an electric current passing through a semiconductor removes heat from a component towards another heat sink. Thermoelectric coolers are used for fine temperature control of smaller components. Their lifetime and structural integrity can be problematic for some space missions; however, they have been used successfully in several space missions.

Louvers. Louvers are a mechanical means of changing the surface properties by exposing different reflectivity values, depending on the position of movable external surfaces. Think of these devices as being similar to the louvered blinds on your window. You open them to let more light in, close them to keep light out. The reflectivity is high when the external surface is closed. If these surfaces are exposed to solar illumination and the temperature increases above some desired level, the external surfaces open.

Open and Closed Loop Systems. If the thermal control requirements for a mission are very demanding, such as very low temperatures or high heat loads, the thermal engineer has to consider more sophisticated methods; sometimes these are heavier, less reliable and more expensive in terms of complexity. An open loop system is an expendable cooling method, normally used when temperatures below -70°C are required. With these types of systems a certain amount of liquid (helium or ammonia) is carried in a tank within the spacecraft. This liquid absorbs the heat from a spacecraft component such as an infrared camera; the heated gas is then vented to space. When the cooling liquid is exhausted, the mission obviously comes to an end. The Space Infrared Telescope Facility (SIRTF) carried a tank of 360 liters of liquid helium to maintain the camera temperature at 5.5K.

Closed loop systems are in many ways similar to a home refrigerator. A cooling liquid is circulated by a pump, the cooled part being attached to an evaporator (the cooling compartment of your refrigerator). The heat rejection part is attached to the spacecraft radiator (the external condenser of your refrigerator). Crewed space vehicles, such as the Space Shuttle and the International Space Station are two examples of where closed loop cooling systems are used. These systems can handle large heat loads; they also have other advantages, such as better control for varying mission requirements. However, they are heavier, more complex, more expensive, and less reliable due to the various moving parts in active systems.

Keeping it all Together – Spacecraft Structures

We've covered almost all of the main subsystems that together form a spacecraft. Now we discuss the elements of the spacecraft that hold everything together – the structural components. The spacecraft structure provides the mechanical support to all the subsystems within the spacecraft (and the launch system). This support must be provided from the time that manufacturing begins until the end of the mission. We define the *primary structure* as the load-carrying part of the structure. The primary structure is the spacecraft frame, or skeleton, and functions much like the large bones in your skeleton – your legs, hips, and backbone – supporting your weight. Examples of primary structures in a spacecraft include the thrust cone and cylinders, support bulkheads, longerons, launch vehicle adapter, and propellant tanks. We also define *secondary structure* as the non-load carrying part. Secondary structures comprise brackets, panels, and deployable items like antennas and solar panels. Secondary structures are not as massive as primary structures, in general.

There are a number of different kinds of structures found on a spacecraft. These include skin panel assemblies, trusses, ring frames, fittings and brackets, pressure vessels, component boxes, and deployable structures like solar arrays, astromasts, and antennas. Can you identify the different kinds of structures used in the Magellan spacecraft, shown in Figure 7.10? Which are primary structures? Which are secondary structures?

There are three main properties which characterize structures, namely their *strength, stiffness* and *stability*, and which play a key role in how spacecraft structures are designed. Strength is defined as the amount of load a structure can carry without experiencing either complete failure or sufficient deformation to jeopardize the mission. Stiffness is a measure of the load required to cause a given deflection. And stability is the ability of a structure to maintain its configuration within a certain tolerance in order to support the correct orientation of the payloads or subsystems. In essence, stability is the ability of a structure to resist buckling. Buckling is the phenomenon that happens when the structure collapses. A good way to demonstrate buckling is to stand an empty soda can on the floor (flat end on the floor) and then carefully try to stand on it. When the soda can collapses under your weight – that's buckling! A structure is *sized* by the largest of the strength, stiffness, and stability requirements.

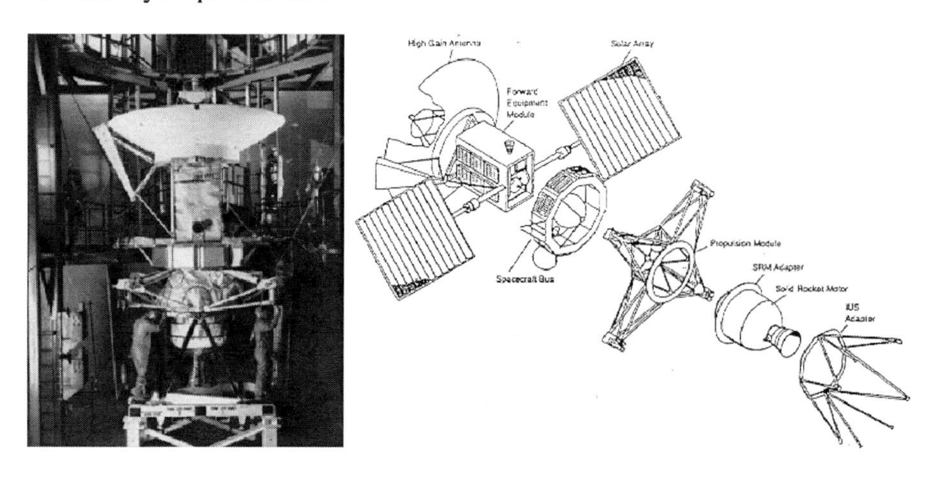

Figure 7.10. Examples of primary and secondary structures (Courtesy of NASA).

Other structural properties can be defined as well. The *structural life* is the number of loading cycles a structure can withstand before it fails, or the duration of a sustained load that a structure can withstand until its materials fail. Damping is a characteristic inherent to the structural material, and refers to the dissipation of energy during vibration. Without damping, an excited structure would continue to vibrate forever.

Structural requirements are derived from all of the environments in which the structure must function, including ground, launch, and space environments. Each of these environments imposes their own particular requirements. The structural elements of a spacecraft must retain their integrity after exposure to these environmental conditions with minimum mass, minimum cost, and maximum reliability. No small task, this!

During launch, the spacecraft structural requirements are dictated by the launch vehicle, which creates major mechanical loads at liftoff, including thrust, acceleration, acoustic noise, and shock. The launch environment is generally the main spacecraft structural design driver. During launch, the spacecraft is subjected to:

- Low-frequency sinusoidal vibration that is generated by aerodynamic forces acting on the launch vehicle, and transient events like staging, fairing jettison, and engine cutoff,
- High-frequency random vibration that is produced by mechanical parts in motion like engine turbo-pumps, fuel combustion, and structural elements excited by acoustic noise,
- Acoustic noise that is reflected off the ground and pad facilities at liftoff and peaks with the maximum dynamic pressure being experienced during ascent, and
- Shock that occurs during pyrotechnic events such as booster separation, staging, fairing jettison, and spacecraft separation.

During the launch phase, structural *strength* and *stability* are the most important! Table 7.4 summarizes structural requirements by mission phase.

Mission Phase	Source of Requirements
Manufacture and assembly	• Handling fixture or container reactions • Stresses induced by manufacturing processes (welding)
Transport and handling	• Crane or dolly reactions • Land, sea, or air transport environments
Testing	• Environments from vibration or acoustic tests • Test fixture reaction loads
Prelaunch	• Handling during stacking sequence and pre-flight checks
Launch and ascent	• Steady-state booster accelerations • Vibro-acoustic noise during launch & transonic phase • Propulsion system engine vibrations • Transient loads during booster ignition & burn-out, stage separations, vehicle maneuvers, propellant slosh, and payload fairing separation • Pyrotechnic shock from separation events
Mission operations	• Steady-state thruster accelerations • Transient loads during pointing maneuvers and attitude control burns or docking events • Pyrotechnic shock from separation events, deployments • Thermal environments
Reentry & landing (if applicable)	• Aerodynamic heating • Transient wind and landing loads

Table 7.4. Structural requirements defined by all mission phases.

When the spacecraft is in orbit, the internal components, i.e. all of the subsystems we've discussed, dictate the structural requirements. On-orbit mechanical loads are substantially smaller in magnitude. Vibration is the main problem when the spacecraft is in orbit. Examples of vibration sources include payload pointing and slewing, turbine-driven machinery, thrusters, cryogenic coolers, and even the motion of the crew. Vibrations are transmitted through the structure; therefore, structural stiffness, or resistance to deflection, is the most important characteristic. Structures vibrate at their natural frequencies, which are the frequencies at which the structure vibrates when excited by a given transient load and then allowed to move freely. The fundamental frequency of a structure is the lowest natural frequency.

Something design engineers strive to avoid is structural *resonance*, which is the condition in which an object experiences large amplitude vibrations in response to an external stimulus acting at or near a natural frequency of an object. The natural frequencies of spacecraft structural elements must be kept separated to avoid resonant coupling caused by excitations from vibrating sources and with the control system. Otherwise, catastrophic situations can occur.

Spacecraft structures, and all spacecraft components and subsystems, undergo rigorous testing procedures at every step of the fabrication process. Examples of the kinds of tests include vibration, acoustic, shock, vacuum and thermal cycling, and mechanical cycling tests. There are a host of other tests that can be reviewed on the web page accompanying this book. Of paramount importance in all testing is the development of a detailed and thorough test plan. At a minimum, a test plan includes the following elements.

- Test objectives
- Test methodology or protocols
- Step-by-step procedural details
- Test equipment required
- Input/output value limits
- Expected results
- Pass/fail criteria.

Following the test plan faithfully is crucial to ensure accurate results and to protect expensive hardware!

Designing a Spacecraft

A television commercial made by the Holiday Inn hotel chain claims that, if staying at one of their hotels is "smart", then merely staying there one night will make you an expert on virtually any subject—from being a doctor to a rocket engineer. We do not expect you to be able to design and build a spacecraft from reading this chapter, nor to have mastered the systems design skills to know how to optimize the design. We do hope that you now understand the major components or subsystems and their overall importance to the proper function of a spacecraft.

† Unless otherwise indicated, the authors developed the graphics for this chapter.

1 J. Wertz and W. Larson, Space Mission Analysis and Design, 3rd edition, Microcosm, 1999.
2 W. J. Larson and L. K. Pranke (eds.), Human Space Flight: Mission Analysis and Design, McGraw-Hill, 1999
3 See http://www.j-cds.nl/index.php?site=solutions&search=alle&block_id=66
4 D. G. Gilmore, Spacecraft Thermal Control Handbook: Fundamental Technologies, 2nd edition, Aerospace Press, 2002.
5 P. Fortescue, J. Stark, and G. Swinerd, Spacecraft Systems Engineering, 3rd edition, 2003

Chapter Eight

Space Missions and Programs – Why and How we Go to Space

"Ships and sails proper for the heavenly air should be fashioned. Then there will also be people, who do not shrink from the dreary vastness of space."

— Johannes Kepler, 1609

The engineering pieces of space missions are many, and include much more than just spacecraft and their associated subsystems discussed in the previous chapter. We also need to know about astrodynamics, launch vehicles and mission operations that are covered in this chapter. Most importantly, we will examine how all these pieces fit together to form working "systems" to support space missions and programs.

Space Missions

A spacecraft alone cannot perform a space mission. It must be combined with a launch vehicle, payloads, communication and navigation infrastructure, mission operations, and an understanding of orbital dynamics to assemble a working space mission.[1] The second shell of the three-level model introduced earlier illustrates how the spacecraft exists within the overall mission, and is shown in Figure 8.1. Space missions may also involve other specialties such as other cooperating and supporting spacecraft, human crewmembers, ground command, testing and processing facilities, and planetary astrodynamics.

As was the case with the design of the individual spacecraft, an expert Mission Designer who understands each of the above specialties as well as the interconnections among them must make difficult choices. Those in charge of the mission must choreograph the design of the overall space mission. Mission designers must also understand how the choices they make in mission performance affect the costs, the risks, and the mission goals and objectives. Missions are rarely flown because they are the lowest cost, lowest risk, and best designed to meet program options. Seasoned designers and managers will choose mission designs that often have the best balance among all three of these prime factors.

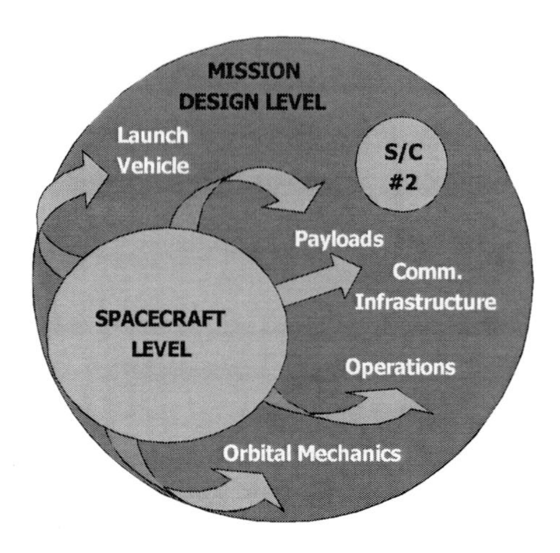

Figure 8.1. The second level of the space model – the Space Mission.†

It is hard, for instance, to obtain a meal replete with filet mignon and a fabulous dessert at the price one pays at a hamburger stand. Further, it is hard to ensure that such a meal will also provide the healthiest possible intake of calories (with the lowest risk of food poison) and nourishes your body to the greatest possible extent. Mission designers clearly have to make tradeoffs and seek the best overall result based on those "balanced decisions."

Just what is an Orbit? Astrodynamics 101

We have already discussed spacecraft orbits and introduced the idea of LEO, MEO and GEO orbits back in Chapter 6. We now want to go back and explain the physics behind these orbits, and indeed the planets and their various moons as well.[2]

Let's begin with some history. *Astrodynamics* is the branch of engineering related to the motion of natural and artificial objects (satellites) in space. It is a subject that some people find difficult and intimidating, but this should not be the case. In reality, the basic concepts of astrodynamics are quite straightforward, and the subject has a very rich and fascinating historical foundation.

For much of recorded human history, it was believed that the other planets revolved around the Earth. This *geocentric* ("Earth-centered") model of the solar system is commonly attributed to the Egyptian astronomer and mathematician Ptolemy (87–150 AD). In Ptolemy's elaborate model, the planets moved about in small circles called epicycles.

Under this theory the centers of these epicycles moved around the Earth in perfect circles called deferents. This system was used for 1400 years and was considered a true interpretation of planetary motion. The Ptolemaic system was useful in at least describing the motions of the planets as seen from Earth.

In 1543, the Polish astronomer Nicolaus Copernicus (1473–1543) broke with Ptolemy's theory and advocated a *heliocentric* ("Sun-centered") model of the solar system. Mathematically, the heliocentric model is simpler than the geocentric one and it explained better many astronomical observations like the "retrograde motion" of Mars. By the principle of Occam's razor, if two theories explain the observations equally well the simpler theory is more likely to be true. However, acceptance of the heliocentric model was slow because at first it did not seem to explain the observations any better than the geocentric model. This was because of the measurement errors inherent in the naked eye measurements of the day. In addition, the Copernican model still had the planets orbiting in perfect circles, mostly for philosophical and religious reasons, which was later shown to be incorrect.

Much of the foundation of astrodynamics can be credited to the Danish nobleman and astronomer Tycho Brahe (1546–1601). While a student, he lost part of his nose in a drunken sword fight, and wore a prosthetic nose for the rest of his life. Tycho Brahe did not support Copernicus' heliocentric model of the solar system, but he was open-minded and was determined to catalogue the planets with enough accuracy to determine which of the two competing theories was correct. From the observatory in his castle, he collected very accurate observations of planetary motion over the span of 20 years. Of equal importance to the observations themselves was the fact that Tycho Brahe recorded the *measurement errors* associated with the data. This would eventually become standard scientific practice, but it was revolutionary at the time. His achievement is all the more remarkable considering it was done without the use of a telescope. He could not explain his observations mathematically, but his apprentice, Johannes Kepler, would eventually accomplish this.

The Laws of Kepler

Upon Tycho Brahe's death in 1601, the German astronomer and mathematician Johannes Kepler (1571–1630) inherited his mentor's job and, more importantly, his astronomical data. Kepler analyzed Brahe's observations to look for patterns, with particular emphasis on trying to mathematically explain the orbit of the planet Mars. Like Copernicus before him,

Kepler originally believed the orbits of the planets to be perfect circles, again because of the philosophical and religious notions of the day. But there was no way Kepler could fit Brahe's data to a perfect circle, and thanks to Brahe's meticulous recording of observational errors he knew the data was accurate. Eventually, Kepler concluded that the orbits of Mars and the other planets are *ellipses*, not circles. It was a triumph of the scientific method, in which a preconceived belief, however appealing, must yield to empirical evidence.

Around 1605, Kepler presented his *three laws of planetary motion*. These laws form the basis of our understanding of satellite and planetary motion.

Kepler's First Law – The orbit of each planet is an ellipse, with the Sun at one focus as shown in Figure 8.2.

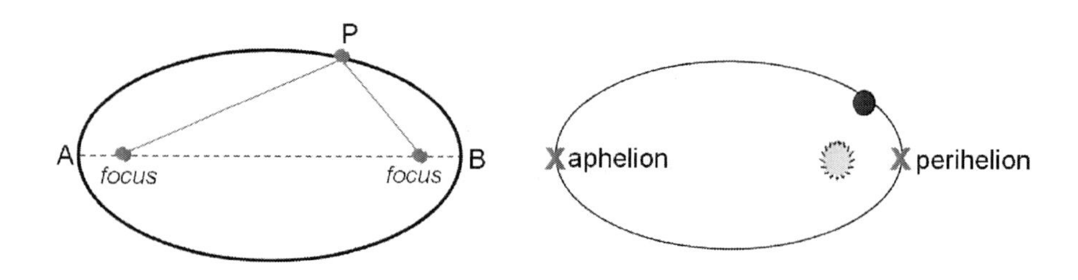

Figure 8.2. Planetary orbits are ellipses.

Kepler's Second Law – A line that joins a planet to the Sun sweeps out equal areas in equal times, as shown in Figure 8.3. For satellites in Earth orbit, Kepler's Second Law could be restated as follows: "A satellite orbits such that the line joining it to the center of the Earth sweeps over equal areas in equal time intervals." In other words, a satellite moves faster when it is close to the Earth and more slowly when it is further away. This is an illustration of a principle called *conservation of angular momentum*, which will be discussed later.

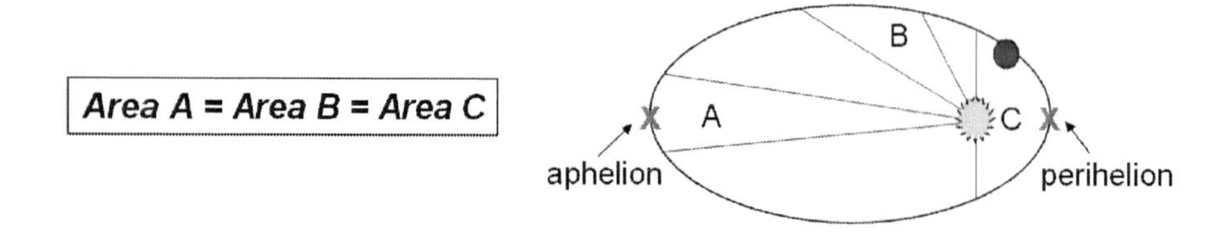

Figure 8.3. Illustration of Kepler's second law.

Kepler's Third Law – The square of the period of a planet's orbit (***P***) is proportional to the cube of its mean distance from the Sun (***a***). In other words, the further a planet is from the Sun (or the higher a satellite is from the surface of the Earth), the longer it takes to move along that part of the orbit. This is illustrated in Figure 8.4.

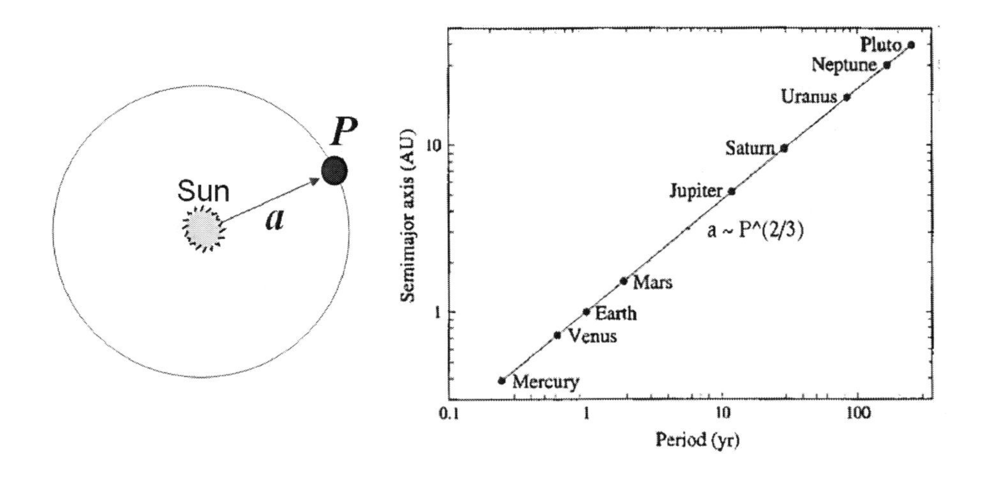

Figure 8.4. Illustration of Kepler's third law.

Classical Mechanics and Newton's Laws

While Kepler's Laws describe planetary motion, it was the English mathematician and physicist Sir Isaac Newton (1642–1727) who would discover the underlying physics behind the observed motion. Newton's most famous book was called *Philosophiae Naturalis Principia Mathematica*, more commonly known as simply the *Principia*. This incredible book transformed the world's understanding of almost everything—starting with gravity. This amazing work was published in 1687 and contained his Three Laws of Motion. This is the basis of what is now known as "Newtonian" or "classical" mechanics.

Before discussing Newton's Laws, two basic terms need to be defined: *mass* and *inertia*. In everyday conversation, the words "mass" and "weight" are often used interchangeably, but in fact they are quite different. Mass is the quantity of matter in an object, whereas weight is the force that gravity exerts on a mass. This means that if you were to go to the Moon or another planet in the solar system, your weight would change but your mass would remain the same.

The Italian scientist Galileo Galilei (1564–1642) was the first to define inertia as "the tendency of an object to stay at rest or stay in motion unless acted on by an outside force." In other words, unless some outside influence forces a change, objects will tend to keep doing what they are already doing. The amount of inertia an object has is related to its mass. A massive object like an ocean liner has lots of inertia, so a large force is required to change its motion, while a far less massive object like a sailboat has little inertia and can be moved with little force.

Newton's Laws of Motion are as follows:

> ***Newton's First Law*** – An object at rest will tend to stay at rest, and an object in motion will tend to keep moving in a straight line, unless acted upon by some outside force.
> ***Newton's Second Law*** – The acceleration of an object is proportional to the force applied, and is in the same direction as that force.
> ***Newton's Third Law*** – For every action there is an equal and opposite reaction.

The latter is probably the best known of Newton's three laws, and is really what makes spaceflight possible. A spacecraft is fixed to a rocket engine that propels exhaust gases out the back. Because every action has an equal and opposite reaction, the spacecraft is propelled forward. Newton's Third Law is an expression of a physical principle known as the *conservation of momentum.* Momentum is the "quantity of motion" that an object has. *Linear momentum* is the quantity of motion in non-spinning objects, while *angular momentum* is the quantity for an object that is spinning or revolving about another object.

Linear momentum is defined mathematically as the product of mass and velocity. If there are no external forces acting on a system, its total linear momentum does not change. This is called the conservation of linear momentum. The recoil of a cannon is an example of this phenomenon. Before the cannon is fired both it and the ball are at rest, so the total linear momentum is zero. Upon firing, the ball shoots out at some velocity. For the system to maintain a net linear momentum of zero the cannon must go in the opposite direction at its own velocity to compensate, i.e. it recoils.

Angular momentum is defined mathematically as the product of *angular velocity* (the speed of rotation) and a quantity called *moment of inertia.* Moment of inertia is the rotational analog of mass. It is measured with respect to an axis of rotation and depends on the shape of an object and the distribution of its mass about that spin axis. Mathematically, moment of inertia is calculated by multiplying the mass by the square of its radius from the spin axis. If there is more than one mass, these terms are added up to form the total moment of inertia about that axis. The greater the radius or the larger the mass, the greater is the moment of inertia about that axis. Just as it takes more force to start or stop moving an object having a large mass, it requires a lot of torque to start or stop the spinning of a body with a large moment of inertia.

If no external torques act on a system, total angular momentum remains unchanged. Think of an Olympic figure skater spinning on the toes of her skates as an example of the *conservation of angular momentum*. Once she starts spinning, she has a certain amount of angular momentum. When she puts her arms out sideways, the distribution of mass has changed such that the moment of inertia (normally denoted as ***I***) increases. Because the total angular momentum must be conserved, the spin rate (normally denoted as **?**) must therefore decrease. The opposite occurs when the skater pulls her arms in. Mass is now distributed closer to the spin axis, so the moment of inertia ***I*** decreases. To keep the same angular momentum, the spin rate (or angular velocity) **?** goes up to compensate, and she spins much faster.

Along with being the primary way in which attitude control of a satellite is accomplished, as discussed previously, conservation of angular momentum is the physical basis behind Kepler's Second Law of planetary motion. The speed of a satellite in Earth orbit (or of a planet in orbit around the Sun) varies in manner similar to that of the spinning figure skater in the previous example. A satellite's speed is greatest at *perigee*, when it is closest to the Earth, and is slowest at *apogee*, the furthest point in its orbit from the Earth.

Universal Gravitation and the Basic Principles of Orbits

Why doesn't the Moon fall out of the sky, the way the apple fell out of the tree and hit Isaac Newton on the head? The answer is…the Moon actually *is* falling! Imagine firing cannon balls from the top of a hypothetical "extremely high" mountain as shown in Figure 8.5. According to Newton's First Law, if there were no external forces, the cannon balls would just keep going in a straight line. But in this case, there *is* an external force – gravity. The Earth's gravity (actually air resistance too, but we will blithely ignore that) would bend the path of the cannon ball and cause it to fall back to Earth (1). The faster the cannonball is fired, the farther it goes before falling back to Earth (2). If the cannonball is going fast enough, it will not hit the surface because the Earth is curving away from the cannonball at the same rate as the cannonball is falling. Like the Moon, the cannon ball would be in orbit around the Earth (3).

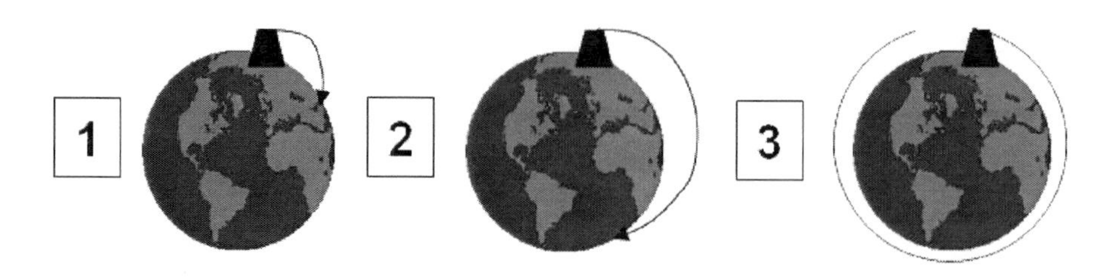

Figure 8.5. If a cannon ball is fired with a high enough velocity, it will go into orbit.

The key point here is that *velocity in a particular gravitation field determines the characteristics of the orbit.* If you remember this key point and understand the concept of the mountain and cannon balls diagram, you already understand most of the theory behind astrodynamics.

According to *Newton's Law of Universal Gravitation*, the force of gravitational attraction between any two objects (such as a cannon ball and the Earth) is directly proportional to the product of their masses and is inversely proportional to the square of the distance between their centers. The more massive the objects and/or the closer they are together, the greater the force of gravitational attraction between them. This law is called "universal" because as far as we know the same principle applies everywhere in the Universe. The motion resulting from this gravitational attraction is called *two-body motion.*

In two-body motion, the shape of an orbit is determined by the speed at which the object is moving. The "slowest" launch produces a circular orbit, because this is the simplest shape that can be achieved without the object crashing back to Earth. As the launch speed increases, the orbit becomes more and more elliptical, or *eccentric*. Parabolas and hyperbolas are fully open trajectories that result when the object is launched so fast that it "escapes" the gravity of the planet. These four types of orbits – circle, ellipse, parabola and hyperbola – are called *conic sections* because they are produced when a circular cone is sliced at various angles.

Describing an Orbit

Recall from the previous chapter that there are exactly six parameters (or degrees of freedom) required completely and unambiguously to describe the position and orientation of any object in space – three fixing positions in three-dimensional space (3D), (x, y, z), and three describing attitude (roll - ϕ, pitch - θ, yaw - ψ). The most common notation for describing satellite orbits uses *Keplerian* or *"classical" orbital elements*, which define the following:

• The size of the orbit

• The shape of the orbit

• The attitude of the orbit (tilt, orientation and rotation)

• The location of the satellite in the orbit.

These orbital elements, together with a time stamp (or epoch), make up a *state vector*, which is simply a set of parameters (numbers), which describes the location of a satellite in space at a particular instant in time. The six Keplerian orbital elements are summarized in Table 8.1.

CLASSICAL OR KEPLERIAN ORBITAL ELEMENTS		
Name	**Symbol**	**Description**
semimajor axis	a	Size of the orbit.
eccentricity	e	Shape of the orbit.
inclination	i	Tilt of the orbit.
longitude of the ascending node	Ω	Orientation of the orbit.
argument of perigee	ω	Rotation of the orbit.
true anomaly	θ	Satellite position in the orbit.
epoch	t	Time stamp of the elements.

Table 8.1. The key elements that describe an orbit.

Orbital Perturbations

In the real world, satellite orbits around the Earth (or other heavenly bodies for that matter) are not the perfect conic sections dictated by Newtonian two-body motion. This is because they experience orbital *perturbations,* or disturbances, which result from a variety of external forces. The major perturbation forces are:

• Atmospheric drag

• The Earth's equatorial bulge, or oblateness

• The Earth's triaxiality, or non-spherical equator

• Third-body effects from the gravitational attraction of the Sun, the-Moon and other planets in the solar system

• Solar radiation pressure

• Unexpected thrusting, either from out-gassing or erroneous thruster activity.

For Low Earth Orbit satellites, atmospheric drag and the Earth's mid drift bulge are key factors. These are lesser factors for orbits that are further from the Earth, such as GEO orbits that are nearly a tenth of the way to the Moon.

Types of Earth Orbit

Table 8.2 summarizes the various orbits and their primary usage. Figure 8.6 shows, not quite to scale, the relative spacing of these orbits in relation to the Earth. It is clear that the orbital characteristics, the sizes of the gravitational forces acting on the satellites, and virtually all of the orbital state vectors are quite different for the satellites in these different orbits.

Type of Orbit	Description	Examples of Missions
Low-Earth Orbit (LEO)	Corridor between about 200 km and 830 km altitude in which 90% of the artificial objects in orbit reside.	Human spaceflight, atmospheric physics, astronomy, remote sensing.
Sun-Synchronous Orbit (SSO)	Uses the Earth's equatorial bulge to "twist" the orbit at a rate of one revolution per year so that the orbital plane always has the same angle with respect to the Sun.	Remote sensing, weather monitoring, reconnaissance.
Molniya	High inclination elliptical orbit with a period of 12 hours, in which the apogee remains fixed above one hemisphere.	Communications, reconnaissance.
Semi-Synchronous	Circular orbit with a period of 12 hours.	Global positioning, navigation.
Geosynchronous (GEO)	Circular orbit above the equator with a period of 24 hours, such that a satellite appears "stationary" when viewed from the Earth.	Communications, weather monitoring, early-warning surveillance.

Table 8.2. Common types of orbits and typical missions associated with each are described in the table above.

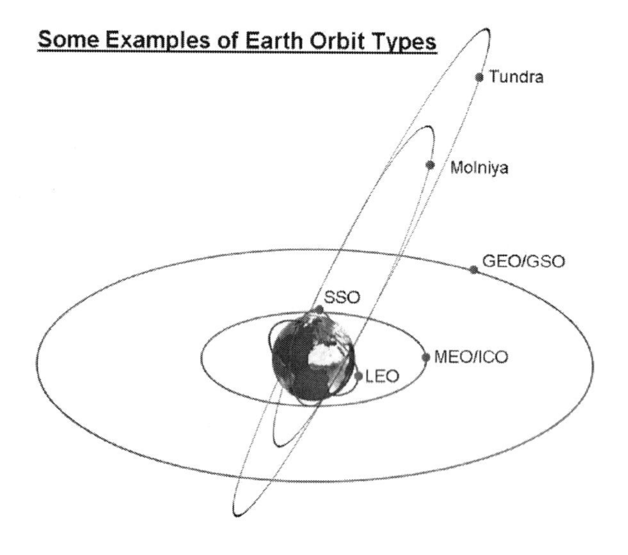

Figure 8.6. Most important types of Earth orbits.

Maneuvering in Space

On nearly every space mission, it is necessary to change, or adjust, the orbit by making *orbital maneuvers.* There are several reasons for making orbital changes, including the following:

- When a satellite is launched, it is usually not deployed directly into its final orbit.
- The satellite's orbit must be maintained against orbital perturbations, especially atmospheric drag for satellites in Low Earth Orbit and for gravitational effects of the Moon and the Sun for GEO orbits.
- The satellite may need to be sent to a new orbit in order to accomplish its mission.

As we have previously discussed, velocity determines the characteristics of the orbit; therefore, changing the orbit requires changing the satellite's velocity. For this reason, an orbital maneuver is often called a *delta-V* (ΔV). The Greek letter "delta", or Δ, is used in mathematics to indicate a "change." So ΔV denotes a change in velocity V. If the velocity is increased along the direction of flight, the altitude is raised. The orbital period lengthens, and by Kepler's Laws the satellite "falls behind" satellites in lower orbits. Similarly, a velocity decrease reduces the satellite's altitude. The period gets shorter and the satellite "pulls ahead" of satellites in higher orbits. To change the apogee (highest point) of the orbit, the satellite must change its velocity at perigee. On the other hand, to change perigee (lowest point in the orbit) the satellite must change its velocity at apogee. *Out-of-plane maneuvers* are those that change the orientation of the orbit, thus altering the inclination. To change the inclination only, the spacecraft's thrusters must be fired at either the ascending or descending node. Firing the thrusters at any other point in a generic orbit alters both the inclination and the ascending node. GEO satellites must maintain their inclination to minimize their North-South drift. Out-of-plane maneuvers require a tremendous amount of energy because of the high velocities involved.

It is often necessary to move a satellite from one orbit to another. The most fuel-efficient method of doing this is called the *Hohmann transfer*, which consists of two separate thruster firings as shown in Figure 8.7. At maneuver point 1, the spacecraft fires thrusters to increase velocity and move from the initial orbit into a transfer orbit. At maneuver point 2, on the opposite side of the Earth, a second firing again increases the velocity for insertion into the target orbit. This is sometimes called the *circulariza-*

tion burn because the target orbit must be less eccentric than the transfer orbit. If the circularization burn does not occur, perhaps due to a malfunction, the spacecraft gets stuck in the transfer orbit.

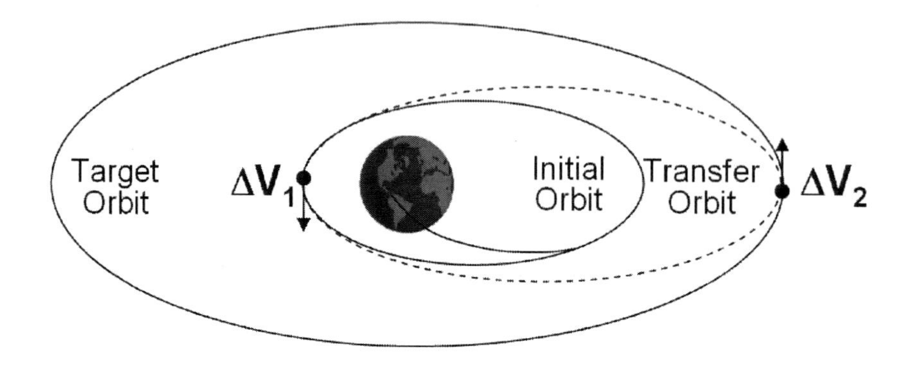

Figure 8.7. Delta-V's are applied at two different maneuver points in a Hohmann transfer.

Interplanetary Travel

The Hohmann method is the most fuel-efficient technique for *interplanetary transfer* to leave the Earth's gravity and go to other planets in the solar system. As an example, consider a Hohmann interplanetary transfer for sending a spacecraft from Earth to Mars. To achieve the transfer, the Earth at launch and Mars at arrival must be directly opposed. The launch has to be timed so that the transfer trajectory takes the spacecraft to the point where Mars will have "caught up" by the time of arrival. For Earth-Mars transfers, this optimum alignment occurs every 26 months. This period defines the *launch opportunity.* Launching at other times requires greatly increased fuel consumption, or it may be impossible to achieve the mission with current propulsion technology.

However, using the Hohmann transfer to reach other planets generally takes a long time. For example, it would take 40 years for a spacecraft to reach Pluto using a Hohmann trajectory. Faster, non-Hohmann trajectories are possible, but these require either a tremendous amount of chemical fuel or entirely different technologies, such as the use of solar sails or nuclear energy. Another approach is for the spacecraft to get a "slingshot" boost through a "gravity assist" maneuver, in which the gravitational field of a planet or the Sun is used to change the spacecraft's velocity (its speed and direction) while en route to another planet.

Changing the velocity of a spacecraft through gravity assist is called "pumping". As a spacecraft flies close to a planet or the Sun, it is pulled towards it by gravity and its velocity increases. If it passes "behind", as shown in the upper diagram of Figure 8.8, without the gravity assist the spacecraft would continue on the dotted trajectory. But thanks to its encounter with the planet, the trajectory has been "bent" so that, in effect, its velocity has increased with respect to the Sun. Gravity assist can also be used to decrease velocity, as illustrated in the lower diagram in Figure 8.8. If the spacecraft passes "inside" or "in front" of a planet, it is pulled against its direction of flight, decreasing its velocity with respect to the Sun. Mariner 10 mission planners were the first to use this technique in 1974. In this case a gravity assist maneuver at Venus was used to slow the spacecraft towards a rendezvous with Mercury.

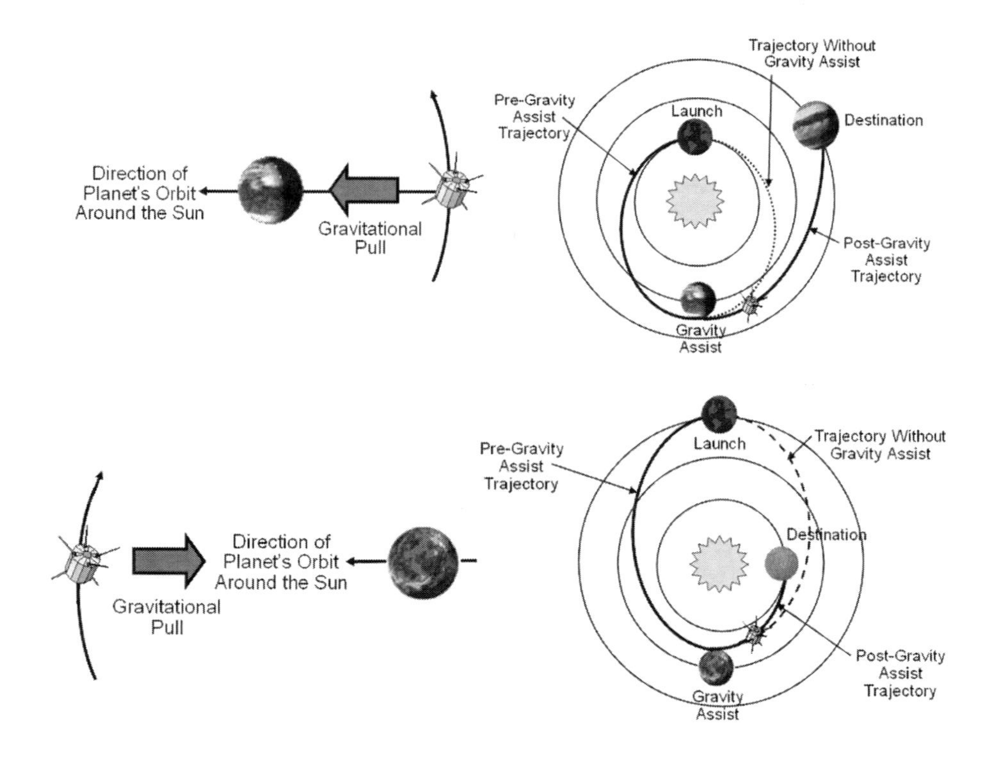

Figure 8.8. Gravity assist.

Getting There – Let's take a Rocket Ride!

Getting the spacecraft into orbit is accomplished with a space launch system or rocket. Once in orbit, the spacecraft may be required to perform maneuvers like attitude changes, or even change orbits as was just discussed. Sometimes, as in the case for interplanetary spacecraft, the vehicle must propel itself out of Earth orbit and on to its final destination. To accomplish this, space propulsion systems are employed.[3] Space propulsion is defined as a class of propulsion that produces a propulsive force in (empty) space; it includes chemical and nuclear rocket propulsion, and solar sails. It *excludes* those means of propulsion that require:

- air to provide the propulsive force, e.g., an air-breathing jet engine, and
- direct contact with the Earth, like wheeled or tracked propulsion.

Space propulsion is needed to do many things including[4]

- Launch a space vehicle from the Earth,
- Accelerate a spacecraft to a velocity that will take it from one orbit to another (depending on the velocity, the spacecraft will either remain within the influence of Earth's gravity or go beyond its influence, on to a distant planet, or even beyond the influence of the Sun and into interstellar space),
- Compensate for the influence of disturbing forces like drag, solar wind, or gravity (e.g., due to the dynamics of a rotating, non-spherical Earth) on the orbit or trajectory,
- Control the attitude of the vehicle (e.g., to orient antennas or sensors, or thermal control surfaces) or to compensate for disturbing torques.

Other functions that are accomplished using space propulsion include controlling the spin-up or spin-down of spacecraft, "unloading" reaction or momentum wheels on a satellite, separating vehicle stages, or settling propellants (sometimes required by liquid propellant systems).

Most spacecraft (including space launchers, satellites, space probes, and human-crewed systems) are currently propelled by expelling mass from within the spacecraft in a direction opposite to the direction of travel. This is referred to as rocket propulsion. All of the hardware and software needed to generate thrust is referred to as the propulsion system or subsystem. Most rocket propulsion uses chemical propellants both to generate the energy needed for thrust and as the mass expelled. Some spacecraft, such as the Pegasus rocket, the X-43 A and C hypersonic scram jet experimental planes and the now famous Spaceship One, use air-breathing propulsion for their lower stages. This has several advantages. The principal one is that there is no need to carry the oxidizer on the vehicle.

There are also incipient attempts to build a "space elevator" that could lift mass up to GEO orbit, and there have been a few experiments with "solar sail" systems. However, to date we remain heavily dependent on chemical rocket propulsion. In time this may well pass since chemical rockets are expensive, not all that "environmentally friendly", difficult and expensive to build, and not the safest of vehicles.

Most satellites have simple and generally reliable chemical rocket thrusters to keep them in position, termed "station keeping". Recently spacecraft and planetary probes have started to use non-chemical means, such as electrical rockets (e.g., ion rockets and "resistojets") for some tasks.

Despite enhancements over the years, space propulsion systems are really inefficient propositions. For instance, to put a 5 to 10-ton (5000 to 10,000 kilogram) payload into space, a giant space rocket, like the European Ariane 5 or Atlas 5, is needed. These rockets are essentially an empty tank that houses several rocket motors and the required propellant, which can be compared to the fuel for your car. The total mass of a fully fueled Ariane 5 rocket is over 746,000 kg, of which more than 80% (or 642,000 kg) is propellant. Given that it costs in the range of about €7,000-20,000 to launch one kilogram of mass into space, one can readily estimate the high cost of having to carry so much propellant along for the ride.

We should mention system failures. Investigations have shown that over 40% of all failures of space systems are due to propulsion system failures. These are often a catastrophic failure, an explosion. They can also be a rocket engine that does not start, or an engine that does not function properly.

Before getting further into the details of propulsion systems, there is one term that requires a definition, as it describes how each is characterized; the term is *specific impulse*. Specific impulse, denoted I_{sp}, is a rough measure of how fast the propellant is ejected out of the rocket nozzle. It is also a term that is used to describe the efficiency of the rocket. Specific impulse, given in seconds, represents the impulse, or change in momentum per unit of propellant consumed. The higher the specific impulse, the less propellant is needed to gain a specific amount of momentum, or to gain a certain increment of velocity. High I_{sp} is good! Solid fuel rocket motors typically have a specific impulse around 250 seconds, while a bipropellant liquid fuelled rocket might have a specific impulse of over 400 seconds.

Options for Space Propulsion

Space propulsion methods generally fall into one of two classes:

1. Rocket propulsion wherein the propulsive force is generated by expelling mass (initially stored in the vehicle) from the vehicle, at a

high velocity,
2. Non-expulsion (non-rocket) propulsion through forces associated with the solar wind, a plasma beam, a magnetic field or the gravitational field of a celestial body.

The majority of rocket propulsion systems today are (thermo-) chemical or cold gas-based rocket systems, wherein the propellant both acts as the matter expelled and as the energy source needed to accelerate the expelled matter to a high velocity. They include liquid fuel, solid fuel, and hybrid rockets, and cold gas rockets.

In addition to these chemical systems, there are also non-chemical, or separately powered, rocket systems using nuclear and electrical power sources. The amount of power needed for thrust generation is so high, however, that such systems are limited to relatively modest thrust levels. In short, chemical rockets give you a short and highly energetic source of thrust. Nuclear and electrical systems give you much lower accelerative force but for much longer durations. Solar sails provide the least thrust of all but the fuel (i.e. photons from the Sun) is continuously available, an inexhaustible source of available thrust for years or even decades.

A further distinction is made as to how the propellants are accelerated. In cold gas and chemical systems the propellant is accelerated to a high velocity by expanding a hot high-pressure gas through a nozzle (we refer to these engines as using thermal expansion for acceleration). Typical systems of this type include the resistojet, arcjet, solar/laser-thermal and nuclear-thermal systems. In addition to these systems, other methods of accelerating the propellant include by applying electrostatic and/or electromagnetic forces. These can generate much higher exhaust velocities.

Chemical Rocket Propulsion

In chemical rockets the energy necessary for heating the propellant is derived from a chemical reaction. The energy required for the thrust generation is stored in the chemical reactants themselves. The expellants are also formed by the chemical reactions taking place in the rocket motor. Therefore, we refer to both reactants and expellants using the general term propellants. Along with the propellants, chemical rockets essentially consist of four components:

1. a chemical reactor, where the propellants are heated at a high pressure to a high temperature,
2. a thrust generation device, wherein the hot propellant gases are accelerated to a high exhaust velocity,

3. a propellant feed and storage system that stores the propellant prior to its use, and also pressurizes and feeds the propellant to the thrust generation device, and
4. a control system.

We distinguish different types of chemical rocket motors according to the state of aggregation of the propellant under storage conditions. In *liquid* propellant rockets the propellant components are stored in a liquid state. In *solid* propellant rockets the propellant is stored in a solid state, usually in the form of a pre-shaped charge. *Hybrid* propellant rockets carry a solid and a liquid propellant component. *Cold gas* rockets carry with them a gas under high pressure to provide the propulsive force.

The Solid Rocket Motor

In a solid propellant rocket, the propellant is entirely stored within the combustion chamber in the form of one or several shaped blocks, which are called grains and are supported by the walls or by special grids, traps or retainer walls. Solid rockets produce thrust by creating an internal explosion and directing the resulting hot gases through the narrow nozzle. Once a solid rocket is ignited, it cannot be "turned off" - it just keeps burning until all the propellant is expended. The one exception to this rule is the case of a hybrid system such as used in Space Ship One and now in Space Ship Two. This is a system that literally "burns rubber". The fuel is neoprene and the oxidizer is laughing gas (i.e. nitrous oxide). In this engine one can turn off the laughing gas and the propulsive system shuts down.

Solid rockets were the first rockets; the Chinese invented them around 1150 A.D. Today, solid rockets are mostly used when it is necessary to maintain a high thrust for a short period of time. For example, this is the case for fireworks, model rocket engines, booster rockets for space rocket launchers, military missiles, and kick motors to provide orbit insertion or the final boost into circular orbit at apogee. In the past, solid rockets have also been used as auxiliary power plants for aircraft take-off, with varying degrees of success! Recent motor developments include the development of the VEGA P80 motor for Europe's envisaged Vega Small Launch Vehicle, and the GEM solid rocket booster for Boeing's Delta-type of launchers. In these designs composite materials are used for the motor case whose mass is substantially less than for a metallic case.

Perhaps the most well-known and "notorious" solid rockets are those used for the Space Shuttle. Although solid rocket motors can be quite reliable, many space safety experts consider the fact that they cannot be shut down on command as a good reason for their not being used for human space flight. (Again the "throttle-able" neoprene and laughing gas engine for Space Ship One and Two represents a different type of solid rocket engine.) The need for foam insulation to shield the liquid fueled Space Shuttle from the solid boosters thrust might have been eliminated with a different design. That could have increased the reliability to that achieved by the use of all liquid fueled rocket engines.

Liquid Propellant Rocket Motors

The first use of liquid propellant dates back to 1926, the rocket pioneering days of Robert Goddard.[5] In the 1940s this was followed by the illustrious V-2, also known as the A-4, designed by Werner Von Braun and his team, which clearly demonstrated the high performance capability of liquid propellant rocket propulsion systems. This was followed, in 1957, by the first true space launch when the Soviets launched Sputnik I. This event spurred a surge in rocketry in the United States that culminated in 1969 with the first manned lunar landing, made possible by the design and development of the powerful Saturn V all liquid propellant rocket system. Today, liquid propellant propulsion systems launch most rockets for crewed spacecraft, so allowing humanity to expand its presence into space. Typical examples are Ariane 5 core stages, the Space Shuttle main engines (Figure 8.9), and the Cassini, Galileo, and NEAR spacecraft.

An important distinction for liquid rockets is with regard to application. We distinguish between *launcher propulsion* and *spacecraft propulsion*. The prime focus for launcher propulsion is to achieve a high thrust (to overcome gravitational acceleration) and a high specific impulse to reduce propellant mass. For space launchers, because of the short thrust duration and because of the limited importance of short launch preparation time, cryogenic propellants may be used. For spacecraft propulsion, the focus is on storable and hypergolic propellants (i.e. propellants that "explode" when coming into contact with each other) to attain high reliability and a longer mission life.

Figure 8.9. Space Shuttle Main Engine (Courtesy NASA).

Hybrid Rockets

In a hybrid rocket motor the oxidizer and fuel are stored in different physical states. In most cases, a solid fuel is stored in the combustion chamber into which a liquid or gaseous oxidizer is injected. The solid fuel is stored in the shape of a shaped block referred to as the fuel grain. The liquid oxidizer is either stored at room temperature or at very low (cryogenic) temperatures in a storage tank. Pressurized gas is used to feed the liquid into the combustor, gas generator or pump. To start motor operation, an ignition device is needed. After full combustion is accomplished, the motor burns until the oxidizer flow is halted, and the motor shuts down.

The first investigations into hybrid rocket propulsion took place in the 1930s. It was the former USSR that first flight-tested a hybrid rocket motor, reaching an altitude of 1500 m. In this test motor, liquid oxygen was used as oxidizer and colloidal benzene as fuel. At the same time, the Germans tested a hybrid rocket motor using laughing gas (N_2O) as oxidizer and coal as fuel. In the 1960s, ONERA in France developed a hybrid sounding rocket using amine fuel and nitric acid oxidizer. At the same time, United Technologies (USA) developed a high-altitude supersonic target (HAST) drone using a polybutadiene fuel and nitric acid oxidizer. HAST flew for the first time in 1968. It demonstrated a range in excess of

150 km. In the 1980s and early 1990s AMROC (USA) designed and hot test fired a wide variety of hybrid rocket motors of all sizes, utilizing non-toxic, storable propellants.

To date, only few applications have been found for hybrid rocket engines. One of the reasons is that during the 1960s solid rocket systems were studied in depth for military purposes. This resulted in a large amount of knowledge on solid propellants and how they burn. Hybrid rocket motors are sometimes used for drones, meteorological rockets and educational purposes. The motor developed by James Benson's company for Burt Rutan, that uses neoprene as the propellant source and nitrous oxide as the oxidizer, is a unique rocket engine design in that a "throttle" can be applied to the propulsion system; it can shut down in emergency situations. This safety feature developed for Space Ship One and Two is of great interest to the space tourism industry. The problem with this propulsion system is that the noxious gases released are much more toxic than the exhausts of conventional liquid fuel systems.

Cold Gas Systems

Cold gas systems use an inert cold gas as the propellant. In addition to the propellant, cold gas systems consist of one or more engines, a propellant storage tank, sometimes a gage, and tubing. Any gas may be used as a propellant; nitrogen and helium are generally used. This is because these gases are highly inert and have low molecular mass. Helium offers the advantage that its specific impulse is about 2.5 times higher than that for nitrogen, giving a significant reduction in propellant mass, but at the expense of an increased storage volume or a pressure about seven times higher, plus a higher cost. The thruster essentially consists of a nozzle and a valve; the nozzle accelerates the cold gas to a high velocity, whereas the valve regulates the thrust generation and on/off switching. The energy required for propulsion is fully contained in the pressurized gas; no heating is needed. Cold gas systems were used on many early spacecraft as attitude control systems. Today, these systems are mostly used in cases requiring a low total impulse, or where extremely fine pointing accuracy or thrust levels must be achieved, or when the use of chemical propellants is prohibited for safety reasons.

Non-Chemical Rocket Propulsion

Non-chemical thermal rockets use separate power and energy sources to heat up a working fluid to a high temperature, after which the working fluid (i.e. the propellant) is discharged through a nozzle. Based on the type of heat or energy source used, we distinguish three types of non-chemical thermal rockets.

1. ***Electro-Thermal Rocket Systems.*** These use electrical energy to heat a propellant, which is then discharged through a nozzle. Heating can be accomplished by resistance heating or by an arc discharge. Such systems use either an external energy source like the Sun, or an internal energy source like a radioisotope generator, nuclear fission (or a nuclear fusion reactor should we ever build a viable one).

2. ***Nuclear-Thermal Rockets.*** These use the heat produced by a nuclear reactor to heat the propellant, using either a solid heat exchanger or particulate absorption.

3. ***Laser-Thermal*** **and** ***Solar-Thermal Rockets.*** These use laser energy or solar energy to drive a heat exchanger, which then heats the expellant. Because the heating is direct, no losses occur due to the conversion of thermal energy into electrical energy, and so the thrust levels are higher than for electro-thermal rockets.

So, what does it do? The Payload

Earlier, we learned to think of a spacecraft as collections of subsystems, each of which has a specific task, working together to perform a specified mission. Each of the subsystems introduced previously does not perform the tasks that directly result in completing a mission's objectives, but they perform tasks to maintain and monitor the spacecraft they provide the spacecraft with a frame, or with power, or allow it to navigate properly. The collection of the subsystems that support the spacecraft is referred to as the spacecraft bus. Each spacecraft consists of a spacecraft bus as well as one more subsystem: the payload.

The payload subsystem consists of all the tools in the spacecraft that directly achieve the objectives of its mission. This definition is very broad when we consider the wide range of objectives that space missions can have. For example, there is the payload of an Earth observation satellite, which might consist of photographic, radar or multi-spectral instruments to image the Earth. This is quite different from the payload of, say, the European Space Agency's Automated Transfer Vehicle (ATV), which consists of cargo that is to be delivered to the International Space Station, or the waste materials removed from the space station; as it descends from orbit it burns up in the Earth's atmosphere. In general terms, we can group different payloads into one of three categories, as shown in Table 8.3.

Spacecraft Type	Example Spacecraft	Payload Components	Example Payloads
Launch Vehicle	Ariane launcher, Falcon 1, Atlas V, Taurus, HIIA	Satellites and Spacecraft	Communication Satellites, Space shuttle
Applications Satellite	Landsat, Hubble, Galileo, Intelsat X, Eutelsat Hotbird	Hardware and Software	Imaging instruments, communication systems, navigation and positioning systems
Other	Space Shuttle, ATV	Humans and Cargo for Crewed Space Missions	Crew, food, scientific tools

Table 8.3. Different payload categories.

With all the various types of payloads, it can be difficult to choose what the best payload is to use for any mission – that is one of the key tasks of a spacecraft designer. The wide range of payload options is greatly reduced by the objectives and constraints of the mission. There are only a handful of payloads that can perform a task such as observing the Earth in the visible range of light. First, a spacecraft designer must analyze all the mission objectives and constraints to determine all the possible payloads that would work. Then the designer examines several aspects of the payload to figure out which best meets the mission objectives, at the lowest cost, and yet entails the lowest acceptable risk of failure.

Because of the high cost of launching objects into space, one of the most critical factors in choosing a payload is ensuring that it is as light as possible. As the mass of the payload increases, the mass of the structural subsystem that supports it also increases, which greatly increases the overall mass of the spacecraft. This example shows that choosing the best payload not only depends on the characteristics of the payload itself, but also on the payload's effect on the spacecraft bus.

In some cases a smaller payload may need more power. The designer must consider how much power the payload requires, and its impact on mass and risk factors. As the amount of power that a payload requires increases, the mass of the power subsystem has to increase, leading to a more massive satellite. That leads to higher costs, not only to construct and test the satellite but also to launch it as well.

Robots in Space – Payloads that do Real Work

The word *robot* comes from the word *robota* meaning serf labor, or hard work, in Slavic languages (Czech, Slovak and Polish), and was introduced to the public by Czech writer Karel Capek in his play R.U.R. (Rossum's Universal Robots), premiered in 1920. In the play, the robots are described as artificial creatures, or *androids* that can be mistaken for humans.

Today, the word robot is used for a mechanical, or artificial, agent that can behave interactively with environments or humans in some coordinated ways. A *humanoid*, a human looking robot, attracts the public attention, though the robots used in industry are automated or programmable handling devices, which do not look much like humans. *Industrial robots* are successfully working on the mass production lines of industrial factories, conducting repetitive tasks such as welding or assembling motor vehicles. But most research efforts are now put into robots to work *outside the factory*, such as in offices, homes and hospitals, or in fields, or even inside human bodies (nano-robots), or in outer space. *Robotics* is a discipline of *system integration* making the most of our knowledge of many different disciplines, including mechanics, electronics, computer technology, bio-engineering, and even anthropology and sociology.

Artificial Intelligence (AI) is one major aspect of a robot, but not all of it. Physical motion control is an indispensable basis in many cases. For the robots that function in the real world, *intelligence* means an ability to negotiate (or perform appropriate behaviors) in the physical environment, but not merely mental or psychological ability. For example, *dexterous manipulation* is achieved by a good combination of sensing (such as vision, or force or tactile sensing) and the motion control of the arms, hands, and fingers, under the coordination of the *brain* for perception, recognition, planning and execution. From a roboticist's viewpoint, the *instinct* of primitive animals is the first level of AI that we should implement in an *intelligent* robot, even though the instinct was previously considered to be opposite to the concept of intelligence.

Autonomy is another aspect of the robot. Any unmanned spacecraft that is under an automated control sequence may be referred to as a robotic satellite. However, when we use the term *Space Robot* it means a more capable mechanical system that can facilitate manipulation, assembling, or servicing tasks in orbit, to assist astronauts. In time space robots may become more capable and able to explore remote planets as surrogates for human explorers. The most advanced concept we now have of a space robot is what is called a *von Neumann Machine*. This is a robotic device that could go to another planet, create a processing plant, replicate itself and send additional probes to explore even further out into the cosmos. In theory a von Neumann Machine could not only reproduce itself multiple times but could even improve on its original design. This would demonstrate a "thinking" mechanism.

Key issues in space robotics are characterized as follows:

Manipulation — Although manipulation is a basic technology in robotics, microgravity in the orbital environment requires special attention to the motion dynamics of the manipulator arms and the objects being handled. Reaction dynamics, that affect the base body, impact dynamics when the robotic hand contacts an object to be handled, and vibration dynamics due to structural flexibility are included in this issue.

Mobility — The ability of locomotion is particularly important in exploration robots (rovers) that travel on the surface of a moon or a remote planet. These natural surfaces can be rough, and thus challenging to traverse. The robotic technologies of sensing and perception, traction mechanics, and vehicle dynamics, control and navigation must first be demonstrated in a natural untouched environment on Earth.

Tele-Operation vs. Autonomy — There can be a significant time delay between a robotic system at a remote worksite in space and the human operator in a control room on the Earth. In earlier orbital robotics demonstrations, the tele-operation delay (termed latency) was typically only a few seconds, such as for a mission on the Moon. But the delays could be several tens of minutes, or even hours for planetary missions. Tele-robotics technology is therefore an indispensable ingredient in space exploration, and the introduction of some autonomy is a reasonable consequence of such protracted transmission delays.

Extreme Environments — In addition to a microgravity environment that affects the motion dynamics of the robots, there are many other issues related to extreme space environments that are challenging. They must be solved in order to enable practical engineering applications. Such issues include extremely high or low temperatures, high vacuum or high pressure, corrosive atmospheres, ionizing radiation, and very fine dust.

Versatility — The ultimate space robot would be not only autonomous but also versatile, doing all the tasks by itself using its own resources. There have been communications satellites designed with autonomous control that function close to this goal. A space robot, therefore, should be versatile in such extreme space environments as mentioned before, and versatile with regard to many different situations and scenarios that it might encounter. The ultimate versatility would be the so-called von Neumann machine described above.

Orbital Robotic Systems: Past and Present

The first robotic manipulator arm used in the orbital environment is the Space Shuttle Remote Manipulator System, a.k.a., the "Canadarm," designed and built in Canada. Successfully demonstrated on the STS-2 mission in 1981, it is still operational today. This success opened a new era of orbital robotics and inspired a number of mission concepts in the research community. A long-term goal that has been discussed intensively since the early 1980s is to rescue and service malfunctioning spacecraft using a robotic free-flyer or free-flying space robot. Later, manned service missions were conducted for the capture-repair-deploy procedure of malfunctioning satellites (ANIK-B, Intelsat 6, for example) and for the maintenance of the Hubble Space Telescope (STS-61, 82, 103, 109 and 125). For all of the above examples, the Space Shuttle, a manned spacecraft with dedicated maneuverability, was used. However, unmanned servicing missions have not yet become operational. Although there were several demonstration flights, such as the ETS-VII of Japan and the Orbital Express, an experiment of the US Defense Advanced Research Projects Agency (DARPA), the practical technologies for unmanned satellite servicing missions await the outcomes of future tests and experiments.

Space Shuttle Remote Manipulator System

The Canadarm is a mechanical arm that maneuvers a payload from the payload bay of the Space Shuttle orbiter. It can also grapple a free-flying payload and maneuver it to the payload bay. The Canadarm has been employed more than 100 times to deploy or retrieve payloads as well as assist human extra vehicular activities (EVAs), better known as "space walks". Servicing and maintenance missions to the Hubble Space Telescope, and construction tasks of the International Space Station, have also been successfully carried out by the Canadarm in conjunction with humans on EVAs. As noted in the "story" told by Astronaut Jeff Hoffman earlier, the repair of the Hubble would not have been possible without this capability. The Canadarm is pictured in Figures 3.1-3.3.

ISS mounted manipulator systems

The International Space Station (ISS) is the largest international technology project ever, with 15 countries making significant cooperative contributions. The ISS is an outpost of the human species in space, as well as an orbiting laboratory with substantial facilities for science and engineering research. To facilitate various activities on the station, there are several robotic systems that often work in tandem. The Space Station Remote Manipulator System (SSRMS), Canadarm 2, shown in Figure 8.10, was launched into space in 2001 during STS-100. The Canadarm 2 has played a key role in the construction and maintenance of the ISS both by assisting astronauts during EVAs and using the SRMS on the Space Shuttle to hand over a payload from a Shuttle to the SSRMS.

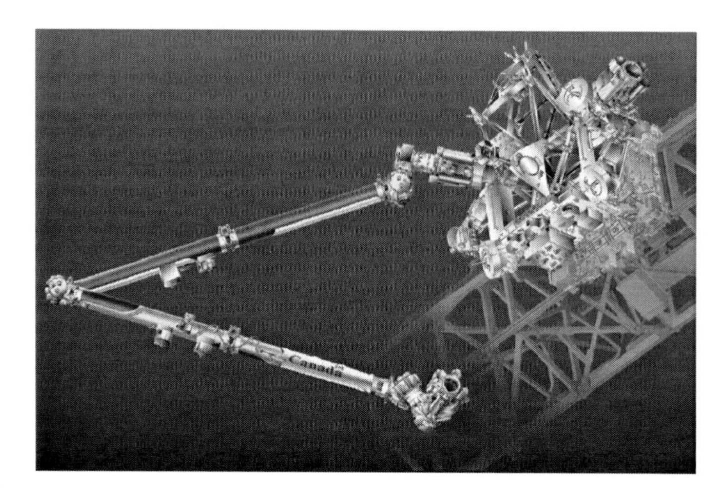

Figure 8.10. Space Station Remote Manipulator System (SSRMS) on the Mobile Base System (MBS) (Courtesy Canadian Space Agency).

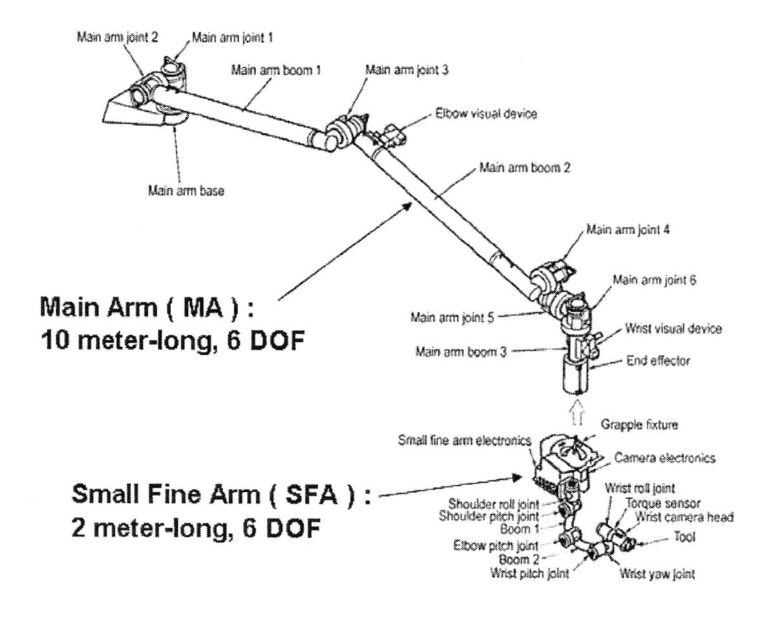

Figure 8.11. This figure shows a schematic drawing of the JEMRMS manipulator system (Courtesy JAXA).

The Special Purpose Dexterous Manipulator (SPDM), or Dextre, which was attached at the end of the SSRM in 2008 by STS-123 (ISS assembly flight 1J/A), is a capable mini-arm system to facilitate the delicate assembly tasks currently handled by astronauts during EVAs. Dextre is a dual arm manipulator system, where each manipulator having seven degrees of freedom is mounted on a one degree-of-freedom body joint. Each arm has a special tool mechanism. The arms are tele-operated from inside the ISS. The European Space Agency (ESA) is also providing a robotic manipulator system for the ISS, the European Robotic Arm (ERA), which will be used mainly to work on the Russian segments of the station.

In Japan, the Japan Space Exploration Agency (JAXA) has developed the Japanese Experiment Module Remote Manipulator System (JEM-RMS), shown in Figure 8.11. The arm will be attached to the Japanese module of the ISS. After installation, the arm will be used to handle and relocate the components for the experiments and observations on the Exposed Facility.

ETS-VII: "Orihime" and "Hikoboshi"

The Japanese Engineering Test Satellite VII was another historical milestone in the development of robotics technology in space, particularly in the area of satellite servicing. ETS-VII was an unmanned spacecraft developed and launched in November 1997 by the National Space Development Agency of Japan (NASDA), now named the Japanese Aerospace eXploration Agency (JAXA). Several experiments were successfully conducted using a two-meter-long, six-degree-of-freedom manipulator arm mounted on its carrier satellite.

The mission objective of ETS-VII was to test free-flying robotics technology and to demonstrate its utility in unmanned orbital operation and servicing tasks. The mission consisted of an autonomous rendezvous/docking and a number of robot experiments. For the rendezvous experiments, the spacecraft was separated into two sub-satellites in orbit, one is named "Orihime" and behaved as a target, and the other is named "Hikoboshi" and behaved as a chaser. The meaning of these names in ancient mythology was mentioned in Chapter 2. The robot experiments included a number of tasks that included testing of tele-robotic operations with up to a seven seconds transmission delay, and a simulated refueling of a satellite.[6,7]

Orbital Express

The Orbital Express Space Operations Architecture program is a US Defense Advanced Research Programs Agency (DARPA) program developed to validate the technical feasibility of robotic on-orbit refueling and reconfiguration of satellites, as well as autonomous rendezvous, docking, and manipulator berthing. The system consists of the Autonomous Space Transport Robotic Operations (ASTRO) vehicle, developed by Boeing Integrated Defense Systems, and a prototype modular next-generation serviceable satellite, NextSat, developed by Ball Aerospace. The ASTRO vehicle is equipped with a robotic arm to perform satellite capture and fuel exchange operations, as illustrated in Figure 8.12.

Figure 8.12. On-orbit spacecraft servicing: artist's rendition of Orbital Express during unmated operations (Courtesy of Boeing).

After the launch of the Orbital Express in March 2007, various mission scenarios were conducted, including visual inspection, fuel transfer, fly-around, rendezvous, and docking with NextSat. These scenarios were successfully completed by July 2007, with ASTRO's onboard autonomy robot employing the onboard cameras and advanced video guidance system.

Surface Robotic Systems: Past and Present

Research on surface exploration rovers began in the mid-1960s, with an initiative for an un-manned rover for the Surveyor lunar landers and a manned rover ("*moon buggy*") for the US human missions on the Moon. During the same period, research and development began for a teleoperated rover named *Lunokhod* in the Soviet Union. Both the Apollo manned rover and the Lunokhod unmanned rover were successfully demonstrated in early 1970s.[8] In the 1990s the exploration target had expanded to Mars; in 1997, the Mars Pathfinder Mission successfully deployed a micro-rover named *Sojourner* that safely traversed the rocky field adjacent to the landing site by autonomously avoiding obstacles.[9] Following up on this success, today, autonomous robotic vehicles are considered indispensable technology for planetary exploration. The Mars Exploration Rovers, *Spirit* and *Opportunity*, launched in 2003, have had remarkable successes, remaining operational in the harsh environment of Mars for over three years. Each has traveled more than 7.5 km, and has made significant scientific discoveries using on-board instruments.[10]

Lunokhod

The first remotely-operated robotic space surface vehicle was Lunokhod, a drawing of which is shown in Figure 8.13. Lunokhod 1 landed on the Moon on November 17, 1970 as a payload on the lander Luna-17, and Lunokhod 2 landed on the Moon on January 16, 1973. Both were 8-wheeled vehicles having a mass of about 840 kg. Almost all components were in a pressurized thermal enclosure. The Lunokhod looked like a *bathtub*, with a lid that closed over the tub. With the temperature on the surface of the Moon reaching around 120 degrees Celsius during the daytime and -170 Celsius degrees during the lunar night, the day and night each lasting 14 Earth days, it was a challenge to design a robot that could survive both the extreme heat and cold. The design of the bathtub and lid aided the thermal protection of the vehicle's controls. On the lid were solar arrays that recharged batteries during the day.

Figure 8.13. Drawing of a Lunokhod (Courtesy NASA).

Lunokhods were remotely driven by a driving team in the U.S.S.R., consisting of a driver, a navigator, a lead engineer, an operator of a communication antenna, and a crew commander. There was approximately three seconds of delay with commands and return signals traveling at the speed of light. The driver viewed a monoscopic television image from the vehicle, and gave the appropriate commands such as "turn", "proceed", "stop", or "back up". The vehicles had both gyroscope and accelerometer-based tilt sensors that could automatically stop the vehicle in the event of excessive tilt of the chassis. Typical mobility commands specified a time for which the motors would run, and then stop. Precision turning commands specified the angle through which the vehicle was to turn. A ninth small wheel, that was un-powered and lightly loaded, determined the distance covered.

Lunokhod 1 successfully operated for 322 Earth days. It traversed over 10.5 km during that period, and returned over 20,000 TV images and 200 high-resolution panoramas. It provided the results of more than 500 soil penetrometer tests and 25 soil analyses using its X-ray fluorescence spectrometer. Lunokhod 2 operated for about 4 months. It traversed more than 37 km before the mission officially terminated on June 4, 1973. Lunokhod 2 was lost prematurely when it began sliding down a crater slope.

Apollo Moon Buggy

The Lunar Roving Vehicle (LRV), *Moon Buggy* was an electrically powered surface exploration vehicle deployed during the Apollo program so that the astronauts could extend their range of surface extravehicular activities. Three of the Apollo missions brought LRVs to the moon, namely Apollo 15, 16 and 17. A brief summary of the distances traveled by the various Moon Buggy missions is listed in Table 1.

Mission	Total Distance Traveled	Total Time	Longest Single Traverse	Maximum Range Traveled from the Lunar Module
Apollo 15	17.25 miles (27.76 km)	3h 02 m	7.75 miles (12.47 km)	3.1 miles (5.0 km)
Apollo 16	16.50 miles (26.55 km)	3h 26 m	7.20 miles (11.59 km)	2.8 miles (4.5 km)
Apollo 17	22.30 miles (35.89 km)	4h 26 m	12.50 miles (20.12 km)	4.7 miles (7.6 km)

Table 8.4. A Summary of journeys made on the various Apollo LRV missions

Mars Pathfinder Rover - Sojourner

The first successful Mars rover was *Sojourner*, shown in Figure 8.14, was part of the NASA Mars Pathfinder Mission launched on December 4, 1996. After a 7-month voyage it landed on Ares Vallis, in a region called Chryse Planitia on Mars, on 4 July 1997. The Sojourner rover was then deployed from the lander to execute many experiments on the Martian surface.

The Mars Pathfinder mission carried a series of scientific instruments to analyze the Martian atmosphere, climate, geology and the composition of its rocks and soil. It was a project from NASA's Discovery Program, which promotes the use of low-cost spacecraft and frequent launches under the motto "cheaper, faster and better." The goal of the mission was regarded as a "proof-of-concept" for various technologies, such as airbag-assisted touchdown and automated obstacle avoidance, both later exploited by the Mars Exploration Rovers.

Figure 8.14. The Pathfinder Rover, "Sojourner", on the Mars in Front of the Rock Nicknamed "Yogi." (Courtesy NASA/JPL)

For rough terrain mobility a unique suspension arrangement called the Rocker-Bogie system was developed. The term "rocker" comes from the design of the differential, which keeps the rover body balanced, enabling it to "rock" up or down depending on the various positions of the multiple wheels. The term "bogie", on the other hand, comes from the old railroad systems, and refers to the train undercarriage with six wheels that can swivel to curve along a track. The Rocker-Bogie is a suspension system that connects six wheels using a passive linkage mechanism, performing as a type of a mechanical equalizer. Thanks to this the rover can go over a rock obstacle larger than the diameter of the wheel.

The Sojourner rover had a semi-autonomous navigation capability, which was successfully demonstrated during the remote operation under the communication round-trip delay of 10 to 15 minutes during the mission.

Mars Exploration Rovers - Spirit and Opportunity

NASA's Mars Exploration Rover (MER) Mission is a robotic mission using two rovers — *Spirit* and *Opportunity* — to explore the Martian surface and geology. The MER-A rover, *Spirit,* was launched on June 10,

2003, and MER-B, *Opportunity*, on July 7, 2003 (see Figure 8.15). Spirit landed in Gusev crater on January 4, 2004; Opportunity landed in the Meridiani Planum on the opposite side of Mars from Spirit, on January 25, 2004. The hardware design concept of these twin rovers follows that of the Mars Pathfinder mission which was successful in 1997, but with larger rovers. The Sojourner was about the size of a microwave oven, but the Spirit and Opportunity are about the size of a golf cart. The recent rovers used an air-bag system for landing and a six-wheel system with Rocker-Bogie suspensions. They have on board stereo cameras and hazard cameras for long-range navigation, and a manipulator arm for scientific operations. At the end tip of the arm are a number of instruments to perform scientific investigations on the rock and soil targets of interest. The total distance traveled by the two rovers is more than 19 km, as of sol 1524 (April 16, 2008; Spirit's total odometry remained at 7,528 meters and as of sol 1497 (April 9, 2008), Opportunity's total odometry was 11,689 meters).

Figure 8.15. The Mars Exploration Rovers, Spirit and Opportunity, each have a manipulator arm in front (Courtesy NASA/JPL).

Keeping in Touch – Ground and Mission Operations

Having introduced the various subsystems that make up a spacecraft and the different types of payloads, a designer knows the capabilities and requirements of the subsystems and payloads to meet a particular mission's objectives; however, there is one more aspect of a space mission that greatly affects its chances for success as well as its overall costs. This missing aspect is referred to as "mission operations." This is not a physical component of the spacecraft, but the collection of activities and tasks involved with controlling and operating the spacecraft.

The people who focus on mission operations are called "mission operators;" their activities begin once the spacecraft is launched. These activities include such tasks as controlling the spacecraft, monitoring the spacecraft's health and processing the data gathered during the mission. If the mission involves a crew, mission operators must also monitor the health and plan the activities of the crew. Mission operators not only need to determine what operations must be done, but also when the operations must occur. Improper scheduling of operations can lead to increased costs and complexity as well as affecting the likelihood of a successful mission.

During the mission, there are several different aspects of a mission on which operators must focus. One of their key jobs is to perform the duties of "mission control". These members of the team focus on the daily activities of the spacecraft that are necessary to achieve the mission. These are usually laid out in the Mission Operations Plan, but there are often times when the operation of the spacecraft requires real-time decision making because of an unforeseen incident. The operators may need to receive data from the spacecraft's instruments before developing the appropriate plan.

In order for the mission control team to perform its duties, there are several other teams that work together. There is always a team in charge of monitoring the health and performance of the spacecraft bus and passing that information on to mission control. Similarly, the health and performance of the payload is monitored to ensure that it is working according to plan. Another team focuses on ensuring that the positioning and navigation of the spacecraft is performing as required. If one of the payloads is a human crew, there will also be a team focusing on their health and activities. The final area where a supporting team is required is to deal with communications with the spacecraft. All these teams work together to ensure that the mission is achieved as smoothly and successfully as possible.

The number of people required to form these teams can vary greatly. The number of operators can be either a few individuals or a large team depending on the complexity of the mission. For very complex missions, each area can have its own team or, for simple missions, several areas can be the responsibility of a single person. The members of the mission operations team are housed in facilities called "mission control centers." Linked to a mission control center is a ground system which houses all the electronics and antennas required to communicate with the satellite. The location and number of ground stations depends on several factors, one of which is how often the mission operators need to be in communication with the spacecraft. For any communication the ground station's antenna must be in line-of-sight to the spacecraft. If operators require constant communications with the spacecraft, they will need to have ground stations spaced around the globe to ensure that at least one antenna can always "see" the spacecraft.

Mission Control Centers are generally rather large and expensive installations, not only because of their size, the equipment involved and the 24-hour a day staffing, but also because of their redundant communications links to ground control facilities. For deep space installations these centers are positioned around the world so as to receive signals continuously as the Earth rotates on its axis. For example, the European Space Agency maintains a facility for ground tracking, telemetry, control and command center located in Kiruna, Sweden.

Mission Design – Putting all the Pieces Together

Space mission design begins with an understanding of the requirements of the mission, and concludes with the assembly of mission components that accomplish those requirements. In order to design a space mission, all of the following elements may need to be integrated into a unified mission: payload(s), spacecraft, science instruments, launch vehicles, orbit and trajectory selection, human crews, in-space transportation stages, entry vehicles, landers, planetary and transfer orbit selection, ground (Earth) elements, communications infrastructure, and ground facilities. Space missions are divided into two categories, robotic and human. Robotic missions comprise the largest number of space missions and include communication satellites, navigation/positioning satellites, remote sensing missions, science missions, technology flight experiments, and military space missions.

Once the designer has reduced the number of mission options, the mission design process becomes an exercise in producing performance, risk and cost metrics with which to compare the many mission options. A first-order analysis that produces only performance (mass) estimates is usually sufficient to further down-select the number of mission options, since to first order cost and mass are linearly related. After eliminating those options that are much more massive than other options, the designer calculates the relative risk (probability of mission success) and cost of each option and evaluates those remaining. A combination of performance, cost and risk, but not necessarily the "best" solution in any one of these areas, will ultimately decide the mission design chosen.

The final steps in mission design are to document the design "drivers", and then compare the design against the original mission requirements. "Drivers" are the decisions necessitated by the type of mission, given mission constraints, or payload performance constraints. Design drivers are important to understand because they shape, and perhaps limit, the number of design options that a mission designer may consider. After considering the performance, cost and risk metrics for the initial set of mission design options, it may be possible that none is a viable option – if this is the case, then the design drivers must be re-examined to reassess other possibilities for better mission options.

Once a mission design choice is made, it should again be compared against the original objectives and constraints of the mission to verify that it accomplishes the intent of the original mission statement. For each mission design, a number of products are created, including a technical mission design (which meets the original objectives, goals and constraints of the mission), a cost estimate (including launch, operations, spacecraft and instrument costs), risk analyses (identifying the segments of the mission having the highest probability of failure), and a schedule (starting with technology development, and continuing through design, fabrication, testing, launch, operations and eventual disposal).

Integration at the space program level requires a specialized skill set that combines an understanding of engineering, politics, business and law. At the highest level a mission such as the International Space Station involves international cooperation and diplomatic skills as well as program management skills (see Figure 8.16).

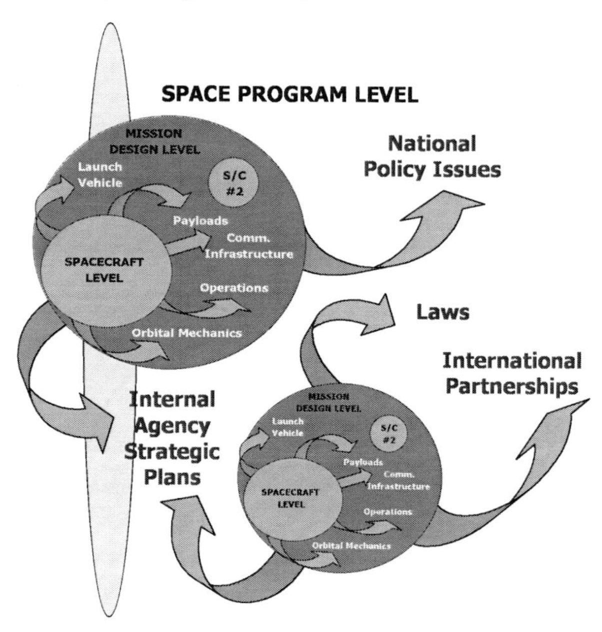

Fig 8.16. The highest level of the space model: the space program.

Successfully Designing Systems to Go Into Space

In the previous two chapters we have learned a great deal about how and why to design spacecraft. We have learned about both the supporting spacecraft "buses" as well as the prime payload. The design of these systems is challenging, involving a great deal of technology and management skills. Good communications are vital to the success of space programs—and we mean communications at all levels. There must be communications with regard to requirements related to all of the subsystems, and how they support the overall mission and the payload design. There must be good communications with regard to how the launch system, and the orientation and control thrusters, are compatible with and adequately support the mission objectives. There is a need to communicate accurately as to when a satellite can operate remotely as a robot and when a human crew is required. There must be clear communications as to what the mission objectives are in terms of schedule, acceptable risk, program budget and whether existing or new technology must be applied to the program. There is also need to communicate as to whether international cooperation is appropriate and what degree of public-private partnership might be required. In the design of launch systems[11], there must be careful communications as to what type of rockets or propulsion system is needed and for what tasks. Many and, in fact, most space programs that ultimately fail probably break down due to communications failures. It is our hope that we have managed to communicate, at least to some reasonable degree: "Why and How We Go to Space."

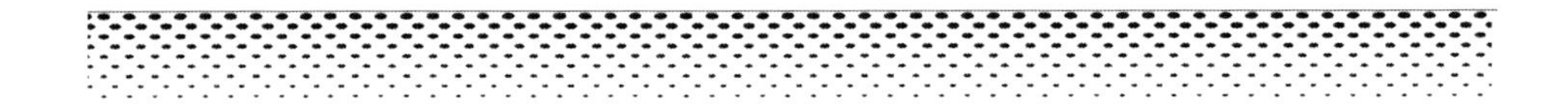

† Unless otherwise indicated, the authors developed the graphics for this chapter.

1 J. Wertz and W. Larson, Space Mission Analysis and Design, 3rd edition, Microcosm, 1999
2 R. R. Bate, D. D. Mueller and J. E. White, Fundamentals of Astrodynamics, Dover, 1971
3 G. P. Sutton and Oscar Bilbraz, Rocket Propulsion Elements, 7th Edition, Wiley, 2001
4 R. W. Humble, G. N. Henry, and W. J. Larson, Space Propulsion Analysis and Design, McGraw-Hill, 1995
5 D. K. Huzel and D. H. Huang, Design of Liquid Propellant Rocket Engines, Revised Edition, AIAA, 1992
6 M. Oda et al, "ETS-VII, Space Robot In-Orbit Experiment Satellite," Proc. 1996 IEEE Int. Conf. on Robotics and Automation, pp.739–744, 1996
7 K. Yoshida, "Engineering Test Satellite VII Flight Experiments For Space Robot Dynamics and Control: Theories on Laboratory Test Beds Ten Years Ago, Now in Orbit," International Journal of Robotics Research, Vol. 22, No. 5, pp. 321–335, 2003
8 "Lunokhod 1 Mobile Lunar Laboratory, USSR" published by the Joint Publications Research Service (JPRS identification number 54525.), US Department of Commerce, a translation of a Russian language monograph: Peredvizhnaya Laboratoriya na Lune Lunokhod-1, responsible Editor Academican A. P. Vinogradov, Moscow, 4 June 1971, 128 pages
9 A. Mishkin, *Sojourner: An Insider's View of the Mars Pathfinder Mission*, Berkley Books, 2004
10 M. Maimone, C.Y. Yang, L. Matthies, "Two years of Visual Odometry on the Mars Exploration Rovers," Journal of Field Robotics, Vol. 24, No. 3, pp. 169–186, 2007
11 M. J. L. Turner, Rocket and spacecraft propulsion: Principles, practice and new developments, 3rd Edition, Springer-Praxis, 2009

Chapter Nine
Space Policy, Law and Security

"Opportunities like this come rarely. The human migration into space is still in its infancy. For the most part, we have remained just a few kilometers above the Earth's surface – not much more than camping out in the backyard."

— Global Exploration Strategy Document, 2008

Space Policy, Law and Security

Dynamic Elements Sometimes In Conflict and Sometimes in Concert

Space policy, space law and space security all go together. They go together not like peas in a pod, but rather like members of a family that are mutually dependent on one another but sometimes in conflict as to goals and objectives. These three elements, each in their own way, derive from national interest and national defense on one hand, and yet from global opportunity and worldwide commercial enterprise on the other. Space policy, space law and space security essentially boil down to attempts at the level of the nation state, and sometimes within international organizations, to solve various types of problems and address realistically various interests that derive from global interactions around the world and in outer space.

Thus, space policy, space law and space security all derive from the desire of various countries of the world (often to varying degrees) to derive benefit from what is seen as a common resource—namely outer space. In many ways outer space has been likened to the high seas or the continent of Antarctica—a resource that no nation "owns", but in which many countries of the world have a strong economic, political and even military interest.

As we have been more and more able to access outer space, launch satellites into Earth orbit and send machines, probes, and even people, beyond the grasp of our planet's gravity, the need to coordinate space activities have obviously grown. Space policy, space law and space security pacts, and even space treaties and conventions, help us to coordinate how countries view space and, to a degree, regulate its usage.

This is a complicated and sometimes difficult task because there are more and more space faring nations, with growing and particular interests to protect, and thus more and more problems of coordination to solve. In order to utilize various types of satellites for communications, remote sensing, navigation, meteorology, or even defensive applications, for instance, there is a need to use the electromagnetic spectrum for sensing and telecommunications. Without coordination of how the various frequencies are used there would be interference and chaos. Likewise, if the various orbits with satellites deployed in them were not coordinated, there would be difficulties of interference and possibly even collisions.

For these reasons and more, the specialized agency of the United Nations known as the International Telecommunication Union (ITU) takes a leading role in providing for the registration of radio frequencies used in outer space, for assigning the usage of various orbits, and for helping to set standards for the provision of various types of space services. This is but one example of the international coordination process related to outer space.

Maturing Space Law and Policy Systems

As space exploration, space sciences and space applications have matured over the last half-century key elements of space policy, space law and space security have matured with the increasing number of space programs and missions. More and more countries have become involved. International organizations, like the ITU, as well as regional organizations, that address various elements of space activities have also increased their sway. Through these entities nations, to varying degrees, have agreed to various "rules of the road", adopted guidelines for cooperation in space, such as procedures for the deployment and operation of the International Space Station (ISS). They have even formally ratified treaties governing how, when, why and how to go into space, processes for deploying and operating satellite networks, and processes for protecting astronauts. They have even considered how we might go into space to exploit the resources of the Moon and other extra-terrestrial bodies for the common benefit of humankind.

Today, as there is more and more commercial interest in exploiting the resources of space, and as there are more and more entities seeking to operate commercial vehicles taking people into space as "tourists", the need for regulations on safety and common oversight provisions is expanding. To understand these things more clearly, we begin with some historical background.

International Law, International Coordination and the Nation State

The two Treaties that ended the Thirty Years War and resulted in the Peace of Westphalia in 1648 established for the first time on a systematic basis that the "nation state" would be the prime actor in establishing who is "in charge" around the world.[1] The setting of rules and laws covering matters such as taxes, criminal behavior, social and religious practices, the licensing of businesses and so on, would henceforth be the duty and responsibility of the nation state. This concept has been more or less accepted now for three and a half centuries. There have been wars and "interventions" and attempts to use religious or social or political influence to undermine the authority of the nation state, but nevertheless countries and their governments establish the law of the land.

The problem is that the world, now composed of some 200 nation states, is increasingly international in scope when it comes to trade, finance and capitalization, transportation, communications and broadcasting. In order to work out problems related to international interactions, various types of international organizations have evolved. Some are universal public international organizations. Some are private international or regional organizations. Some are non-governmental organizations that carry out such tasks as international assistance or the coordination of professional standards and practices. There are even what are sometimes called "BINGOs", or business international non-governmental organizations. Dealing with trade and business practices, these represent the most rapidly growing type of international organization. Indeed the Yearbook of International Organization records tens of thousands of various types of international organizations, ranging in import and size from the globally scoped United Nations Organization to the quite small International Institute of Communications (IIC) or the Arthur C. Clarke Foundation. Even the International Space University (ISU) is one form of non-governmental international organization. All are actors on the international stage, but the major actors are of course, the nation states, the major international organizations and worldwide commercial interests.

The world's first entity that might be called an international organization was the Rhine River Commission. This was formed long ago to address a problem of international pollution. Printers all along the river were throwing their inks into the Rhine and not only discoloring it but also killing the fish, making the water undrinkable and hazardous. The countries along the Rhine joined together and formed the Commission, which eventually solved the problem through regulations and fines. Most of today's major international organizations, and particularly the specialized international organizations of the United Nations, are still largely concerned with solving a variety of economic, social, technical, cultural, environmental or security issues and problems.

International Regulation of Outer Space

The international laws, treaties, and regulations related to outer space that have evolved over time are closely related to, and often derived from, the principles of the open seas. Legal scholar Myres McDougal in his book about the "laws of the seas" and "laws of outer space" explains that the historical development of such "international laws" relates very much to those areas that a nation can "protect" from outside incursion by a hostile force. Originally international waters were defined as a 3-mile limit from a nation's shoreline. With improved armaments, navies and surveillance equipment a nation's international waters were extended to 12-miles. A book by David Goldsmith Loth, entitled *How High Is Up?: Modern Law for Modern Man*, takes a similar view about a nation's claim to airspace over its skies.[2] The regulation of air space has increased to higher and higher altitudes as planes, rockets and missiles were developed to fly higher and higher, and as anti-aircraft guns and rockets were able to protect a country's airspace further overhead.

The famous U2 incident in the 1950s in which American U-2 pilot Gary Powers was shot down over the airspace of the Soviet Union seemed to confirm the principle that a nation controlled its airspace for as high as its defense systems can defend. Today the end of the Earth's atmosphere and the start of outer space are generally defined to be at a height of 100 kilometers, or 62.5 miles. When one applies such a definition of outer space and calculates how many people have flown in outer space and thus become astronauts or cosmonauts, one finds that the number is almost 500. In future years, with the advent of space tourism, that number should increase substantially. It may well be then that "outer space" is redefined to be at an even higher level.

Exactly who should regulate commercial and governmental air space and who controls operations in outer space is not clearly established. Recently, with the advent of high altitude platform systems (HAPS) that can fly above commercial air space, there have been new questions about who should control so-called "sub-space", or "proto-space". This proto space is above the normal "air space" where commercial aircraft fly, i.e. to a ceiling of about 60,000 feet (or some 21 kilometers), on up to 100 kilometers.

The Chicago Convention that establishes the international regulatory authority of the International Civil Aviation Organization (ICAO)—a U.N. specialized agency for the regulation of aviation, headquartered in Montreal, Canada—establishes ICAO's authority for the control of aircraft and winged vehicles, but its authority (or lack of it) over rockets, missiles or space planes has yet to be firmly addressed and resolved. (See "An ICAO for Space? www.iaass.org/.) Today, national entities such as the Federal Aviation Administration (FAA) in the United States or regional entities such as the European Aviation Safety Agency (EASA) are responsible for regulating the safety of space planes, but this could change with new national regulations or even some form of new international charter.

The Basic Question in Space Policy: Why Go to Space?

Sending machines and people into space is technically difficult, hard to manage successfully, risky, and expensive. Why then do governments and private sector firms undertake space activities? What objectives, purposes, or goals provide the rationale to pay the costs and accept the risks of doing things in space?

The answer to this question differs for the public and private sectors. The fundamental rationale for the private sector to engage in space activities is the potential for profit, which, after all, is the basic goal of business. For the public sector – governments and intergovernmental organizations – there is a changing mix of rationales, differing in priority among different actors and at different times. The reasons that governments support space activities can be found in various places, such as speeches by government leaders, decisions taken by executives or parliamentarians, budget proposals, or formal statements. Sometimes they must be inferred from the activities undertaken. In some nations, governments develop and announce comprehensive space policies, which usually combine both a statement of goals and objectives and a statement of the paths of action - the policies - that governments will pursue to achieve those purposes.

As one example, the most recent United States National Space Policy, announced in October 2006, set as an objective to "strengthen the nation's space leadership and ensure that space capabilities are available in time to further U. S. national security, homeland security, and foreign policy objectives." One of the means identified for achieving these objectives was to "encourage international cooperation with foreign nations and/or consortia on space activities that are of mutual benefit and that further the peaceful exploration and use of space." Similarly, the European Space Policy approved in May 2007 set as a strategic objective "a competitive European space industry." To create and sustain such an industry, the policy declared: "it is essential that European public policy actors define clear policy objectives in space activities and invest public funds to achieve them."

Early Rationales. From the start of their space activities, the former Soviet Union and the United States pursued three overriding rationales for their space programs. Similar rationales also later motivated the decisions of other countries to become active in space. These three rationales were:

- Protecting national security
- Increasing national prestige and demonstrating international leadership
- Increasing scientific knowledge.

Other Rationales. Other rationales have emerged in the years following the start of space activity. They include:

- Enhancing military capabilities
- Creating the basis for space commercialization
- Providing tangible benefits to society
- Assisting in social and economic development

Each of these rationales is discussed in more detail below.

Space Policy and Space Exploration

Exploration – the travel to distant destinations to learn more about them – has been an objective of space activity since the beginning. Robotic spacecraft have visited the Earth's Moon, every planet in the solar system and many of their moons, asteroids, and comets. A U.S. spacecraft is on its way to Pluto for a 2015 encounter with this controversial body – is it a planet, or not? However, human exploration of the solar system has not until recently been accepted as a rationale for space activity. Although U.S. astronauts did visit the surface of the Moon between 1969 and 1972, their journeys were motivated primarily by the U.S. quest for international prestige, not by a desire to explore the Earth's nearest neighbor. In 2004, U.S. President George W. Bush proposed that the United States undertake "a sustained and affordable human and robotic program of the solar system and beyond." Whether the United States or any other country will pursue such an ongoing program of human exploration is a major space policy issue, and will be discussed in more detail below.

Space Policy and Enhancing National Security

Space systems provide information to national leaders that can be used to understand what is happening around the world, and particularly information relevant to the activities of potential adversaries. This information can then be used to support planning for national and homeland security strategies. Among the space capabilities used for national security purposes are optical and radar observations, electronic intelligence, and measurements and signals intelligence.

The United States was the first to employ satellites for national security purposes; this action was motivated both by memories of the surprise attack on Pearl Harbor that began its participation in World War II and by the Cold War competition with the Soviet Union, which made its military preparations in secret. The U.S. leadership of the 1950s was determined not to be caught by surprise again, and saw in intelligence satellites a means of obtaining information not otherwise available about potentially threatening Soviet activities.

A question was raised as the United States began its reconnaissance satellite efforts was about the status in international law of satellite overflight. Was a satellite passing over a country's territory without that country's permission a violation of its national sovereignty? The precedent that this was not the case was first set by the Soviet launch of Sputnik 1 in October 1957; however, the Soviet Union protested that the operations of U.S. intelligence satellites were illegal until it also began to operate its own "spy satellites" in the early 1960s. Using space systems for national security purposes has become attractive to other countries in recent years, particularly those countries with global or regional security interests, such as China, India, France, Germany, Italy, Israel, Japan, South Korea, and member countries of the European Union. Either through dedicated national security or space systems called "dual-use" systems, which serve civilian and defense purposes alike, these actors have acquired the capability to obtain national security information from space. In response to the new Basic Space Law legislation enacted by the Diet in 2008, Japan is seen as being likely to acquire dedicated capability in this area in the coming years.

In addition, space systems have become an important means for verifying arms control and other international treaties. It can be argued that national security satellites have been a stabilizing force in international politics, avoiding miscalculations or misunderstandings about the actions and relative capabilities of potential or actual adversaries.

For many years, the kind of imagery produced by national security photoreconnaissance satellites was highly classified and available only to select government officials. Recently, however, commercial operators have provided, or are deploying, high resolution (i.e. 35 to 40 centimeter) imagery that rivals that provided by past security satellites. The implications of such global transparency are not clear; is it a stabilizing factor, or can it allow more countries and non-state actors to plan military or terrorist activities?

Space Policy and National Prestige

In the fifteen years after World War II ended in 1945, the United States and then Soviet Union engaged in a global rivalry for influence over international affairs known as the Cold War. Each country claimed its form of political and economic organization was a model for the rest of the world to emulate. In this competition, national prestige – a country's reputation for excellence that was admired by others – became an important factor. When the Soviet Union was first to launch a satellite in 1957 and followed this achievement with a series of other space "firsts," the Soviet leadership claimed that its space capabilities were evidence of its overall superiority. There was a perception in the United States and around the world that such achievements contributed to a significant increase in Soviet national prestige vis-à-vis the United States. American President Dwight Eisenhower, who was in office until January 1961, rejected the idea of a "space race" with the Soviet Union, believing that any prestige gained from space that was not based on solid scientific achievement would not be of lasting value.

On April 12, 1961, soon after John F. Kennedy came to the White House, the Soviet Union was first to launch a person into space. Kennedy reversed the Eisenhower policy and decided that the United States should enter a space competition with the USSR. He asked his advisors to identify a "space program which promises dramatic results in which we could win." The response was that the first such undertaking, that the United States had more than an even chance of accomplishing first, was landing humans on the Moon. Kennedy's advisers argued: "This nation needs to make a positive decision to pursue space projects aimed at national prestige. Our attainments are a major element in the international competition between the Soviet system and our own . . . '[C]ivilian' projects such as lunar and planetary exploration are, in this sense, part of the battle along the fluid front of the cold war." Project Apollo, the successful U.S. effort to send astronauts to the Moon, was thus driven primarily by considerations of national prestige; it was a very visible, and peaceful, way of demonstrating to the world

U.S. technological and organizational superiority.

Apollo was a unilateral demonstration of U.S. space leadership, and maintaining that leadership has been an important consideration in U.S. space policy ever since. It was one of the rationales cited in the 1972 decision to develop the Space Shuttle and the 1984 decision to develop a Space Station. The Soviet Union did try to compete with the United States in missions to the Moon, but for a variety of factors was not successful. The USSR in the early 1970s shifted its emphasis to activities in low Earth orbit, and by the late 1980s claimed that its accomplishments had earned it leading space status.

In the 1960s, countries other than the Cold War rivals, most notably France, identified space capability as one of the attributes of a leading nation. Although other rationales have been important for the now almost fifty countries operating their own systems in space, the quest for national prestige and leadership influences almost every country's decision to go into space. In 2003, China became the third country to launch one of its citizens into orbit. Like their predecessors in the Soviet Union and the United States, Chinese leaders hailed this achievement as signifying China's success in becoming a modern nation; after the third Chinese spaceflight in 2008, President Hu Jintao, who had met with the three Chinese taikonauts shortly before their launch, suggested that the mission was "another feat on the Chinese people's journey to ascend the peak of science and technology."

After its success in being first to the Moon, the United States shifted its approach to space leadership and the accompanying desire for prestige towards a more cooperative strategy. In both the Space Shuttle and Space Station projects, the United States invited other countries to participate, although on terms set by the United States. The U.S. goal was to demonstrate its leadership by being a respected "managing partner" in a multilateral undertaking. The U.S.-led International Space Station program has been carried out by a coalition of sixteen countries, and represents the largest cooperative technological project ever undertaken. Russian Space Agency (Roscosmos) participation in the International Space Station has not only been critical to its success because of problems in the Space Shuttle launch program, but has been a key means of fostering international cooperation during the past decade.

One issue associated with seeking national prestige through space activities is whether the direct involvement of humans is essential, or whether highly visible robotic programs such as the Hubble Space Telescope, the Cassini-Huygens mission to Saturn, or the Mars Exploration Rovers also make major contributions to national prestige. The 1961 recommendation to the President to undertake Apollo suggested that "it is men, not machines, that captures the imagination of the world." As human access to space becomes relatively more common, it is not clear that the mere fact of human presence contributes to the prestige element of a particular space mission.

Space Policy and the Search for Knowledge

Scientists are motivated in their work by a persistent desire for new understanding of some aspect of the natural world. That understanding is enabled by new data to test existing theories and by new theories based on the ability to extend the reach of observations to new phenomena. The end product of the scientific process is new knowledge. As it became clear in the 1950s that it would soon be possible to send their instruments into the new environment of outer space, scientists were quick to identify unanswered questions about the Earth, the solar system, and the Universe that could be addressed by space missions. Science from, and in, space has become a central feature of the space activities of many countries, and generations of space scientists have created a community that has grown to be one of the significant influences on the directions taken by government space activities. That a primary rationale for space activities should be the increase of scientific knowledge has become a widely accepted belief among both members of this community and the political leaders who provide the funding for their efforts.

If space missions were carried out only for scientific reasons, however, it is not clear that governments would be willing to support their costs, which are high compared to most other areas of scientific inquiry. Links between space science and the other original rationales for space – security and prestige – may account for the willingness of governments to bear the substantial costs of space science.

Some have suggested that many space science missions do not address fundamental scientific questions, but rather primarily reflect the desire of a specialized community to continue its work through a series of incremental advances over previous missions. It is unarguable, however, that results from increasingly complex missions have revolutionized humanity's understanding of the cosmos. The Nobel Prize in physics was, for the first time in 2006, awarded for a space-based mission, the Cosmic Background Explorer, which provided convincing evidence of the validity of the Big Bang theory of the origins of the Universe and a clearer estimate of its age.

Space Policy and Enhanced Military Power

The use of space systems to enhance the effectiveness of land, sea, and air forces was initiated during the Cold War by the United States and the Soviet Union, and both the United States and Russia continue to have significant military space programs. More recently, developing military space capabilities has become a course of action of interest to other countries with substantial armed forces. The use of space for "force enhancement" is traditionally separated both conceptually and organizationally from the use of national security space systems for strategic intelligence, although there has been movement towards using some space systems for both intelligence and military purposes.

Space systems can serve a variety of traditional military purposes, such as providing early warning of attack, offering secure communications to forces deployed at distance from military headquarters, understanding current and future weather patterns in areas of potential or ongoing combat, locating targets on the battlefield, and providing precision navigation information for deployed forces and for weapons aimed at specific targets. While a few countries have developed space capabilities dedicated to military purposes, to date only the United States has made a significant commitment to such capabilities as a central element in its approach to the use of military force.

So far, outer space itself has been a "sanctuary," free from armed conflict between opposing forces; there have been no "space wars." The United States and the Soviet Union developed and tested anti-satellite (ASAT) weapons between 1960 and1985, but these weapons were never used. Although an attempt in the late 1970s to negotiate a bilateral treaty to ban the use of ASATs failed, the two countries reached a tacit agreement not to deploy ASAT capabilities against each other's intelligence and military satellites.

As the United States has become increasingly dependent on space systems to confer military advantage, there has been a corresponding recognition that those systems can be vulnerable to attack, and a concern regarding how best to protect them from hostile threats. In 2007, China demonstrated its ASAT capability against one of its own satellites, calling attention to such vulnerability, and in 2008 the United States used a ship-based missile to destroy a reentering U.S. intelligence satellite.

These recent incidents and earlier perceptions that the United States, and perhaps other countries, intended to develop "space weapons" – a term very difficult to define – have led for more than two decades to calls for negotiating a treaty to prevent an arms race in outer space. To date, there has been very little progress towards such an agreement. How best to maintain outer space as a secure environment where all can operate for peaceful purposes free from threats of disruption is a major space policy issue.

Space Policy and Space Commerce

As discussed elsewhere, revenues from private sector uses of space have continued to grow in recent years. The use of space systems to obtain and transmit various forms of information is a well-established commercial industry.

Governments have nurtured, and continue to foster, the commercial uses of space in a variety of ways. These include supporting research and development with commercial applications, paying the cost of needed infrastructure such as launch ranges and tracking and data relay systems which are used by both governments and the private sector, serving as a major customer for commercial space products and services, partnering with the private sector in specific space projects, and creating a policy and regulatory environment favorable to commercial activities. An explicit objective of European space policy has been to create and support a space industry that can compete with the industries of other countries in the global space marketplace. Other governments have a similar objective, although it may not be so explicitly stated. However, some actions, such as U.S. export control regulations, may act counter to commercial interests.

Space entrepreneurs continue to seek new commercial opportunities. Among them are providing cargo and perhaps even crew transportation services to the International Space Station. A particularly promising opportunity in the view of many is private space travel – "space tourism." Further into the future, some suggest that using solar energy collected in space to power networks on Earth and using the resources of the Moon and near-Earth asteroids for revenue-producing purposes can be shown to be technologically and commercially feasible. Governments will be challenged regarding how best to assist the private sector in determining such feasibility.

Space Policy and Mission to Planet Earth

Space systems can produce services that have many of the characteristics of "public goods," i.e., use by one person of such a good does not diminish its availability to others, and the good is available to all who choose to use it. Examples are observing meteorological conditions by orbiting satellites, providing navigation and timing services, disaster warning and mitigation, and monitoring environmental security and global change. Because it is difficult to provide these services on a commercial basis, but also because the services have high social value, it has fallen to governments to develop and operate the space systems providing such services. For example, the signals from the U.S. Global Positioning Satellite System are used on a global basis for many different purposes and by many different users; GPS thus approaches the status of a global utility, provided without charge by the U.S. government. Attempts in Europe to develop a similar system, called Galileo, were first made on the basis of a public-private partnership, with the objective of charging specific users for some of Galileo's services. Private sector skepticism regarding the viability of this approach led European governments to decide to use public funds to pay for the development of the Galileo satellites. It is unclear whether the eventual operation of the Galileo system will also be publicly financed, or whether the original idea of charging users for some Galileo services can be put into practice. By contrast, there is no controversy over the public good character of environmental monitoring satellites planned by Europe in its Kopernikus program, previously known as Global Monitoring for Environment and Security. Those satellites will be totally funded by European governments.

Space Policy and Social and Economic Development

Space capabilities can address important issues associated with national development, such as improving education of the population, providing more robust communication capabilities, delivering services such as health care and weather forecasting, and helping to locate and exploit in a sustainable manner a country's resources. For countries with large territories, large populations, and limited infrastructure, developing their own space systems can have major payoffs. An example of this reality is India, which has made societal development the top priority of its space program. Other large countries such as China and Brazil have developed a variety of indigenous space capabilities for application to their development problems. China and India have tens of millions of its citizens participating in satellite-based educational and health care services.

Some smaller countries also have either developed or purchased space systems or leased capacity on other space networks, particularly communication satellites, as part of their development efforts. At the end of 2008, almost fifty countries operate at least one satellite and over 100 either have a space system or are leasing space segment capacity. Many of those countries are still classified as "developing." The availability of simpler, relatively inexpensive space systems has helped poorer countries gain access to space capabilities that address their development problems. Even so, the full spectrum of space capabilities has not yet been applied to the world's social and economic development problems; this has not been a top priority issue for leading space countries, although some countries such as Canada have been true leaders in the space education and health services area. The United Nations has for several decades promoted space applications for development, but limited funding has hampered its efforts.

Space Policy and Exploring the Solar System

As mentioned above, sending robotic spacecraft to various destinations in the solar system to carry out close up or *in situ* exploration has been taking place since the early years of space activity. In the 1960s spacecraft flew by Venus and Mars and explored the Earth's Moon. In the years since increasingly complex spacecraft have journeyed to many destinations in the solar system. Twelve Americans walked on the Moon between July 1969 and December 1972, but their missions were driven by U.S.-USSR geopolitical rivalry, not as the initial steps in a long-range effort of human exploration. Since 1972, no human has traveled more than a few hundred miles away from Earth.

In January 2004, the United States Space Policy took a new turn with the proclamation of a new Presidential mandate to undertake a "sustained and affordable human and robotic program to explore the solar system and beyond." This proposal came to be known as the "Vision for Space Exploration." It called for a return of humans to the Moon by 2020 and eventual human missions to Mars and beyond.

The U.S. space agency NASA has been working since 2004 on implementing this vision. It has also joined with thirteen other space agencies – the European Space Agency, the French CNES, the Italian Space Agency, the German Space Center DLR, the British National Space Center, the Canadian Space Agency, the Australian Commonwealth Industrial and Scientific Organization, the Japanese Aerospace Exploration Agency, the Chinese National Space Agency, the Korean Aerospace Research Institute, the Indian Space Research Organization, the Russian Space Agency Roscosmos, and the Ukrainian Space Agency - to coordinate exploratory plans and to prepare a "Global Exploration Strategy" that spells out the rationales for carrying out a long-term program of human space exploration.

Whether the new U.S. Congress and President Barack Obama's Administration that assumed office in January 2009 will maintain a commitment to human space exploration is not known. If it does, it will recognize the wisdom of this statement contained in the Global Exploration Strategy document: "Opportunities like this come rarely. The human migration into space is still in its infancy. For the most part, we have remained just a few kilometers above the Earth's surface – not much more than camping out in the backyard." It is possible that human exploration of the solar system could become the compelling rationale for 21st century space activities (see Figure 9.1).

Figure 9.1. The exploration of the solar system—21st Century Space Policy Challenge (Courtesy of NASA).

Basic Concepts of International Space Law and Key Space Treaties

Some rules are necessary for the smooth functioning or avoidance of chaos, in a society, be it a nation or the international community. Basic rules of behavior and laws are sometimes needed even to ensure a society's very survival. For the most part, however, laws are adopted only when it is necessary, both at national and international levels. As in the field of economics, laws are usually imposed only when problems, or even a crisis, indicate that action is necessary. In short, in market economies, political systems, and human society in general, there is a tendency toward what is sometimes called the "let it be" (in French the *laissez faire*) approach.

Outer space is an international sphere; thus, it is a difficult arena for firm laws or regulations to be agreed and imposed through global consensus among nations. To the extent there are regulations that are observed, it is through what is called international space law, although the number of national laws related to space has recently started increasing. Human laws regulate human behavior and those under the jurisdiction of the states that make these laws. International treaties to regulate international behavior are more difficult to negotiate and agree, and even more difficult to enforce, since there is often no clear-cut enforcement mechanism. Since humans are not perfect, their laws are not perfect either. National self-interest plays a major role in the formulation and application of international law.

During over fifty years of the space age, some general, but fundamental, principles of international space law have been elaborated mainly through the following international treaties that have been laboriously negotiated and finally agreed upon by many nations:

- The 1967 Outer Space Treaty.[3]
- The 1968 the Rescue and Return Agreement.[4]
- The 1972 Liability Convention.[5]
- The 1975 Registration Convention.[6]
- The 1979 Moon Agreement.[7]

Currently the launching and operation of all types of space objects are space activities that are governed by the provisions of the 1967 Outer Space Treaty, which is considered to be virtually the "constitution" of outer space. These principles were intentionally drafted in general terms in order to regulate a field of activity that was still in its infancy and, for the most part, as of 1967, confined to research. Specific rules to define, clarify, develop and elaborate the principles enunciated in the Outer Space Treaty have been formulated in various other agreements, and continue to be formulated by the UN and its specialized agencies.

Basic Principles of International Space Law [8]

Benefit and Interests of all Countries

The most basic provision of the Outer Space Treaty is that the exploration and use of outer space, including the Moon and other celestial bodies, must be carried out for the benefit and in the interests of all countries, irrespective of their degree of economic or scientific development, and shall be the province of all mankind (Article I, paragraph 1 of the Outer Space Treaty). The term "province of all mankind" implies that outer space is within the domain and concerns all mankind as opposed to an individual or a group of states. This common interest principle is of a legally binding nature and thus operates in what lawyers would call a predominantly obligatory fashion. Other specific principles incorporated in the Outer Space

Treaty come as a natural derivation and extension of this principle. The initial statement of principle implies that the exploration and use of outer space must be in some way beneficial to mankind. This does not, however, imply that all outer space activities must be undertaken only if they are exclusively beneficial to all mankind.

This statement of the basic principle of the Outer Space Treaty was included on the insistence of developing countries that believed that they should be the recipients of the benefits in the exploration and use of outer space along with the space faring nations.

Freedom of Exploration and Use

According to Article 1 (2) of the Outer Space Treaty, all states are entitled to launch and operate their space objects without discrimination of any kind. However, such freedom is not unlimited. States must carry out their space activities in accordance with international law and the United Nations Charter.[9] They are prohibited from "appropriating" outer space by any means.[10] States Parties to the Treaty are obliged to conduct all their space activities "with due regard to the corresponding interests of all other states Parties to the Treaty."[11]

Non-Appropriation

What exactly does non-appropriation mean in the Treaty? Specifically it means that outer space, including the Moon and other celestial bodies, is not subject to national appropriation by claim of sovereignty, by means of use or occupation, or by any other means. When a used car salesman from Australia began "selling" tiny plots of the Moon over the Internet many people participated as a lark, but such activity had absolutely no legal validity whatsoever.

This principle of the non-appropriation of outer space, like the freedom principle, implements the common interest principle, i.e. if some states are allowed to appropriate outer space it cannot be used for the benefit and in the interests of all countries. A broad prohibition against appropriation applies to all states as well as to their private individual/natural persons or legal persons (companies). Private persons cannot do what their states are prohibited from doing. Moreover, the states are under obligation to ensure that the space activities of their private entities are in conformity with the 1967 Outer Space Treaty. These treaty provisions following the concept and the spirit of the principles found in the Treaty with regard to the exploration and "non-appropriation" of the continent of Antarctica.

Non-Weaponization and Non-Militarization of Outer Space

Under Article IV of the Outer Space Treaty, states are obliged not to place in orbit around the Earth any objects carrying nuclear weapons or any other kinds of weapons of mass destruction, install such weapons on celestial bodies, or station such weapons in outer space in any other manner. In addition, the Moon and other celestial bodies must be explored and used by all states …... exclusively for peaceful purposes. The establishment of military bases, installations and fortifications, the testing of any type of weapons and the conduct of military maneuvers on celestial bodies are forbidden. However, anti-satellites and anti-ballistic missiles, not being nuclear weapons or weapons of mass destruction, are presumably allowed. The use of outer space for military purposes is not explicitly prohibited.

Ownership, Jurisdiction and Control

Ownership of objects launched into outer space, including objects landed or constructed on a celestial body, and of their component parts, is not affected by their presence in outer space or on a celestial body.[12] Similarly, jurisdiction and control over a space object remain with the state in whose registry that object is registered.[13] This provision of the 1967 Treaty was elaborated by the follow-on 1974 Registration Convention. Article II (1) of the Registration Convention obliges the launching state to register its launched space objects in an appropriate registry which must be maintained by that state. Furthermore, the launching state is required to provide information concerning each space object carried on its registry to the Secretary-General of the United Nations. Such information is incorporated in the International Register maintained by him.[14]

National space registers are established under national policies and laws. International registration is carried out after the actual launch - often after substantial delays on the part of some nations and with very general information about objects launched. In short, military-related satellite launches are typically registered with a very minimum of description.

International Responsibility for the Activities of Private entities

According to Article VI of the 1967 Outer Space Treaty, states are responsible for the space activities of their private entities. States are also obliged to ensure that all space activities of their non-governmental entities are carried out in compliance with the provisions of the Outer Space Treaty. [15]

The activities of non-governmental entities in outer space, including the Moon and other celestial bodies, require authorization and continuing supervision by the appropriate state. The level and legal mechanism for "authorization and continuous supervision" is to be determined, under appropriate national licensing systems, by the relevant state through its national laws and policies.

When activities are carried on in outer space, including the Moon and other celestial bodies, by an international organization, responsibility for compliance with the Outer Space Treaty then becomes the responsibility of both the international organization and all the nations participating in such organization.

International Liability

Each state that launches or procures the launching of an object into outer space, including the Moon and other celestial bodies, and each State from whose territory or facility an object is launched, is internationally liable for damage caused by its space object to another nation or to its "natural or juridical persons" by such object or its component parts on the Earth, in air space or in outer space, including the Moon and other celestial bodies.[16]

Under the 1972 Liability Convention, which expanded the 1967 Outer Space Treaty, the launching state assumes absolute liability if damage is caused on the surface of the Earth or to an aircraft in flight (i.e. no need of proof of fault or negligence). However, if damage is caused elsewhere (i.e. outer space) then the launching state shall be liable only when its fault is proved. The term "damage" means loss of life, personal injury or other impairment of health, or loss of or damage to property of States or of persons, natural or juridical, or property of international intergovernmental organizations.[17] Thus "damage" covered under this Convention is limited to physical damage only.

The Treaty specifies that the "launching state(s)" includes the nation that either launches or procures the launching of a space object or from whose territory or facility a space object is launched.[18] Whenever two or more states jointly launch a space object, they are jointly and severally liable for any damage caused.[19] The principle of joint and several liability also applies to intergovernmental organizations and their member states when a space object belonging to such organizations causes the damage.[20]

Under the Liability Convention, the victims of damage caused by a space object of another state cannot, on their own, get redress. All claims must be made to the concerned state on behalf of its nationals or permanent residents who have suffered any damage. The Liability Convention can be invoked only when the claimant state and the launching state are different and both are parties to the Convention. States to a dispute are required to settle their disputes though diplomatic negotiations. However, if they are unable to reach a satisfactory solution, the matter could be referred to a Claims Commission that "shall be composed of three members: one appointed by the claimant state, one appointed by the launching state and the third member, the Chairman, to be appointed by both parties jointly" (Art. XIV). "The decision of the Commission shall be final and binding if the parties have so agreed; otherwise the Commission shall render a final and recommendatory award, which the parties shall consider in good faith" (Art. XIX. 2). The Liability Convention has been used in a dispute settlement in the case of COSMOS 954 accident (i.e. the claim of Canada vs. the U.S.S.R.).

The Soviet space object, Cosmos 954, carried on board a nuclear reactor using radioactive uranium-235 as its fuel source. On January 24, 1978, the satellite entered the Earth's atmosphere intruding into Canadian air space. On re-entry and disintegration, debris from the satellite fell on Canadian territory. This intrusion created immediate damage, including nuclear damage, to persons and property in Canada. In accordance with international space law, the U.S.S.R. was obliged to compensate Canada for the deposit on Canadian territory of hazardous radioactive debris. Canada claimed from the U.S.S.R. a sum of about CAN$6 million but, during negotiations, finally accepted CAN$3 million in a full and final settlement. (This amount would represent about US$30 million in today's financial markets.)

International Cooperation and Respect for Interests of Others

States must conduct all their activities in outer space with due regard to the corresponding interests of all other states. According to Article IX of the Outer Space Treaty, in the exploration and use of outer space, including the Moon and other celestial bodies, states must be guided by the principle of co-operation and mutual assistance. If a state has reason to believe that an activity or experiment planned by it or its nationals in outer space, including the Moon and other celestial bodies, would cause potentially harmful interference with activities of other states in the peaceful exploration and use of outer space, it must undertake appropriate international consultations before proceeding with any such activity or experiment. A state that has reason to believe that an activity or experiment planned by another state in outer space would cause potentially harmful interference with activities in the peaceful exploration and use of outer space may request consultation concerning the activity or experiment.

Protection of the Earth-Space Environment

States are required to conduct exploration of outer space and celestial bodies so as to avoid their harmful contamination and also adverse changes in the environment of the Earth resulting from the introduction of extraterrestrial matter and, where necessary, must adopt appropriate measures for this purpose.[21] States must have due regard to the corresponding interests of all other states; therefore, they must not intentionally create hazards like space debris, which could adversely affect the safe conduct of space activities by other states.

Contamination of outer space could be caused *inter alia* by nuclear radiation, i.e. by (i) nuclear weapons tests and (ii) the use of nuclear reactors to provide electric power for satellites. As noted above, states are obliged not to place in orbit around the Earth any objects carrying nuclear weapons or any other kinds of weapons of mass destruction, install such weapons on celestial bodies, or station such weapons in outer space in any other manner.

Moreover, under the 1963 Partial Test Ban Treaty,[22] states agreed to prohibit, to prevent, and not to carry out any nuclear weapon test explosion, or any other nuclear explosion, at any place under its jurisdiction or control in the atmosphere, beyond its limits, including outer space, or under water, including territorial waters or high seas.

With respect to the use of nuclear reactors to provide electric power for satellites, the states have agreed to several principles,[23] which recognize that for "some missions in outer space nuclear power sources are particularly suited or even essential owing to their compactness, long life and other attributes." However, "the use of nuclear power sources in outer space should focus on those applications which take advantage of the particular properties of nuclear power sources."[24]

States launching space objects with nuclear power sources on board must endeavor to protect individuals, populations and the biosphere against radiological hazards. Any state launching a space object with nuclear power sources on board must, in a timely fashion, inform all states concerned, in the event such space object might malfunction with a risk of re-entry of radioactive materials to the Earth.

The launching state is responsible for such a spacecraft and its components after its re-entry into the Earth's atmosphere. The launching state must promptly offer and, if requested by the affected state, provide promptly the necessary assistance to eliminate actual and possible harmful effects. All states with relevant technical capabilities must, to the extent possible, provide necessary assistance upon request by an affected state.

Astronauts as Envoys of Mankind

Under Article V of the Outer Space Treaty, states are required to regard astronauts of all nations as envoys of mankind in outer space and must render to them all possible assistance in the event of accident, distress, or emergency landing on the territory of another state or on the high seas.

These provisions of the Outer Space Treaty were further elaborated in the 1968 Rescue and Return Agreement. Prompted by sentiments of humanity, states are expected under this Agreement to render all possible assistance to astronauts of other states in the event of accident, distress or emergency landing, and the prompt and safe return of astronauts. If a space object or its component parts are found beyond the limits of the state on whose registry they are carried, they must be returned to that nation, on furnishing identifying data prior to their return.

Exploration and Use of Natural Resources [25]

It has been recognized that significant benefits could be derived from the exploitation of the natural resources of the Moon and other celestial bodies. In order to define and further develop the provisions of the 1967 Outer Space Treaty in relation to the Moon and other celestial bodies, the 1979 Moon Agreement was adopted. Article 4 (1) of this Agreement specifies that the "exploration and use of the Moon shall be the province of all mankind and shall be carried out for the benefit and in the interests of all countries, irrespective of their degree of economic or scientific development. Due regard shall be paid to the interests of present and future generations as well as to the need to promote higher standards of living and conditions of economic and social progress and development in accordance with the Charter of the United Nations." The Agreement specifies that states, parties to the Moon Agreement, have the right to collect on and remove from the Moon samples of its mineral and other substances. Such samples would remain at the disposal of those states that caused them to be collected and may be used by them for scientific purposes. States may in the course of scientific investigations also use mineral and other substances of the Moon in quantities appropriate for the support of their missions.

Article 11 (1) of the Moon Agreement specifies that the Moon and its natural resources are the common heritage of mankind (CHM), which finds its expression in the provisions of the Agreement and in particular in its Article 11 (5). Article 11 (5) in turn states that when exploitation of the natural resources of the Moon is about to become feasible, states, parties to the Agreement, must establish an international regime, including appropriate procedures, to govern such exploitation. The main purposes of the envisioned international regime must include (a) the orderly and safe development of the natural resources of the Moon, (b) the rational management of those resources, (c) the expansion of opportunities in the use of those resources, and (d) an equitable sharing by all states parties in the benefits derived from those resources, whereby the interests and needs of the developing countries, as well as the efforts of those countries which have contributed either directly or indirectly to the exploration of the Moon, shall be given special consideration. Therefore, the principle of CHM would not apply to the exploitation of the natural resources of the Moon, unless implemented through the envisioned international regime.

Overview of Space Law Principles

The various legal principles as outlined above were drafted intentionally in general terms in order to regulate a field of activity, which was still in its infancy. Specific rules to define, clarify and elaborate these principles have been formulated in various other agreements.

The most fundamental international law principle is that all parties to agreements and contracts must respect their obligations. In international legal jargon it is called the doctrine of *pacta sunt servanda*; the provisions of the agreement must be observed and implemented. Implementing that legal doctrine, all applicable laws, both at national and international levels, require that the parties must observe their duties and perform their tasks in good faith. Those who do not do so are normally made to face some consequences. In other words, the law prescribes some kind of redress or penalty for the breach of one's contractual obligations both in private contracts and intergovernmental agreements. The opinion of the international community is a very strong deterrent against non-performance by a state of its international obligations. The International Court of Justice may be asked to decide a dispute; its decisions are normally followed. They could be implemented through the U.N. Security Council, which could impose sanctions or also could initiate some sort of a joint military action.

The problem with the principles being indicated in very broad terms allows for various interpretations and for nations to observe the provisions in only a general way. Registrations of spacecraft, for instance, have contained very little practical or precise information. The interpretation of what is meant by a space weapon or provisions related to orbital debris is generally left to a nation to define as they themselves wish.

A Review of the Most Important Actors in International Space Law

The U.N. Committee on the Peaceful Uses of Outer Space (COPUOS)

The United Nations is the major forum for the formulation of international space law. The U.N. General Assembly has assumed overall responsibility for all outer space matters, which it discharges primarily through its Committee on the Peaceful Uses of Outer Space (COPUOS).[26]

The COPUOS was first established in 1958 as an *ad hoc* Committee with eighteen member states. A year later it was re-established as a permanent body and its membership has since been increased periodically to the present number of sixty-nine. The COPUOS has two Sub-Committees: the Scientific and Technical Sub-Committees, and the Legal Sub-Committee. These Sub-Committees perform the function related to their respective fields of expertise. The Legal Sub-Committee drafts treaties and agreements regarding outer space and presents them to the General Assembly. The General Assembly, in turn, adopts them as resolutions and recommends them for adherence and ratification by its member states. This is how all above-mentioned space treaties and agreements have been formulated. In addition, the COPUOS has adopted five important declarations and resolutions,[27] which are not strictly legally binding, but influence the international governance of several space activities.

The membership of COPUOS is based on the principle of equitable representation of developed and developing countries, space powers and non-space powers as well as all the regions of the world. The decisions in COPUOS (as well as in its Sub-Committees) are taken on the basis of consensus. Such a procedure ensured the participation of all member states. However, such a consensus process slows the decision making process to such a degree that it may take over a decade to conclude any agreement of substance. The recently adopted guidelines with regard to orbital debris took many, many years to conclude. Further the guidelines now adopted are still marred by many “loopholes” and ambiguous provisions that undercut their effectiveness.

In addition to the COPUOS, there are several UN specialized agencies (international organizations) that have been involved in the international space law making process. These organizations have been established under inter-governmental agreements with a variety of responsibilities in the economic, social, cultural, scientific, educational, and other fields. In order to bring them within the U.N. family, an appropriate agreement is generally reached between the prospective candidate organization and the U.N. Thus they enjoy a special relationship with the U.N. and co-ordinate their activities with the U.N. The following is a list of some of the specialized agencies that deal actively with outer space matters:

International Telecommunication Union (ITU)

After COPUOS, the ITU is the most important organization that is extensively involved in outer space affairs. The ITU, which is headquartered in Geneva, Switzerland, adopts treaties to regulate the use of the geostationary orbit and the radio frequency spectrum, registers frequencies and orbital locations, and adopts recommended technical standards for communications satellite operation and executes many other functions.

The ITU is the oldest inter-governmental organization in the world, originally created in 1865. It was reorganized in 1932 and became a U.N. Specialized Agency in 1947. It was again restructured in 1989, and the current version of its organizational instruments were adopted in 1994 and then revised in 1998, 2002 and 2006. It has been created in order to (a) maintain and extend international co-operation for the improvement and rational use of telecommunications of all kinds, including satellite communications, (b) promote the development and most efficient operation of telecommunications technical facilities, and (c) harmonize the actions of nations in the attainment of these common goals.[28] At present there are 190 states members of the ITU. In addition, over 600 private companies and telecommunication operating agencies participate in the activities of ITU.

The ITU legal principles provide that (a) states must not assign to a radio station any frequency in derogation of the ITU regulations, (b) radio stations must not cause harmful interference to the radio services of others that are legally operating, (c) for international protection against harmful interference, co-ordination through, and registration with, ITU must be followed, and (d) states are obliged to require their private operating agencies to observe the ITU regulations; thus, they require them to obtain radio licenses under their respective domestic laws.

United Nations Educational, Scientific and Cultural Organization (U.N.E.S.C.O.)

U.N.E.S.C.O., which is headquartered in Paris, France, was created in

1946 in order to (a) broaden the base of education in the world, (b) bring the benefits of science to all countries, and (c) encourage cultural exchanges. The U.N.E.S.C.O. Constitution provides that "Since wars begin in the minds of men, it is in the minds of men that defenses of peace must be constructed." In 1972, UNESCO adopted the Declaration of Guiding Principles on the Use of Satellite Broadcasting (DBS) for the Free Flow of Information, the Spread of Education and Greater Cultural Exchange. Article IX of the U.N.E.S.C.O. Constitution provides that "In order to further the objectives set out in the preceding Articles, it is necessary that States, taking into account the principle of freedom of information, reach or promote prior agreements concerning DBS to the population of countries other than the country of origin of the transmission... with respect to commercial advertising, its transmissions shall be subject to special agreement between the originating and the receiving States."

World Intellectual Property Organization (W.I.P.O.)

The World Intellectual Property Organization (W.I.P.O.) which is headquartered in Geneva, Switzerland not far from the ITU headquarters, was established in 1967, by restructuring the 1893 International Bureau for the Protection of Intellectual Property in order to (a) promote the protection of intellectual property throughout the world, and (b) ensure administrative co-operation in the enforcement of various international agreements on such matters as trademarks, industrial design, patents, the protection of literary and artistic works (copyright), the protection of performers, producers of phonographs and broadcasting organizations, etc.

W.I.P.O. encourages the conclusion of new international treaties and the harmonization of national intellectual property laws and policies. It has initiated the conclusion of the 1974 Brussels Convention relating to the Distribution of Program-Carrying Signal Transmitted by Satellite. Radio signals transmitted by a satellite are generally available over wide geographical areas, and thus their unauthorized interception and distribution may occur easily in many countries, which are within the footprints of that satellite. This could give rise to a problem of piracy of satellite programs. In order to address this problem, the Convention in its Article 2 (1) specifies that "each Contracting State undertakes to take adequate measures to prevent the distribution on or from its territory of any program-carrying signal by any distributor for whom the signal emitted to or passing through the satellite is not intended. This obligation shall apply where the originating organization is a national of another Contracting State..."

The W.I.P.O. Convention does not apply to signals emitted by Direct

Broadcast Satellites. However, distribution of pirated signals obtained from Direct Broadcast Satellite (DBS) is prohibited. Piracy of DBS signals not intended for a country is also prohibited. Less developed countries are allowed to intercept and distribute satellite signals for teaching (including adult education) and scientific research purposes.

World Trade Organization (W.T.O.)

The W.T.O., which is headquartered in Geneva, Switzerland, and supersedes the General Agreement on Trade and Tariffs (G.A.T.T.), came into effect in 1994. The W.T.O. (a) addresses world trade both in commodities (products) and services, (b) seeks to reduce barriers to international trade, (c) promotes competition, and (d) works for the protection of intellectual property and of foreign investment, etc.

The W.T.O. has become an important forum for the discussion of space-related products and services, e.g., the 1997 Agreement on Basic Telecommunications, including satellite telecommunications. The Agreement was concluded on 15 February 1997 and involves over 90 countries, which account for more than 90 percent of global telecommunications service revenues. The Agreement is based on a worldwide commitment to opening markets, promoting competition, and preventing anti-competitive conduct. The Agreement ensures that, as soon as the countries begin to open their markets (including satellite communications), they will be available to all W.T.O. Members on a non-discriminatory basis. Furthermore, the W.T.O. dispute settlement mechanism ensures that individual countries will effectively enforce the commitments they have undertaken. The W.T.O. is one of the few U.N. specialized agencies that has enforcement powers at its command and can fine member nations for anti-competitive behavior and other trade infractions.

International Civil Aviation Organization (I.C.A.O.)

The I.C.A.O., which is headquartered in Montreal, Canada, was created in 1947 in order to (a) foster the planning and development of international air transportation, (b) ensure the safe and orderly growth of international civil aviation throughout the world, and (c) develop principles and techniques of international air navigation. ICAO is active in efforts to implement the communications, navigation and surveillance and the air traffic management (CNS/ATM) systems, which are today largely based on satellite technology. I.C.A.O. in the future may also possibly be involved in the regulation of aerospace vehicles, particularly by setting up international safety standards and global procedures for their navigation.

Non-Governmental organizations (NGOs)

Non-Governmental Organizations (NGOs), like the International Institute of Space Law (IISL) and the International Association for the Advancement of Space Safety (IAASS) actively, but indirectly, contribute to the initiation and clarification of several aspects of international space law.

The International Institute of Space Law (I.I.S.L.), founded by the International Astronautical Federation (I.A.F.) in 1960, carries out activities for fostering the development of space law.[29] The I.I.S.L., as the prominent global professional body of space lawyers, comprises of about 500 individual and institutional elected members from all over the world. A large majority of them are those distinguished for their contributions to space law development. Both the I.I.S.L. and I.A.F. are officially recognized observers at the U.N. Committee on the Peaceful Uses of Outer Space (C.O.P.U.O.S.) and its two subcommittees. They are thus able to make contributions to the U.N. deliberations.

The International Association for the Advancement of Space Safety[30] was established in 2004 in the Netherlands and its membership is open to anyone having a professional interest in space safety. The main goals of the I.A.A.S.S., as its website states, *inter alia* include (a) to improve understanding and awareness of the space safety discipline, and (b) to advocate the establishment of safety laws, rules, and regulatory bodies at national and international levels for the civil use of space. The I.A.A.S.S., (a) holds every 18 months international conferences on important issues, including legal aspects, and aspects related to space safety, (b) publishes books for space safety professionals, and (c) organizes training sessions for all those who are interested in achieving and applying space safety standards and procedures (including regulatory procedures).

Key Issues of Space Law and Space Policy

Now that we know that nations and international organizations have created an increasing set of treaties, laws and regulations related to outer space and adopted space policies for a number of reasons related to national security, image, economy opportunity and even international development the question remains: "So what?" What good do all these policies and regulations actually accomplish? The answer is a great deal.

There are serious concerns with the increasing amount of orbital debris and the possibility that the orbital debris itself could start a cascade, or avalanche, making lower Earth orbits almost unusable. The International Telecommunication Union (ITU) has now established due diligence rules for each nation to police, to eliminate explosive bolts, and to create de-orbit capabilities for satellites. The U.N. Committee on the Peaceful Uses of Outer Space has now adopted a systematic set of guidelines to limit orbital debris. While these measures could be made more effective, this is a good start on this issue. And positive regulatory measures for the effective regulation of space go well beyond orbital debris.

There are concerted efforts to use outer space to help protect the biosphere and combat global warming and climate catastrophe. There are efforts now emerging through international collaborative efforts to regulate the safety of space planes through NGOs like the I.A.A.S.S. and through systematic collaboration through national and regional organizations such as the F.A.A. in the U.S. and the E.A.S.A. in Europe. There are also efforts to coordinate the process of exploring outer space that are reflected in the efforts of nearly twenty space faring nations and expressed in the Global Exploration Strategy Document, 2008, as discussed earlier.

It is particularly useful to note that some of these coordinating efforts are carried out within the treaty-making and regulatory processes of the United Nations and its specialized agencies. Other efforts are achieved through the coordinating work of national and regional space agencies, while yet others are carried out through the efforts of NGOs, such as by the I.A.A.S.S., the I.I.S.L., the Secure World Foundation and other such groups around the world. At this time these complementary efforts have produced generally positive results. In the future, more systematic and more closely coordinated or integrated processes may evolve, but we are still in the early days of human space exploration and applications.

The Next Frontiers

Space law and regulation, space policy and national security considerations interact today to create a process for the oversight of commercial, national, regional and international activities in space. These processes are still in their infancy, but they have evolved a good deal in the last half-century since the start of the Space Age. These processes will evolve further as new space commerce, such as space tourism and private space habitats, evolves in the coming decades. The high cost of space exploration and economic necessities may well provide important new incentives for close international cooperation in space, and new (and more precise) space law and regulations. The current regulatory environment in space is relatively weak, since there are no overriding enforcement processes or major penalties for not observing "the rules of the road". The space law that exists is very broadly stated, with a great deal of opportunity for different forms of interpretation. Nevertheless, current progress in such areas as orbital debris, coordinated space exploration, the operation of the International Space Station, and the use of space to combat global warming and to aid human development are all very hopeful and encouraging signs.

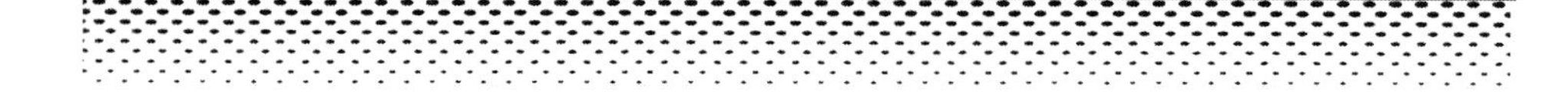

1 Treaty of Westphalia http://www.encyclopedia.com/doc/1O48-WestphaliaTreatyof.html
2 David Goldsmith Loth, *How High Is Up? Modern Man and Modern Law* (1954)
3 The *Treaty on Principles Governing the Activities of States in the Exploration and Use of Outer Space, including the Moon and Other Celestial Bodies*, adopted by the General Assembly in its resolution 2222 (XXI), opened for signature on 27 January 1967, entered into force on 10 October 1967, 98 ratifications and 27 signatures.
4 The *Agreement on the Rescue of Astronauts, the Return of Astronauts and the Return of Objects Launched into Outer Space*, adopted by the General Assembly in its resolution 2345 (XXII), opened for signature on 22 April 1968, entered into force on 3 December 1968, 90 ratifications, 24 signatures, and 1 acceptance of rights and obligations.
5 The *Convention on International Liability for Damage Caused by Space Objects*, adopted by the General Assembly in its resolution 2777 (XXVI), opened for signature on 29 March 1972, entered into force on 1 September 1972, 86 ratifications, 24 signatures, and 3 acceptances of rights and obligations.
6 The *Convention on Registration of Objects Launched into Outer Space*, adopted by the General Assembly in its resolution 3235 (XXIX), opened for signature on 14 January 1975, entered into force on 15 September 1976, 51 ratifications, 4 signatures, and 2 acceptances of rights and obligations.
7 The *Agreement Governing the Activities of States on the Moon and Other Celestial Bodies*, adopted by the General Assembly in its resolution 34/68), opened for signature on 18 December 1979, entered into force on 11 July 1984, 13 ratifications and 4 signatures.
8 For a detailed analysis of these principles, see Ram Jakhu, "Legal Issues Relating to the Global Public Interest in Outer Space", 32 *Journal of Space Law*, (2006), pp. 31-110.
9. Arts. I.2 and 3 of Outer Space Treaty.
10. Art. II of Outer Space Treaty.
11. Art. IX of Outer Space Treaty.
12. Art. VIII of Outer Space Treaty.
13. ibid.
14. Art. IV of Registration Convention.
15. Art. VI of Outer Space Treaty.
16 Art. VII of Outer Space Treaty.
17. Art. I(a) of Liability Convention.
18. Art. I of Liability Convention.
19. Art. V of Liability Convention.
20. Art. XXII of Liability Convention.
21. Article IX of the Outer Space Treaty.
22. *Treaty Banning Nuclear Weapon Tests in the Atmosphere, in Outer Space and Under Water*, 5 August 1963 (125 ratifications and 10 signatures, as of 1 February 2006), 480 UNTS 43.
23. The 1992 *Principles Relevant to the Use of Nuclear Power Sources In Outer Space*, adopted by the UN General Assembly without vote (under General Assembly resolution 47/68 on 14 December 1992).
24. Ibid.
25. For a detailed analysis, see, Ram Jakhu "Twenty Years of the Moon Agreement: Space Law Challenges for Returning to the Moon", 54 *Zeitschrift Für Luft-und Weltraumrecht*, (2005), pp. 243-260; and Ram Jakhu and Maria Buzdugan, 'Development of the Natural Resources of the Moon and Other Celestial Bodies: Economic and Legal Aspects,' *Astropolitics*, (2008), 6:3, pp. 201- 250.
26. For more details, visit: http://www.oosa.unvienna.org/oosa/en/COPUOS/copuos.html
27. They are: (a) The 1963 *Declaration of Legal Principles Governing the Activities of States in the Exploration and Use of Outer Space*, adopted by the UN General Assembly (under General Assembly resolution 1962(XVIII) on 13 December 1963; (b) The 1982 *Principles Governing the Use by States of Artificial Earth Satellites for International Direct Television Broadcasting*; Adopted by the UN General Assembly by 107 votes to 13, with 13 abstentions, on 10 December 1982 (under a General Assembly resolution 37/92: voting results are reproduced from UN document A/37/PV.100 of 17 December 1982); (c) The 1986 *Principles Relating to Remote Sensing of the Earth from Outer Space*, adopted by the UN General Assembly without vote(under General Assembly resolution 41/65 on 3 December 1986); (d) The 1992 *Principles Relevant to the Use of Nuclear Power Sources In Outer Space*, adopted by the UN General Assembly without vote (under General Assembly resolution 47/68 on 14 December 1992); and (e) The 1996 *Declaration on International Cooperation in the Exploration and Use of Outer Space for the Benefit and in the Interest of All States, Taking into Particular Account the Needs of Developing Countries*, adopted by the UN General Assembly without vote (under UN General Assembly Resolution A/RES/51/122 on 13 December 1996).
28. For details, see Ram Jakhu, "Safeguarding the Concept of Public Service and the Global Public Interest in Telecommunications", 5(1) *Singapore Journal of International and Comparative Law*, 2001, pp. 71-102.
29. For details, visit: http://www.iislweb.org/
30. For details, visit: http://www.iaass.org/

Chapter Ten

Life in Space for Life on Earth

"I spent a lot of my time working in the American module, and [the Russian cosmonaut] would stay in the Russian segment working on his things, and we'd meet up at meal times. So it actually worked out very well."

— Leroy Chiao, ISS Expedition 10 Commander

The Challenges of Life in Space

One of the key reasons we go into space is to understand life here on Earth. But to understand life in space we need to know about more than medicine, physiology and biology. When we go into space we find that these traditional life sciences are closely linked with the sciences of physics, chemistry, geology, engineering, and astronomy among others. Space life sciences are not straightforward. A lot of very smart people, with diverse and sometimes unlikely degrees in science and engineering, get involved. We need mechanical engineers to design suits for astronauts, we need electrical engineers to design our experiments, and on it goes.

Space life sciences research helps us to live and work in space. Five decades of experimentation with biosatellites and more than 260 crewed spaceflights with about 500 astronauts from various countries around the world have led us to understand many of the effects that the space environment has on living organisms - from microbes to people. That is the good news! The bad news is the more we understand the more questions we have. The truth is that we still don't know some basic mechanisms concerning life in space. Even so, the field of space life sciences has already benefited many people on the ground. The new challenge of long-duration planetary missions actually helps us to explore fundamental questions about gravity's role in the formation, evolution, and even the aging process of life on Earth.

Objectives of Space Life Sciences

Life sciences are specifically devoted to understanding the living world, from bacteria and plants to humans, including their origins, history, characteristics and habits. You name it, and it is probably in someone's space life sciences research agenda. Through its evolution, life on Earth has experienced only a fairly constant one-g environment. The influence of this omnipresent force, the Earth's gravity, on life is still not well understood. We do know that there is a clear biological response to gravity in the structure and functioning of living organisms. The plant world has evolved gravity sensors. Roots grow "down" and shoots grow "up". Animals have gravity sensors in the inner ear. Many species orient themselves with respect to gravity within a few minutes after fertilization.[1]

To better understand a living system, the scientific method sometimes studies the consequences of changing its environment. Clearly, the removal of gravity is a desirable, even necessary, step toward understanding its role in living organisms. Transition into weightlessness abolishes the stimulus of gravity. This dramatic procedure is physiologically equivalent to shutting off the lights in order to better study vision. What can be accomplished in such an elegant fashion aloft can never be done in Earth-based laboratories. This method has opened a new field of basic scientific interest, known as "space physiology". This field of study deals with fundamental questions concerning the role of gravity in living processes. "Space medicine" is another, albeit more applied, research component concerned with the health and welfare of astronauts. These two objectives complement one another, and constitute key elements of the field of *space life sciences*. We believe that space life sciences open a door to understanding ourselves, our evolution, and the working of our world without the constraining barrier of gravity.[2]

Enhancing Fundamental Knowledge

Access to a space laboratory, where gravity is reduced, aids our research. It allows us to study cellular and molecular mechanisms that are involved in sensing small forces. We begin to understand whether "messages" concerning gravity are transmitted via neural or hormonal signals. How do individual cells perceive gravity? What is the threshold of perception? Gravitational biologists try to answer these questions. Results of flight experiments suggest that a gravity level as low as one one-hundredth of normal gravity (or 0.01g) is perceived by cells through physical changes both in the liquid medium surrounding cells in a culture and in cellular structures that sense and "oppose" mechanical loads (See Figure 10.1). Exactly how the change in gravity then affects cellular functions is yet to be determined. The answer to this question is not only relevant to understanding the fundamental processes in normal cell physiology, but also to the study of certain diseases. This includes the study of such age-related concerns such as bone loss, cancer, or immune system deficiencies.[3]

Figure 10.1. Cell growth experiments in very low gravity fields, using a so-called "glove box" experimental apparatus (Courtesy of NASA).

Although many embryos split along planes in a direct relationship to the pull of gravity, we do not understand the role gravity plays throughout the reproductive cycle. We do not know, for instance, whether gravity, *per se*, is essential to fertilization, cell division and the differentiation between cell function, or whether gravity plays a key role in the formation of various tissues and organs.[4] Thus we would like to know whether any organism can be effectively fertilized and then develop to form viable reproductive cells for the next generation, all in the environment of space.

The formation of the various tissues and organs spans several developmental stages. The formation of specialized cells continues after birth or hatching and continues on well into the post-natal period. In the event that normal development does not occur, the priority is to determine which period of development is most sensitive to microgravity. For each organ system, there appears to be a critical period during which development can be disrupted by relatively small environmental stresses. Experiments on neonate (newborn) rats in space have shown abnormal development of righting reflexes and locomotion.[5] Bone metabolism and the formation of blood cells might also suffer permanent effects if the gravity stimulus were withdrawn during the appropriate stage of fertilization, birth or early development.

Regenerative processes are also fundamental developmental responses to post-natal tissue loss and injury. In many situations, these processes are a response to changes in the environment to which the individual is exposed. Understanding the role of gravity not only in ontogeny (i.e. the development of the individual) but also in phylogeny (i.e. the evolution of species) indicates the need for studies on various species in space for successive generations and indeed over a long period of time.

Such research leads to a host of intriguing questions. If we were to change one or more external factors such as gravity, what difference might this make? Would it be possible to modify the blueprints contained in the genome and change some characters of the species? Might all humans become boneless, jellyfish-like organisms, after many generations in space?[6]

Protecting the Health of Astronauts

For more than four decades crewed spaceflight has proven that humans can survive and function in the space environment for missions longer than one year. Russian cosmonauts hold all of the longer-term endurance records, although they were more than wobbly when they returned to Earth after over a year in orbit.

Future missions to the Moon and Mars, however, will last two or three times longer. While there were many unknown health and performance hazards in the earliest space missions, experience obtained from the aviation medicine community was used to determine that these hazards were not totally uncontrolled and therefore acceptable for spaceflight.

> *"People who were concerned with the future of man in space quickly became aligned with one of two points of view. On the one side, there were the more cautious and conservative members of the medical and scientific community who genuinely believed man could never survive the rigors of the experience proposed for him. The spirit in the other camp ranged from sanguine to certain. Some physicians, particularly those with experience in aeronautical systems, were optimistic. [...] It became the task of the medical team to work toward bringing these divergent views toward a safe*

middle ground where unfounded fears did not impede the forward progress of the space program, and unbounded optimism did not cause us to proceed at a pace that might compromise the health or safety of the individuals who ventured into space."[7]

In the early space missions, physiological and clinical studies were predominantly performed pre- and post-flight. Occasionally in-flight biomedical monitoring provided the means to verify that humans could carry out additional tasks, such as extravehicular activity, rendezvous and docking, robotic manipulation, exercise, and function without significant impairment, as discussed by astronauts Rusty Schweickart and Jeffrey Hoffman in chapter 3.

After a while, human spaceflight seemed to become more commonplace. In truth, however, spaceflight is far from commonplace; it is certainly not free from risk. About 1% of the crewed spaceflights, that have lifted nearly 500 people to orbit - and beyond, resulted in lethal accidents. Indeed 4% of all astronauts and cosmonauts that rocketed into space have died.

In any event, after the early missions were completed the spaceflight goals became less operational and thus more focused on research. A good deal of the research has been to examine the effects of spaceflight on living organisms, including humans. As a result of this research, we now know there are specific impacts on the health and performance of humans both during spaceflight and even several years afterward (See Figure 10.2).

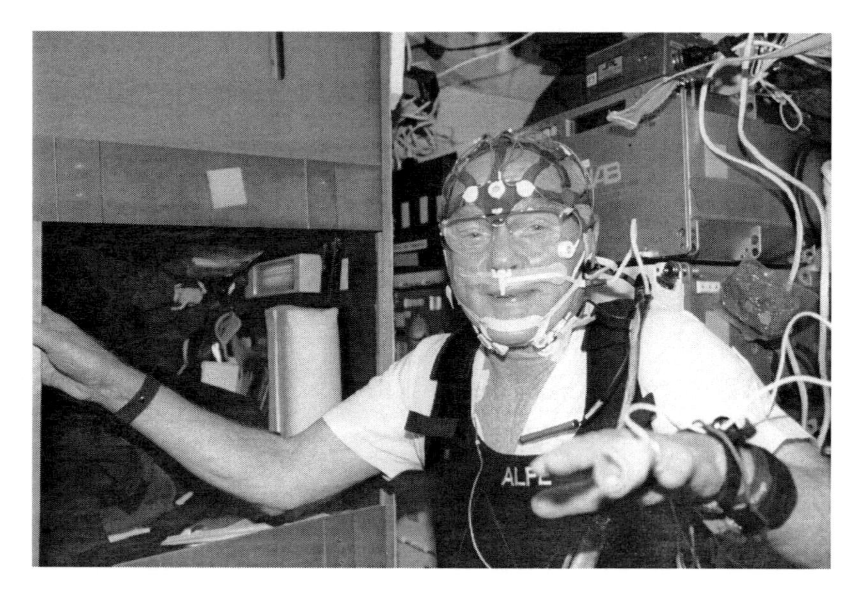

Figure 10.2. Astronaut John Glenn is equipped with biosensors for monitoring his sleep during his mission on board the Space Shuttle (Courtesy of NASA).

As was amply demonstrated by Pasteur, as well as countless successors, basic research is a boon to both medicine and agriculture. In the space life sciences key unknowns abound. The process of bone demineralization in space as seen in humans and animals is a progressive phenomenon. This is not only a serious medical problem but involves questions of biological research. We now know that weightlessness gives rise to questions about abnormalities in the development of bones, shells, and special crystals (such as the otoliths of the inner ear). The study of such abnormalities should provide us with important insights. Research areas that may seem quite arcane, such as the process of biomineralization and the control of gene transcription, will pay off in terms of new understanding. This research may help us to care for bedridden patients, to find new ways to treat a variety of diseases, and to gain many insights into how genetic processes work. Many patients who are bedridden lose muscle mass and suffer bone demineralization, yet many hibernating animals do not. It is important to understand why these animals are protected from bone loss. Even though astronauts exercise daily on the International Space Station (ISS), many still lose a significant amount of bone mass (See Figure 10.3).

In some astronauts this loss of bone mass is never recovered after returning to Earth, while in other cases bone mass is partially restored. Some scientists and physicians are investigating whether drugs used to prevent bone loss in the elderly should be used on astronauts. These drugs are not without their side effects. Thus some scientists believe that we should be examining those animals that seem to be protected from such problems while hibernating.

Figure 10.3. A cosmonaut uses an oscillating device to measure his body mass in weightlessness on board the ISS (Courtesy of NASA).

Developing Advanced Technology

In addition to the scientific need to study basic plant and animal interactions with gravity, there is a practical need to study their responses to gravity as well. These responses are essential to our ultimate ability to sustain humans for a year or more on the surface of extraterrestrial bodies or in spaceflight missions of long duration. On such missions re-supply is not possible, and any new food must be produced in space. Experiments during long-term space missions will determine which plants and animals are most efficient and best suited for our needs. For instance, can soybeans in space germinate, grow normally, and produce an optimum crop of new soybeans for food and new seed for ensuring crops? We believe that new knowledge of this biological cycling, plus the development of equipment for water and atmosphere recycling and management of waste in space, will also bring important benefits for people on Earth.

Research in biotechnology relies on the manipulation of cells of living organisms. The purpose of these manipulations is typically to produce useful molecules, natural or artificial, in useful quantities. Specifically the objective is to develop new biological molecules for specific uses. This may, for instance, be to improve plant yields and enhanced animal products through genetic alteration. Recombinant techniques, for example, make it possible to produce natural or artificially mutated versions of proteins, which can exhibit a wide range of activities and uses, both scientific and medical. In many cases these can be produced in large quantities. The techniques essential to these manipulations are applied in the liquid or aqueous environments within a cell, and are subject to the constraints of fluid dynamics and specific rules that we call "transport processes". Gravity affects biological systems through its influence on the transfer of mass and heat, particularly in the area of fluid dynamics and transport functions. Gravity also impacts cell structure and function. Consequently, we believe that being able to conduct experiments in microgravity can, and will, lead to new knowledge about biological systems. This in turn should lead to improvements in experimental techniques.[8]

Another potential application is in the field of "crystallography". Detailed knowledge of the three-dimensional architectures of biological macromolecules is required for a full understanding of their functions. We also seek to understand the chemical and physical effects of these large organic molecules that allow them to achieve these functions. Protein crystallography is currently the principal method for determining the structure of complex biological molecules. In essence we need relatively large, well-ordered single crystals to achieve a clear and "useful morphology". (This is to say we need to understand the molecules' form and their function). Crystals with these qualities may be difficult to produce for a variety of reasons. Some of these crystals may be influenced by gravity, through density-driven convection and sedimentation. Protein crystal growth experiments conducted in space have provided persuasive evidence that improvements can, in fact, be realized for a variety of protein samples.

However, the continuous production of such biological materials (bioprocessing) on a commercial scale in space has so far proved not to be compatible with the large cost of access to space. Space bioprocessing thus remains marginal today as a viable industrial application, but its potential very much remains. Improved bioprocessing techniques coupled with cheaper access to space may change the business equations in future years. (See chapter 11 that discusses, in greater detail, commercial space opportunities.) Ground-based genetic engineering in mammalian or human embryo cells is now a very strong alternative to space bioprocessing. There continue to be rapid advances in many areas achieving purification methods, including "immuno-affinity chromatography".[9]

The Space Environment and Its Effects on Living Organisms

The deep space environment is really a very hostile and difficult place for people as well as for the other organic matter that we know here on Earth. This environment includes deadly radiation, microgravity, vacuum, and magnetic fields. Then there are the local planetary environments, for the manned missions to the Moon and Mars, that present other major challenges. For more details on the characteristics of these environments, the reader could refer to the excellent review of this subject by Eckart.[10]

Microgravity

The presence of Earth creates a gravitational field that acts to attract objects with a force inversely proportional to the square of the distance between the center of the object and the center of Earth. When we measure the acceleration of an object acted upon only by Earth's gravity at its surface, we commonly refer to it as "1 g" or one "Earth gravity." This acceleration is approximately 9.8 m/s^2 (9.8 meters per second, per second). A microgravity environment is one that imparts to an object a net acceleration that is small (such as 10^{-6} g, or one one-millionth of the Earth's gravity) (See Figure 10.4).

Figure 10.4. Thanks to the weightlessness of space, astronaut Greg Chamitoff easily moves an experiment rack onboard the ISS (Courtesy of NASA).

Microgravity can be created in two ways. Because gravitational pull diminishes with distance, one way to create a microgravity environment is to travel far away from Earth (almost 17 times farther away than the Moon). This approach is impractical, except for automated spacecraft. However, the act of free-fall creates a more practical microgravity environment. Free-fall can be achieved in Earth-based drop towers, parabolic aircraft flights, and Earth-orbiting laboratories. In practice, the residual acceleration will range from about 10^{-2} g on board an aircraft in parabolic flight, to better than 10^{-6} g on board an orbiting spacecraft.

However, after a few seconds, or minutes, of low-g in aircraft, drop tower facilities and small rockets, Earth "gets in the way" and the free-fall stops. To establish microgravity conditions for long periods of time, one must use satellites in Earth orbit, i.e. in a trajectory that arcs above Earth at the right speed to execute a falling path parallel to the curvature of the Earth. When the spacecraft is in a state of free-fall - in orbit - around the Earth, its occupants are in a microgravity environment. However, the acceleration due to gravity at the low Earth orbit (at ~300 km altitude) is only reduced from its surface value.

Other Factors of the Space Environment

Besides microgravity, during spaceflight living organisms are also affected by ionizing radiation (see chapter 5), isolation, confinement, and changes in circadian rhythms (24-hour day/night cycle). In plants, for example, spaceflight offers the unique opportunity to separate the gravitational input from other environmental stimuli known to influence plant growth, such as the effects of light (phototropism), water (watertropism), and electric and magnetic fields. Spaceflight thus gives us the opportunity to distinguish between the various "tropic" responses, i.e. to investigate the mechanisms of stimulus detection and response.

The absence of natural light in spacecraft may have significant effects on humans too. A typical person spends his days outdoors, exposed to light from the Sun (filtered through the ozone layer), including a small but important amount of mid- and near-ultraviolet light. This ambient light also includes approximately equal portions of the various colors of visible light. Beside brightness, this light has numerous additional physiological and psychological effects. For example, light exerts direct effects on chemicals near the surface of the body, such as activating Vitamin D as well as destroying certain compounds such as melanin. This light also exerts indirect effects via the eye and brain to trigger circadian rhythms, secretion from the pineal organ, endocrine operations and even our moods. Many people exhibit major seasonal mood swings, in particular depression, in the fall and winter when the hours of daylight become short. Research is exploring the key question as to whether prolonged exposure to inadequate lighting (i.e., the wrong spectrum, or too low an intensity, or too few hours per day of light) may adversely affect human mood and performance.

The effects of spaceflight on biological specimens might also be related to other factors. Even the gentlest of launch vehicles produces enormous amounts of noise and vibration, plus elevated g forces. This increased "g force" continues until orbital velocity is achieved and also occurs during reentry into Earth's atmosphere. Once in orbit, machines and astronauts continue to produce vibrations and noise that are difficult to control. Biologists who are trying to focus their research on the effects of microgravity try to control other external factors (such as, for example, fluctuations in atmospheric pressure as astronauts enter and exit a spacecraft). For instance, onboard centrifuges can simulate the level of gravity found on Earth's surface in control specimens to create a baseline against which to test the other specimens when they are weightless.

The Radiation Environment

In space there is a broad spectrum of radiation. It varies from extreme ultraviolet radiation, X-rays, and high-energy particles, encountered as electrons, neutrons, protons, and heavy ions with high charges. Additional radiation is created in high-energy collisions of primary particles with spacecraft material. Although all forms of radiation are of significance, we focus on the high-energy ionizing radiation encountered during space missions. Other radiation effects are addressed in greater detail in the book by Eckart.

The highest energy radiation comes in the form of galactic cosmic rays and solar particles. Galactic radiation comes from supernova explosions and is composed of protons, electrons, and fully ionized nuclei of light elements. These high-energy particles interact with the nuclei of the nitrogen and oxygen atoms of the atmosphere, resulting in a highly complex secondary radiation, which irradiates the whole surface of the globe. The solar energetic particle radiation originates from solar flares or in shock waves associated with mass ejections from the Sun's corona (see chapter 5). This radiation consists of 95% protons, electrons and heavy ions of high energy and speed. Space missions that include travel within or through the Van Allen belts also add a third source of radiation. The Van Allen belts consist of protons and electrons trapped by the geomagnetic field. A phenomenon of special importance for missions in low Earth orbit is the South Atlantic Anomaly, in which the charged particles are drawn closer to the Earth than over other regions of the globe, due to the weaker geomagnetic field.

The mechanisms whereby ionizing radiation affects cells are either direct (particles impacting a vital target molecule and directly transferring their energy), or indirect (particles impacting other molecules, e.g., water, to yield longer-lasting, very-reactive free radicals, i.e., unpaired electrons). At the cellular level, when DNA strands break following an impact with a high-energy particle, non-rejoined breaks can lead to cell death, whereas incorrectly rejoined breaks can lead to mutation. The temporal and spatial characteristics of the radiation energy determine the quantity and quality of damage. Single-strand breaks can normally repair themselves. However, double-strand breaks with close single hits or a high-density energy hit do not repair. Cells that are actively dividing (mitosis) are the most vulnerable. High-energy particles, with their large capacity to transfer their energy along the path, can generate a high percentage of double-strand DNA breaks. The effects are widespread and can lead to the death of numerous cells along the track.

The ozone layer, the Earth's magnetosphere and the Van Allen belts block most of the harmful cosmic and solar radiation. Without these key filters above us, radiation would otherwise bombard people and animals on Earth in a disastrous way. These protective shields help to prevent runaway mutation of our genes. These "naturally occurring protective systems" are thus critical to the well being of our species and indeed of all the fauna and the flora across the planet. Extensive holes in the ozone layer observed in the past two decades coupled with the evidence of increased mutation among some species is considered an environmental risk that we must address—and sooner rather than later.

Chromosome damage and abnormalities due to high-energy particle hits have been clearly observed in plants and animals that have flown in space. Some studies showed that standard radiation-protective chemicals didn't stop the damage. It is hard to determine if these effects in lower organisms will lead to tumor induction, life shortening, or chromosome aberration in organisms with longer life spans. On the other hand, experiments with protozoa and bacteria suggest that small doses of radiation may actually be beneficial, because small doses elicit stress responses that have been shown to increase DNA repair.[11]

Micro-lesions (or "tiny cuts") of cultured retina cells from human eyes that were produced by single heavy ions were first discovered via spaceflight experiments. These findings initiated biological investigations using particle accelerators on Earth. However, the results of ground-based and space studies are often conflicting. For example, more mutations were observed in larvae of Drosophila (a type of small fly) exposed to an "artificial" radiation source while in orbit compared to ground controls. This difference suggests the possible existence of a combined effect of radiation and other environmental factors, like microgravity.[12]

Figure 10.5. ISS astronauts hold the Phantom Torso radiation experiment

(Courtesy of NASA).

Clearly there is a difficulty of differentiating between the effects of several factors inevitably present during spaceflight. For this reason, some biological effects, and their protection, can be studied only in space. NASA and ESA developed to this end a "phantom torso" to study the effects of radiation at tissue level, both during activities inside and outside the ISS (See Figure 10.5). Dosimeters are mounted where critical organs are located (head, heart, liver, and kidneys). These dosimeters record the level of radiation received as a function of time. Other instruments mounted on the outside of the ISS measure the energy, the spectrum and the frequency of particles that first hit the ISS shielding. (The higher the frequency of electromagnetic radiation, the smaller is the wavelength, and the higher the energy.) As radiation goes through the station wall and the phantom torso, the radiation is modified. The secondary radiation may have a different effect on tissue than that of the primary radiation. The information gained from the torso experiment will help determine the best types of materials and methods for shielding human crews in space.

All space agencies agree that the effects of space radiation, especially on non-dividing cells of the retina and central nervous system, must be assessed before long-duration human missions beyond the Earth's magnetosphere are attempted. For a human mission to Mars, considerably better quantitative data on radiation dose rates beyond the magnetosphere are still required. In particular, better predictability of the occurrence and magnitude of energetic particles from the Sun is needed, since radiation from the largest solar flares can be life threatening in a relatively short time.

Life Support Systems

Protecting humans from radiation, vacuum, extreme temperatures, and noise requires the use of life support equipment and technologies such as radiation shielding, pressurized and isolated living quarters, and space suits. In addition, certain basic physiological needs must be met in order for human beings to stay alive in space. On Earth, these needs are met by other life forms, in conjunction with chemical processes. These life forms (such as bacteria and vegetation) and associated chemical reactions effectively use human waste products in conjunction with energy from the Sun to produce fresh supplies of food, oxygen and clean water. In the artificial environment of a spacecraft, these natural processes that allow us to live and thrive are not there unless we artificially create them. Food must be available, water recycled (yes, this means reclaiming water from urine), human wastes must be removed, and so on. All this must be done without relying on the natural resources of the Earth's biosphere.

To date, most space missions have largely relied on a simple "open" system. This meant bringing all necessary food and supplies for the crew, and venting waste products to space, or collecting and storing them for return to Earth. When the point is reached where it is no longer cost effective or logistically possible to re-supply the spacecraft or planetary habitat with water, atmosphere, and food, ways must be found to recycle all these components. This recycling of material is referred to as a "closed" system, and can be achieved using physical-chemical systems or, better, using biological systems.

Trying to recreate the cycles of nature in a relatively small volume is a great technical challenge. Critical questions are being addressed during human missions, such as: How far can we reduce reliance on expendables? How well do biological and physical-chemical life support technologies work together over long periods of time? Is a "steady state" condition ever achieved with biological systems? How do various contaminants accumulate, and what are the long-term cleanliness issues? Eventually, in the case of exploration missions, is it possible to duplicate the functions of the Earth in terms of human life support, without having the benefit of the Earth's large buffers, such as the oceans, atmosphere, and land mass? How small can the requisite buffers be and yet maintain extremely high reliability over long periods of time in a hostile environment? We must hope that finding answers to these questions might ultimately bring some solutions to Earth's global environmental problems too.

The Legacy of Space Life Sciences Research

There is today a rich legacy of space life sciences research that is built on the basis of extensive international, intercultural and interdisciplinary teamwork among scientists and engineers who have worked hard to investigate the many unknowns.

Where We Are

Human spaceflight began less than 50 years ago, with Yuri Gagarin's single orbit of the Earth on board Vostok-1. Since then, astronauts and cosmonauts have spent a considerable amount of time beyond the Earth's surface. As we write these lines, the total number of days spent in space is about 31,000 crew days. This number might sound large, but it corresponds to only about 85 years, which can be less than a lifetime for a single individual.

Flight durations longer than six months are limited to about 40 individuals, and only four individuals have experienced spaceflights longer than one year. The record of spaceflight duration is currently held by Dr. Valery Polyakov, a Russian physician who spent 437 days during a single mission aboard the space station Mir in 1994-1995. Five years earlier, he had spent 242 days onboard Mir, so his total time spent in space actually is 679 days, or about 22 months. But this is not the longest duration in space for a single individual. Sergei Krikalev has logged 805 days during stays on board Mir, the Space Shuttle, and the ISS, and he currently holds the all-time cumulative total for days in space. His total time and that of Valery Polyakov corresponds to nearly two years, which is less than the duration of exposure to microgravity that will be experienced during a mission to Mars. Although we know that humans can survive for a long time in space, the data collected on the few individuals who have flown these very long missions is extremely limited.

The Pioneers of Space Life Science

About 120 years before the first airplane flight that lasted only a few seconds, a hot air balloon carried the first animal and human passengers during a flight that lasted more than two hours. This event is traditionally considered as the milestone in aerospace medicine[13] because the first human to fly in a balloon described for the first time the symptoms of hypoxia (altitude sickness, increase in heart rate, fatigue) and also became the first casualty after his balloon accidentally crashed in a later flight. [14]

By the middle of the 20th century, as human spaceflight began to be seriously considered, most scientists and engineers projected that if spaceflight became a reality it would build upon logical building blocks. First, a human would be sent into space as a passenger in a capsule (Projects Vostok and Mercury). Second, the passengers would acquire some control over the space vehicle (Projects Soyuz and Gemini). Third, a reusable space vehicle would be developed that would take humans into Earth orbit and return them. Next, a permanent space station would be constructed in a near-Earth orbit utilizing a reusable space vehicle. Finally, lunar and planetary flights would be launched from the space station using relatively low-thrust and reusable (and, hopefully, lower cost) space vehicles. Today's vision for space exploration is largely based on this approach.[15]

As with balloon flights, animals were sent up in rockets before humans, to test if a living being could withstand and survive a journey into space. The first successful spaceflight for live creatures came after World War II, when the former Soviet Union launched a sounding rocket with a capsule including a monkey and eleven mice. A few attempts to fly animals had been made before in the nose cones of captured German V-2 rockets during U.S. launch tests, but something always went wrong with these tests. These early attempts to study space life sciences sought to study the effects of high acceleration, weightlessness, and exposure to radiation at high altitude.[16]

In 1958, Sputnik-2 carried the first living creature into orbit. This dog named Laika was equipped with a comprehensive array of telemetry sensors. These sensors gave continuous physiological information to tracking stations. Laika's flight demonstrated that spaceflight was tolerable to animals. Other dogs, as well as mice, rats and a variety of plants were then sent into space for longer and longer durations. Monkeys rode U.S. missiles and Mercury capsules as the first step toward putting a human into space (see Figure 10.6). While these animals were in space, instruments also monitored various physiological responses as the animals experienced the stresses of launch, re-entry, and the weightless environment. The results of these animal flights showed that cardiac and respiration functions, as evaluated from pulse rate, electrocardiogram and blood pressure records, remained within normal limits throughout the flight. Animals trained in the laboratory to perform various tasks during the simulated acceleration, noise, and vibration of launch and re-entry were able to maintain performance throughout an actual flight. On the basis of these results, it was concluded that the physical and mental demands that astronauts would encounter during spaceflight "would not be excessive", and the adequacy of the life support system was demonstrated.[17]

Figure 10.6. Chimpanzee Ham and a technician go over the biomedical equipment in preparation for a mission on board a Mercury capsule in 1961 (Courtesy of NASA).

Humans in Orbit

In 1958 and 1959 both NASA and the Soviet Union announced that they had selected fighter pilots for their respective space programs. All candidates had endured stringent physical, psychological, and medical examinations. On April 12, 1961, Yuri Gagarin became the first human to orbit the Earth. According to the press release, Gagarin felt "perfectly well" during the weightlessness phase of his single orbit around the Earth. It was noted, however, that "measures" had been taken to protect the spacecraft from the hazards of space radiation. He was followed by Gherman Titov, who flew 17 orbits and, two years later, by Valentina Tereshkova, who became the first woman in space. She remained in space for nearly three days and orbited the Earth 48 times. Unlike earlier Soviet space flights, Tereshkova was permitted to operate the controls manually. Although her spaceflight was announced as successful, it was 19 years until another woman flew in space; she was Svetlana Savitskaya, aboard

Soyuz T-7 in 1982.

Soviet Cosmonaut Aleksei Leonov and Astronaut Edward White made space walks in March and June 1965, respectively, and both experienced considerable difficulties maneuvering around. Leonov also had difficulty re-entering the spacecraft. (See also the "stories" in Chapter 4 from Astronaut Rusty Schweickart about his spacewalks on Apollo 9 and from Astronaut Jeff Hoffman about his multiple spacewalks to repair the Hubble Space Telescope).

These early flights had made it clear that the body undergoes some real changes during and after a spaceflight of several days in duration. These include such changes to the body as significant cardio-vascular deconditioning (by "deconditioning" we mean the "deterioration" or the reduced state of healthiness experienced by astronauts or cosmonauts during the course of a mission), reduction in red blood cell count, and weight loss. A more complex set of in-flight medical studies was carried out during the subsequent missions, which served as precursors to the lunar missions. In parallel, ground-based studies were initiated to simulate some of the conditions of spaceflight to better understand these changes. These studies utilized bed rest and water immersion as a means of simulating microgravity. In addition, biosatellite missions of longer duration were launched with animals on board.

With the lunar missions, a medical program was developed which would make provision for emergency treatment during the course of the mission in case a serious illness occurs. Indeed, during the orbital flights, it was always possible to abort the mission and recover the astronaut within a reasonable time should an in-flight medical emergency occur. This alternative was greatly reduced during Apollo. The events of Apollo-13 showed that this medical program proved effective. Biomedical findings of the Apollo program revealed decreased post-flight exercise capacity and red blood cell number, a loss of bone mineral, and the impact on the body of extra-vehicular activity. In addition, symptoms of space motion sickness such as nausea and vomiting were experienced by both astronauts and cosmonauts. These observations raised concerns about future human spaceflights, especially if they were of long duration. These results constituted the starting point for detailed life science investigations in the Skylab and Salyut space stations program in the 1970s.[18]

From 1981 the Space Shuttle provided the opportunity to test many more crewmembers. As the first spacecraft designed to be used again and again, the Space Shuttle gave space life scientists the ability to conduct more experiments, as well as to repeat and refine those experiments. However, with the Space Shuttle, other concerns appeared. The Space Shuttle was different from previous spacecraft because it returned to Earth by landing on a runway. Critical issues existed concerning the ability of crews to perform the visual and manual tasks involved in piloting, landing and leaving the Shuttle after a long exposure to weightless conditions. It was later found that the astronauts-pilots were able to pilot and manually land the Space Shuttle, as long as the flight duration did not exceed two weeks. A mission longer than 14 days gave rise to concerns about being able to conduct all of the necessary operations flawlessly.

In April 2001, an American engineer and millionaire, Dennis Tito, paid some US$20 million to fly on a Soyuz and spent eight days on board the International Space Station (ISS). (See more about space tourism and that mission in Chapter 4). His trip precipitated a great deal of controversy when NASA and the other ISS partners objected to a "tourist" visit in the middle of a critical series of assembly operations at the ISS. The ISS partners reluctantly gave their approval for the visit (that was going to take place, with or without their approval), in return for a promise by the Russians that there would be new standards for paying visitors in the future. Since then, several other space tourists have visited the ISS onboard Soyuz. The U.S.-based firm Space Adventures in fact, brokered all of these arrangements for flights on Russian vehicles.

Unofficial "tourists" had already flown on several occasions both on the U.S. Space Shuttle (a senator, a congressman, a teacher, and a prince from Saudi Arabia) as well as aboard the Salyut and Mir space stations (a reporter, an engineer from a chocolate company, and a number of guests from allied countries). Europe took the opportunity of a paying visitor on the Soyuz to allow its astronauts to have regular access to the ISS for one week at a time. These were the so-called "Taxi" missions. Despite less stringent physical requirements and only a short training period, space tourists seem to have adapted well to spaceflights lasting a week or two, once the symptoms of space motion sickness had abated.

Surviving the Odyssey

Early predictions of the response of humans to spaceflight assumed that space adaptation would be analogous to human disease processes rather than to normal physiology. The predictions made by scientists about the ability of humans to endure spaceflight were indeed dire. Despite ground-based studies proving the contrary, there was true concern that the g forces of launch and reentry (6 to 8 g for the earliest rockets) would render human passengers unconscious, severely impaired or even result in their death. The mystique of this alien environment was so great that many feared a psychotic breakdown when separated from mother Earth. Some physicians voiced concerns that body functions such as swallowing, urination, and defecation would be impaired or impossible in the absence of gravity (although anyone who has ever swallowed while standing on their head or hanging upside down could have proven otherwise). They also felt that the bowels would not work without gravity, the heart might cavitate like a pump or beat so irregularly as to cause severe problems, sleep would be impaired (see Figure 10.7), and muscles, including the heart, would become so weakened as to prohibit return to Earth (Wolfe 1979).[19]

Fortunately all of these predictions over-estimated the physiological problems associated with spaceflight.

Figure 10.7. Tucked away in a sleeping bag, an astronaut poses near two extravehicular spacesuits in the airlock of the ISS (Courtesy of NASA).

The first space missions showed, however, that with the proper protection humans could survive a journey into space. Biomedical changes of significance observed during spaceflight were predominantly related to the effects of microgravity. Nevertheless, other phenomena were closely monitored as well. These factors included high launch and re-entry gravitational forces, radiation exposure, and psychological stress. To illustrate these many physiological changes, Dr. Susanne Churchill, in one of her lectures at the International Space University, has used the following "story line" to describe the space journey of a hypothetical space traveler who experiences all of the known problems:

> *"So, let us take a journey with our hypothetical astronaut. She is in excellent health and fully trained for the rigors of her 3-month increment on board ISS. [...] Launch occurs as anticipated and less than ten minutes from lift-off, she finds herself floating in the weightlessness of space. Without warning, however, she suddenly vomits and is overwhelmed with intense symptoms of motion sickness: nausea, a sense of dizziness, and disorientation. Her symptoms become worse when she moves about in the cabin or sees one of her fellow crewmembers floating upside-down. She takes some pills and is getting ready for sleep. However, when looking in the mirror above the sink, she realizes that her eyes seem smaller, her face is round and puffy, and her neck veins bulging. The good news is that her wrinkles have disappeared and she looks younger. When undressing, she notices that her legs look like sticks. [...]*
>
> *Within a couple of days, the motion sickness symptoms begin to subside, though her face and legs remain changed. Rendezvous and transfer to the ISS occur without incident and she starts to settle for a 3-month stay on board with her two fellow crewmates. Personal hygiene is limited to "sponge" bathing; food becomes bland tasting and she must add spices for interest. There are experiments to monitor and several hours of exercise daily on the treadmill or cycle ergometer. After a few weeks, however, the routine is boring and it becomes harder and harder to keep up with the exercise. The more she looks out of the window, the more she longs for the sounds of rain and wind, and the smell of flowers. The crew starts to argue about the smallest things. One planned space walk has to be cancelled because of a persistent irregular heartbeat in one of the crewmembers. Since that incident, this crewmember seems to be withdrawing from the others. The weekly videoconferences with family and friends are eagerly anticipated, but she wonders why there has been no commu-*

nication from her youngest child for several weeks. Has something happened? [...]

But at last the time to return approaches. When donning her reentry space suit, she realizes it is too tight because she has grown a few centimeters. After landing, our traveler reports an unbelievable sense of "heaviness" and finds herself unable to stand up unassisted from her seat, much less walk down the stairs. Her heart is beating fast; she sweats and almost faints. Even after several days of rehabilitation, muscle weakness is very evident; she quickly feels short of breath and is constantly thirsty. Weight loss that occurred in space is rapidly disappearing, but her physician tells her that she had lost much of the bone density in her hips and that her immune system seems to be impaired. Now she is concerned because she remembers that the various bacterial colonies they were studying on board the ISS laboratories showed explosive growth rates! Several months later, though, all her body functions seem to have readapted to Earth gravity.

This story is not meant to discourage anyone from wanting to be an astronaut. In reality, not all people experience all of the adverse effects of spaceflight. It is rather meant to show how little we really know about the human body's response to spaceflight and how very dangerous this new environment can be."[20]

Challenges Facing Humans in Space

Human travel in space is not easy. Some believe that robots should do most of the exploration, a job they can do cheaper and perhaps even better than humans.

The Role of Humans in Space Exploration

The debate over space exploration is often framed as humans versus robots. Some scientists fear that sending humans to the Moon and Mars might preclude the pursuit of high quality science. On the other hand, some proponents of human exploration believe that if we actually undertake as much science as possible using robots this would diminish interest in sending humans. Man vs. machine remains a big question in space exploration. However, space exploration should be thought of as a partnership to which robots and humans each contribute important capabilities. Robots are particularly good at repetitive tasks; they excel at gathering large amounts of data and doing simple analyses. Although difficult to reconfigure for new tasks, robots are highly predictable and can be directed to test hypotheses suggested by the data they gather. However, robots are subject to mechanical failure, design and manufacturing errors, and, of course, errors can be made by their human operators. Certainly, as robots improve in terms of their mobility, functionality and "thinking" capabilities, they will be able to do more.

Robotic operation on the Moon is not a problem, with only a few seconds of delay between command and response. For Mars, radio communication delays between Mars and Earth (on the order of 40 minutes round trip) pose a serious problem for teleoperation maneuvers. However, "self aware machines" and machines that are artificially intelligent should make such problems more manageable.

People, in contrast to today's robots, are capable of integrating and analyzing diverse sensory inputs and of making connections that machines overlook. Humans respond well to new situations and adapt their strategies accordingly. They do better than automated systems in any number of situations, either by deriving a creative solution from a good first hand look at a problem or by delivering a more brainless kick in the right place to free a stuck antenna! For the next decade humans will remain better at "connecting the dots" than our increasingly capable machines.

Today we can still say that humans are more adept at field science, an activity that demands all of these properties. Obviously, humans have a clear role in doing geological fieldwork and in searching for life on Mars. However, humans are less predictable than robots; they are subject to illness, homesickness, and stress from confinement, hunger, and thirst. They need protective space suits and pressurized habitats. Hence, they require far greater and more complicated and expensive support than do robots. In the future space agency planners may have to address questions like: What do we do when it is ten times less costly to send a robot to Mars but the machine is only half as capable as a human? A clear-cut answer is really impossible because there are "apples to oranges" comparisons. Cost, mobility, agility, intuitive capability and more have to be considered.

Automatic probes have indeed returned spectacular results, but it is wrong to compare these directly with the outcomes of human spaceflights. Space life science, a rather recent discipline, suffers from the small number of subjects studied and the many confounding factors that are difficult to control.

> *"But life sciences data obtained in low Earth orbit studies will be practically utilized for going further (such as establishing a Mars base) or for improving our knowledge of clinical and aging disorders on Earth, long before information on the magnetic field of Neptune."* [21]

The opportunities for in-depth studies in space life sciences have indeed been sparse. This is simply the nature of the current space program, with much to do and only a few flight opportunities that must be shared. Experiments that might take weeks on Earth take years to plan and execute in space. Limitations in the spaceflight environment also have limited control experiments and often kept the number of specimens studied far from the statistically ideal number. Often space studies are paralleled by Earth-based simulation studies using centrifuges or clinostats, but the results obtained in actual microgravity conditions can be somewhat different.

Human Missions to the Moon and Mars

As in the ISS program, the political objectives of human exploration missions will presumably focus on a large-scale international cooperation. Beyond cooperation across nations, a human Mars mission can also be regarded as an important cultural task for humankind, with the objective to globalize the view of our home planet Earth, thereby contributing to the solution of local conflicts and global-scale climate changes. In any case, a human Mars mission would meet the natural human need to explore and expand our horizons.

Obviously, for cost-effective Mars exploration, an appropriate combination of unmanned and manned activities supplementing each other in a logical way will be developed. Unmanned precursor missions will certainly help in the selection of a landing site. Such missions will also help assess the local resources available, and seek to answer questions such as whether rocket propellant can be generated from the CO_2 of the Mars atmosphere or water or hydrogen obtained from water ice.

According to conventional spacecraft configurations and current mission designs, the most fuel-efficient trajectory calls for a mission of about 1000 days, with about nine months traveling to Mars, fifteen months on the surface, and nine months returning to Earth.[22] It is likely that the gravity of Mars (equivalent to 0.38 of Earth-gravity) would act to some extent as a countermeasure to the physiological deconditioning that will take place during the trip from the Earth to Mars. However, landing maneuvers on Mars and Earth are characterized by maximum g-loads of up to 6 g due to the atmospheric drag. If the interplanetary cruise is carried out at zero gravity level (i.e. if no artificial gravity is provided within the spacecraft), such high g levels in deconditioned astronauts might prove to be a critical health issue.

Beyond the shield of Earth's magnetosphere, solar and galactic radiation can cause severe cellular damage or even cancer. The crew will need to be protected against the occasional solar flare lasting a few hours. This can be done with a "storm shelter", e.g., with food racks and water tanks packed around the walls to absorb the radiation. Fortunately, most of a solar flare's energy is in alpha and beta particles that can be stopped with a few centimeters of shielding. The ever-present cosmic rays are so energetic as to require several meters of shielding for complete blockage, and a storm shelter is insufficient to protect the crew. However, the permanent habitats of the Mars base can be covered with thick layers of soil to provide full-time radiation protection, so nearly all the crew's radiation exposure would occur during interplanetary travel. Even if such a protective system proves difficult to design, some scientists believe that the cosmic ray doses can simply be endured. Exposure to a continuous stream of radiation does far less damage than an equal magnitude of radiation delivered in one day. There is still the possibility of cancer, but this probability is rather low.

As mentioned above, not much research can be done safely on Earth to investigate these radiation effects, since the highest energy cosmic rays cannot be generated, and no one would consent to being exposed to a theoretically fatal dosage. The ISS could provide a good testing ground, since large numbers of astronauts will be exposed to modest amounts of radiation in their six-month tours of duty, but a full investigation might require waiting decades until these astronauts retire and die either of natural causes or of cancer. Mars mission advocates have no intention of waiting that long. It makes the most sense to accept the radiation risk on the Mars mission, since this is a journey into the unknown, and the risk of radiation is lower than other risk factors.

Astronaut Selection and Training

The minimal medical criteria for the selection of astronauts are different for those astronauts who:

- Pilot the vehicle,
- Support onboard operations,

- Perform extra-vehicular activities,
- Carry out specific onboard experiments, and
- Participate either as politicians, journalists, or tourists.

Knowledge of task-related health risk factors is used in astronaut selection. Specific and proven tests are utilized to make the determinations. Annual medical evaluations are performed to identify and correct medical risks, to help sustain health, provide certification for flight duties, and ensure career longevity. These tests may include further clinical evaluation or fitness assessments in order to prescribe individualized exercise programs and provide pre- and post-flight conditioning activities. Both selection and periodic medical evaluations rely on the accepted ground-based standards of preventive medicine, health maintenance, and medical practice.

During pre-flight training, the primary emphasis of medical support is on prevention. For example, the purpose of the crew health stabilization program is to prevent flight crews from exposure to contagious illness just before launch. A pre-flight quarantine limits access to flight crew for the seven days prior to launch. Even before this period, the health of an active duty crewmember's family is considered to be critical. Factors such as infectious disease and/or stress affecting a crewmember's family are taken into account. Crewmembers are also trained in the use of special countermeasures to spaceflight physical deconditioning, and in medical monitoring and clinical practice procedures (see Figure 10.8).

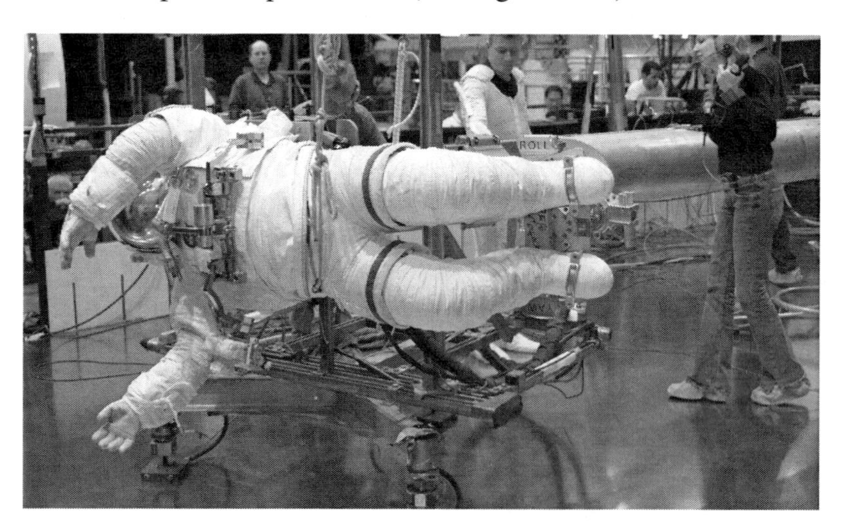

Figure 10.8. An astronaut in training for extra-vehicular activity on the air-bearing floor at the NASA Johnson Space Center (Courtesy of NASA).

Health Risks

The primary emphasis of in-flight medical support is on health maintenance. Health monitoring, countermeasures to body function deconditioning, and environmental monitoring enable a comprehensive program tailored to crew and mission needs.[23]

There are a number of potential health risks to be considered. Risk factors include: (1) levels of acceleration; (2) vibration and noise during launch; (3) exposure to toxic substances; (4) pressure changes and exposure to a vacuum; and (5) cosmic and solar radiation. With the possible exception of changes to the immune system, body changes that occur after entering microgravity represent normal homeostatic responses to a new environment. The body's control systems recognize the lack of gravity and begin to adapt to this unique situation, not realizing that the ultimate plan is to return to 1-g after a transient visit to microgravity. In-flight, adaptive changes typically occur: there is a reduction of blood plasma volume, the muscles shrink in size and strength, the bones begin to reduce in density, there is disorientation that often leads to nausea, and the immune system functions with reduced efficiency, to name just a few. These changes in human physiology are rather "logical" adaptations to the space environment; they are not life threatening, at least for 14 months, which is the longest period that an individual human has been in space. That's the good news. The bad news is that adaptation to space creates problems on returning to Earth. For example, piloting tasks are challenged by the presence of g-forces in deconditioned individuals. After making a nominal landing, an astronaut often has difficulty when standing, a phenomenon known to flight surgeons as "post-flight orthostatic intolerance", or when walking, as well as dizziness and muscle weakness. These complications could prove dramatic in the case of a non-nominal landing where the crew may be required to undertake an emergency escape from the vehicle with no help from ground support personnel. Therefore, appropriate countermeasures must be developed that balance health risk against mission constraints.

Psycho-sociological issues become increasingly more important as space missions become longer, and spaceflight teams become larger and more heterogeneous. The isolated, confined, and hazardous environment of space creates stress beyond that normally encountered on Earth. This remains a problem even after intensive training for a space mission. Extended duration missions place an even greater stress on relationships. There is stress on individuals, on interpersonal relationships, and the group dynamics for astronaut crews. There is stress not only on astronaut crews but also on ground control personnel and on astronaut families. Current countermeasures focus primarily on the individual, mission crew, and to some extent the families of mission crews; psychological training and support through in-flight communications are provided. The movie "Apollo 13" directed by Ron Howard captured many of the stresses on everyone involved in the mission.

Medical and psychological personnel also have an opportunity to review all design considerations early in the design process to ensure that spacecraft design and life support systems meet the medical and psychological requirements. Having a pleasant spacecraft environment is vitally important to crew health, well being, and productivity, especially as the mission duration increases. Thus colors, equipment layout, and hardware design are important issues, as are adequate and ergonomically correct work and living spaces. Having adequate storage space is essential; there must be sufficient personal space for restful sleep, adequate lighting and outside views. Schedules must produce interesting work, with sufficient rest and recreation periods to avoid chronic fatigue (see Figure 10.9).

Figure 10.9. An astronaut is shown inside the Soyuz capsule (Courtesy of NASA).

Flight surgeons ensure that time and resources are set aside for crew personal hygiene and sanitation, and that a variety of healthy and palatable foods and beverages is provided. The daily food supply totals a high 3000 calories, plus snacks. The meals also attempt to compensate for the body's tendency in microgravity to lose essential minerals, such as potassium, calcium, and nitrogen. At the same time, the meals must be attractive, not like the early missions when astronauts had to suck their meals out of "dentifrice" tubes or plastic bags without being able to see or smell the food. Nowadays, attention is given to an individual crewmember's preferences as well as the nutritional adequacy of food available during missions.

Space medicine emphasis is not only on health maintenance, disease prevention, and environmental issues, but also on the provision of medical care to manage possible illnesses and injuries. Being closed compartments, spacecraft standards for air, water, microbiology, toxicology, radiation, noise, and habitability must be established. In-flight environmental monitoring systems prevent crew exposure to toxicological and microbial contamination of internal air, water, and surfaces. Sensors also provide alerts as to elevated radiation sources either from within and external to the spacecraft, as well as to excessive vibration and noise. These systems must have both near real-time and archival sampling capabilities, and also warn crewmembers instantly when measured values exceed acceptable limits.

Astronauts must have career longevity, normal life expectancy, with rehabilitation and recovery capabilities available upon their return from space. After landing, health monitoring and physical rehabilitation therapies accelerate the return of crewmembers to normal Earth-based duties, especially for pilot astronauts. Reported post-flight symptoms are captured in mission medical debriefs after a space mission through interviews between the astronauts and the crew flight surgeon. The information is entered into a database to assess any increased health risks. Such studies are particularly relevant regarding radiation exposure, but have many other applications as well.

Countermeasures and Rehabilitation Issues

The two prime goals for operational space medicine are: (1) to prevent the occurrence of illness or impaired performance in spaceflight, and (2) to rehabilitate or treat impaired function in a manner that does not jeopardize the mission while maximizing crew health and performance.

Long- and short-term exposure to microgravity significantly alters many organ systems. These changes in physiology may eventually show up in many forms of impairment. These problems might include difficulty in standing and walking, decreased exercise capacity, orientation issues, and/or suppression of immunity. The flight surgeon needs to understand these important physiological effects on humans and place them within the operational context of a space mission. Furthermore, crewmembers may have sub-clinical disease or pathology, which may be exacerbated by these adaptive responses to microgravity. Unfortunately, the "sample" represented by just under 500 astronauts is quite small in relationship to the totality of humanity. To have a statistically valid sample concerning the physiology and pathology associated with human space travel it is necessary to cover a range of ages, race, occupation, nationality, culture, occupational exposures, as well as the differences between males and females. The limitation in the number of astronauts prevents the flight surgeon from deriving generalized knowledge about spaceflight. Accordingly, risk data for the astronaut population must be extrapolated from the incidence and prevalence of diseases in similar cohorts such as the military and civilian aviator communities.

The occupational medicine approach of preventing illness and injury is focused on reducing "medical hazards". This is usually accomplished by applying mitigation strategies or hazard controls to known risk factors. In this case "risk" is defined as the likelihood of a medical issue or problem occurring to a certain level of severity. Humans are protected medically in extreme environments by taking a preventative approach to illness or injury, by progressively adding levels of prevention until the risk is considered to be "adequately controlled". Three levels of prevention care are detailed in Table 10.1. Should a medical hazard prove to be uncontrollable yet still pose a significant risk to the mission, a waiver must be filed and approved to permit the mission to proceed. An example of this would be accepting the risk of long-term cancer induced by radiation exposure when the actual risk is unknown or uncontrollable.[24]

Table 10.1. Levels of prevention care for humans in extreme environments. (Definitions are adapted from U.S. Preventative Services Task Forces' Guide to Clinical Preventive Services.)

Prevention	**Definition**	**Rationale**	**Methods**
Primary	Those measures *"provided to individuals to prevent the onset of a targeted condition"*	Eliminate the hazard, e.g., by selecting in crewmembers without disease and who have no symptoms of disease	This is achieved by estimating the incidence and prevalence of pathology in the astronaut cohort. The astronaut cohort is small, and therefore risk data must be derived from observations, likelihoods, and severity of illness or injury in similar cohorts such as the military.
Secondary	Those measures *"that identify and treat asymptomatic persons who have already developed risk factors or preclinical disease but in whom the condition is not clinically apparent"*	Protect against a hazard that could not be controlled by primary prevention alone such as the effects of reduced gravity on bone or chronic low dose radiation on increased cancer likelihood.	During space travel there are root-causes to environmental and operational hazards that are not adequately controlled by mission design or other primary prevention strategies. These root-causes need to be mitigated using secondary prevention such as load bearing exercise to reduce bone loss in reduced gravity environments.
Tertiary	Those measures *"which the care for established disease, with attempts made to restore to highest function, minimize the negative effects of disease, and prevent disease-related complications"*	Tertiary prevention is invoked in most medical systems when primary and secondary prevention has failed. It is the least cost effective means of providing medical care in an extreme environment.	Illness or injury may be due to an uncontrolled hazard harming the crew (fire), an ineffective countermeasure (decompression sickness), or previously undetected disease causing an acute illness requiring treatment. The ability to provide tertiary capabilities can be categorized as advanced life support care, transitional care and ambulatory care.

Countermeasures (also known as secondary prevention) refer to the application of procedures or therapeutic means to maintain health, reduce risk, and improve the safety of human spaceflight. These measures may involve chemical, biological, physical or psychological actions. The countermeasures typically aim at:

- Eliminating or preventing adverse and harmful effects on crew health. (e.g., using artificial gravity to overcome the effects of being in a low-g environment);
- Mitigating the effect of harmful agents or warding off their impact. (e.g., pre- and in-flight exercise to counteract the effects of microgravity, use of medications to prevent space motion sickness, or modifying spacecraft design to minimize radiation exposure);
- Reducing the effect of adverse or harmful agents on the crew once mal-adaptation, disease, or injury has been identified (e.g., fluid loading to minimize post-flight orthostatic intolerance or rehabilitation programs to reverse space mission-induced muscular-skeletal or cardiovascular de-conditioning).

Both the Mir and ISS experience with long duration missions indicates that current countermeasures are far from optimal. There are vivid images of the cosmonauts unable to stand immediately after returning to Earth after a long-duration stay on board Mir. They are helped from the spacecraft and "ceremoniously hauled around like nabobs in sedan chairs" (see Figure 10.10). The situation with the ISS has shown only limited improvement. Astronauts on board the ISS exercise on a cycle-ergometer or a treadmill for two hours per day. Even so considerable muscle and bone loss is observed after landing. We cannot reasonably request astronauts to exercise more than three hours a day. We therefore believe that new countermeasures must be developed for humans traveling to and from Mars. This might involve an on-board centrifuge or even designing the spacecraft to have a reasonable gravity level using spinning modules connected by tethers or a rather large-radius vehicle. Even if this were accomplished there would still need to be improved countermeasures for the time spent on the surface of Mars.[25]

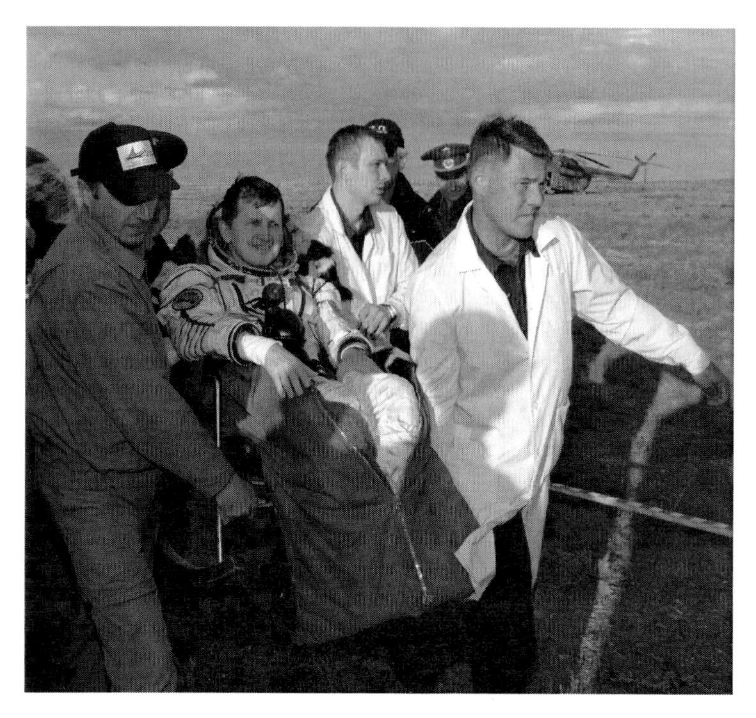

Figure 10.10. An astronaut is carried in a chair from the Soyuz landing site to an inflatable medical tent after an extended mission aboard the ISS (Photo credit NASA/Bill Ingalls).

With spinning spacecraft, increasing levels of artificial gravity can be achieved either by increasing the radius, or by increasing angular rate. The increase in radius serves to create a more "natural gravity", but also increases costs. Increasing the angular rate can be achieved at much lesser cost, but it raises physiological and psychological concerns for the astronaut's health.[26] A drawback of using a centrifuge is that, every time a linear motion is attempted in any plane not parallel to the axis of rotation,[27] the astronaut's sense of balance is affected. Rotation of the astronaut's head out of the plane of rotation generates competing forms of angular accelerations. This in turn results in illusory sensations of bodily or environmental motion. Nausea and vomiting may result after a few head movements, particularly if the angular rate of the centrifuge is high. However, recent studies have shown that humans can adapt to high rotation rates if they are increased gradually.[28]

A rotating spacecraft also presents serious design, financial, and operational challenges for a maneuvering station. From a practical perspective, it is very likely that humans do not need gravity 24 hours a day (or even a significant fraction of it) to remain healthy. If intermittent gravity is sufficient, a permanently rotating spacecraft is not needed. Instead, an onboard short-radius centrifuge may present a realistic near-term opportunity for providing astronauts with artificial gravity; various designs have been proposed. Tests are being conducted by having healthy volunteers exercise on a treadmill or step-machine while lying on their back in a short-radius centrifuge (see Figure 10.11).

The approach will determine whether exercising under increased gravitational forces will decrease the amount of time required to maintain health and fitness. If the results prove positive and the amount of in-flight exercise is reduced through the use of a centrifuge, then such devices could be a good countermeasure for the ISS or the spacecraft en route to Mars.[29]

Figure 10.11. A subject is lying in a short-radius centrifuge for studying the effects of centrifugal force (artificial gravity) on sensory, motor and cardiovascular functions

(Courtesy of CNRS).

Where We Go From Here

Space life science is a young science. This field came into existence with the first studies carried out on animals during the first suborbital flights less than 60 years ago. Since then, people have visited the Moon and have lived in space for about the period planned for a journey to Mars. Still, our understanding of how spaceflight affects living organisms remains rudimentary.

Today's opportunities for carrying out space research are scarce, since the Shuttle is to be retired and the ISS is not yet equipped with all the planned biological research facilities. Some space life science experiments will be performed in the coming months and years, but they are still limited in scope and duration, and the number of test subjects will still remain small. When the ISS is fully operational with a permanent crew, it should provide greatly enhanced knowledge of the effects of microgravity on human beings.

Gravity provides a directional stimulus that plays an important role in basic life processes at the cellular level as well as at the organism level. As of today, we only have short snapshots of how living organisms actually adapt to the space environment. Will these functions develop normally when deprived of the gravitational stimulus for a long time? Fundamental questions raised by the space environment can be addressed in the areas of gravitational biology, developmental biology, and radiobiology.

Most of our central nervous system functions depend on the presence of gravity. These functions include balance, the control of movements, spatial orientation, the regulation of cardio-vascular responses, and control of load-bearing bones and muscles.[30] The "functional hypothesis" theory suggests that what is not used is lost. If this theory holds over multiple generations in space, then gravity-dependent structures may ultimately disappear or assume a very different appearance in space. There is evidence that the load-bearing structures change following acute exposure to space. What will happen over multiple generations is speculative. A strong skeleton becomes useless and legs not only get in the way in weightlessness but also are involved in the fluid shift that pulls the blood away from the heart and brain. Form follows function and, as function changes, so will form. How much change and what form organisms will assume over time in space is unknown.[31]

Carrying out research in space often comes at a considerable cost, sometimes human. The most striking difficulties are the small subject pool available, the lack of adequate controls, and the fact that science is (by necessity) secondary to mission safety when conducting experiments in such a hostile environment. Nevertheless, the success of the manned space program is, in fact, dependent on the success of life science research in microgravity. There is still much research to be done to address the considerable dangers still faced by crewmembers on long-duration missions.

The primary objective in sending humans to Mars is to explore and perform science. This mission, when it will be undertaken, will probably be the longest period of exposure to a reduced gravitational environment, and probably the longest period away from Earth as well. The maintenance of human health and performance, using appropriate medical care, is essential to meet these mission objectives. The medical support will most likely focus on medical prevention and intervention.

Medical planning must address the expected hazards secondary to the mission profile in addition to the stochastic medical events inherent in any population of healthy humans. On the one hand, the 0.38-g environment of Mars enables a much simpler design of medical hardware and procedures when compared to the microgravity of ISS. On the other hand, the voyage to and from Mars will most likely occur in a microgravity environment, unless artificial gravity can be provided in some way on transfer vehicles. Medical systems designed for these disparate gravitational fields may require unique hardware and procedural solutions to handle samples and waste under varied environmental conditions.

Medical hazards for space exploration are diverse in their incidence, severity, and outcome. The ability to maintain up-to-date standards of care for exploration missions over the next 25 years is clearly a challenge. Setting these standards will be complicated by the lead time of 5 to 10 years that it takes to design, build, validate, and certify a medical support system for projected flights to the Moon and Mars. While planning future space missions, terrestrial medical standards of care continue to evolve. Planning the medical support for a manned Moon mission in 10 years time, and a manned Mars mission in 20 years from now, means "shooting at a moving target".

Most medical provider organizations rely on an evidence-based approach to deriving standards of care. Exploration medicine will, to some extent, be able to call on the experiences of past and current spaceflight. It will also depend on advances that have occurred in occupational medicine in extreme environments. However, lunar and Mars missions represent a "new frontier", where we may find ourselves in a situation similar to the lunar missions of the 1960s, inventing a space medical risk mitigation strategy based on "a whole bunch of smart people thinking hard in the absence of experience and data". Yet they have to get it right the first time. Today flight surgeons and researchers in the space life sciences are still striving, with the limited information available, to "get it right" for impending travels to the Moon and Mars.

1 Ingber DE (1999) "How cells (might) sense microgravity". *FASEB Journal* 13: S3-S15.
2 Clément, G (2005) *Fundamentals of Space Medicine.* Microcosm Press, El Segundo, California and Springer, New York.
3 Bouillon R, Hatton J, Carmeliet G (2001) "Space biology. Cell and molecular biology". *A World Without Gravity.* Seibert G (ed) Noordwijk: European Space Agency, ESA SP-1251, pp 111-120.
4 Duprat AM, Husson D, Gualandris-Parisot L (1998) "Does gravity influence the early stages of the development of the nervous system in an amphibian?" *Brain Research Reviews* 28: 19-24
5 Walton K (1998) "Postnatal development under conditions of simulated weightlessness and spaceflight". *Brain Research Reviews* 28: 25-34
6 Souza K, Theridge G, Callahan PX (2000) "Life Into Space". *Space Life Sciences Experiments. Ames Research Center. Kennedy Space Center, 1991-1998.* Life Sciences Division, NASA Ames Research Center, Moffetts Field: NASA SP-2000-534.
7 Berry C (1975) "Perspectives on Apollo", In: *Biomedical Results of Apollo.* Johnston RS, Dietlein, LF, Berry CA (eds). Washington DC: NASA Scientific and Technical Information Office, p 581-582
8 Bonting, SJ, Brillouet C, Delmotte F (1989) "Bioprocessing". In: *Life Sciences Research in Space.* Oser H, Battrick B. (eds) European Space Agency, Paris, ESA SP-1105, Chapter 9, pp 109-117
9 Clément G, Slenzka K (2006) *Fundamentals of Space Biology.* Microcosm Press, El Segundo, California and Springer Press, New York.
10 Eckart P (1996) *Spaceflight Life Support and Biospherics.* Kluwer Academic Publishers, Dordrecht, The Netherlands
11 *Op cit*, Clément and Slenzka (2006).
12 Planel H (2004) *Space and Life. An Introduction to Space Biology and Medicine.* CRC Press, Boca Raton, Lousisana.
13 DeHart RL (1985) "The historical perspective", In: Fundamentals of Aerospace Medicine. DeHart RL (ed) Philadelphia: Lea and Febiger, pp 5-25.
14 Clément G (2005) *Fundamentals of Space Medicine.* Microcosm Press, El Segundo, California and Springer, New York
15 Lujan BF, White RJ (1994) *Human Physiology in Space. Teacher's Manual. A Curriculum Supplement for Secondary Schools.* Universities Space Research Association, Houston, Texas.
16 *Op cit*, Clément and Slenzka (2006).
17 Henry JP (1963) "Synopsis of the Results of the MR-2 and MA-5 Flight". In: *Results of the Project Mercury Ballistic and Orbital Chimpanzee Flights*, Houston, TX: National Aeronautics and Space Administration, NASA SP-39, Chapter 1
18 Clément G, Reschke MF (2008) *Neuroscience in Space.* Springer, New York.
19 Wolfe T (1979) *The Right Stuff.* Farra, Straus, and Giroux, New York.
20 Churchill SE (1999) "Introduction to human space life sciences". In: *Keys to Space. An Interdisciplinary Approach to Space Studies.* Rycroft M and Houston A (eds.) Boston: McGraw Hill, pp. 1813- 1821
21 Barratt M, Pool SL (2008) *Principles of Clinical Medicine for Space Flight.* Springer, New York.
22 Zubrin R, Wagner R (1996) *The Case for Mars: The Plan to Settle the Red Planet and Why We Must.* New York: A Touchstone Book, Simon and Schuster, New York.
23 Barratt M, Pool, SL (2008) *Principles of Clinical Medicine for Space Flight.* Springer, New York.
24 Hamilton D, Smart K, Melton S, Polk JD, Johnson-Throop KJ, Hamilton D (2008) "Autonomous medical care for exploration class space missions". *J Trauma* 64: S354-S363.
25 Clément G, Bukley A (2007) *Artificial Gravity.* Microcosm Press, El Segundo, California and Springer, New York.
26 Diamandis P (1997) "Countermeasures and artificial gravity". In: *Fundamentals of Space Life Sciences.* Churchill SE (ed) Krieger Publishing Company, Volume 1, Chapter 12, pp 159-175
27 Stone RW (1973) "An overview of artificial gravity". *Fifth Symposium on the Role of the Vestibular organs in Space Exploration.* 19-21 August 1970, Naval Aerospace Medical Center, Pensacola, NASA SP-314, pp 23-33.
28 *Op cit*, Clément and Bukley (2007)
29 *Ibid.*
30 *Op cit*, Clément and Reschke (2008)
31 Morey-Holton ER (1999) "Gravity, a weighty topic". In: *Evolution on Planet Earth: The Impact of the Physical Environment.* Rothschild L, Lister A (eds) Academic Press, New York.

Chapter Eleven
Creating Space Markets: Understanding the Economic Rationale

"Clearly our first task is to use the material wealth of space to solve the urgent problems we now face on Earth: to bring the poverty-stricken segments of the world up to a decent living standard, without recourse to war or punitive action against those already in material comfort; to provide for a maturing civilization the basic energy vital to its survival".

— Gerard K. O'Neill, The High Frontier (1976)

The Vision of Space Commerce

Throughout human history, exploration and commercial interest have gone hand-in-hand. From Magellan's circumnavigation of the Earth to Zheng He's oceanic expeditions, one common theme in past exploration missions was the objective of extending the economic sphere of the exploring entity by discovering new resources and establishing new markets. Commerce and trade have been influential not only in the economic realm, but also in the cultural one, by helping to diffuse ideas, cultures and practices.

Exploration of space and utilization of space-based assets are also based on a strong economic rationale. Today, the economic footprint of our civilization reaches beyond Earth's orbit and extends all the way to GEO. Tomorrow, it is very likely that this economic frontier will reach the Moon and other celestial bodies in the solar system. Achieving this expansion will no doubt require technical expertise. However, it will also require mastery of disciplines such as business, management, and economics as they relate to space activities.

Economics and business studies are branches of social sciences, and their main topic of interest is allocating scarce resources for the production of goods and services. These resources include labor, intellectual capital, financial capital, land and natural resources. These "factors of production" are used to transform raw materials into goods and services. These outputs include not only everyday consumption items, but also services such as national defense, education and healthcare.

Space projects can provide many economic benefits and contribute to the production of goods and services on Earth. In fact, we can argue that this production function has already extended beyond Earth, such as in the commercial utilization of space stations. In earlier chapters we have covered a myriad of space applications and emerging markets such as space tourism. These are only some examples of goods and services enabled by our space-based capabilities.

Here we explore the current status of the space industry, various underlying trends, emerging opportunities and the changing nature of the industry. The factors driving this change include a growing space workforce, space entrepreneurship and new collaboration models across the world.

Setting the Stage

The space industry has a number of dimensions and characteristics that makes its imprint, dynamics, and geographic and strategic scope different from those of many other industries.

Current Outlook

If the space industry were a jet airplane, it would have two engines thrusting it forward: commercial services and government spending. Even though reliable industry statistics are a scarce commodity in the space industry, initiatives such as the Global Forum on Space Economics of the OECD, various reports published by the Satellite Industry Association, "Industry Facts & Figures" published by Eurospace and "The Space Report" published by the Space Foundation are providing us with very useful metrics. We should note that the data compiled by these different sources are not always in agreement and a good dose of common sense is required to interpret these metrics.

It is estimated that the global footprint of the public sector space budgets is in the range of US$48 billion to US$77 billion per year, covering both civilian and military uses of space[1]. The budgets of NASA and other space-related U.S. departments and agencies easily dwarf the amounts spent by other governments. OECD reports that, for every dollar spent within the *combined* public budgets of major space faring nations, the U.S. spends three dollars[2].

The estimates vary even more drastically when it comes to the revenue estimates of the commercial sector. The combined annual revenues from the three main types of satellite services (telecommunications, Earth observation and navigation) as well as other space products and services, such as satellite manufacturing, launch services, ground equipment manufacturing and various support services, are estimated to be in the range of US$130 billion to US$173 billion[3].

Therefore, a combined estimate of public space budgets and private sector revenues ranges between US$178 billion to US$250 billion, giving us a glimpse of the direct economic value of space activities for any given year. Indirect benefits, such as technology spin-offs, as well as the book value of space infrastructure, such as launch pads, are not included in this estimate.

Satellite applications constitute the bulk of the commercial products and services in the space industry, and about 70% of the aggregate revenues come from a single market segment, satellite telecommunications. The satellite navigation segment has been growing at a very rapid pace in recent years, as navigation equipment and related services are becoming essentials of our modern economy. Together, direct-to-home broadcast and navigation segments constitute the dynamic duo of the space industry, with high revenue growth rates. Earth observation (EO), a relatively mature segment of the space industry, accounts for about 6% of the total revenues.

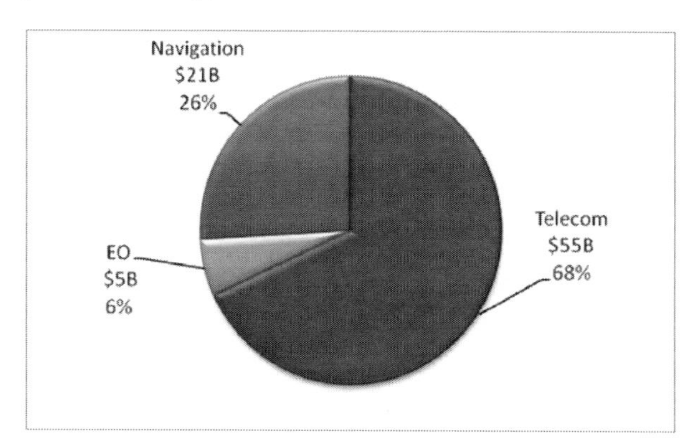

Figure 11.1. Distribution of satellite services revenues (in US$ billions) between the three main types of satellite applications in 2005 (Source: OECD [4]).

Another key indicator of the global space industry is the workforce of space professionals. It is estimated that the global workforce pool of the space sector is composed of around one million individuals. The combined space workforce of Russia (300,000), China (260,000) and the U.S.A. (120,000) account for about 70% of the total workforce[5].

Evolution of the Space Industry

Given that the commercial space activities are dominated by satellite communications today, it is no surprise that the first truly commercial space launch placed an experimental telecommunications satellite, Telstar, into orbit in July 1962. More than 45 years later, today, Earth's orbit is bustling with activity with over 850[6] operational satellites serving both civilian and defense sectors. In fact, contrary to the image of space as an infinitely vast ocean, there is a real estate crunch above our heads: it's getting more and more challenging to acquire orbital rights and frequency allocations due to the increasing demand for various satellite services. The estimated annual value of a slot in Geosynchronous (or Clarke) Orbit (GEO) can run into millions of dollars for the more desirable locations.

The private sector is playing an increasingly important role in the space industry. Given the dominance of governmental and military budgets in the early stages of the space age, this is a relatively new phenomenon. During the last two decades, a major trend in space activities has been increasing levels of commercialization and decreasing levels of public funding. In the early to mid 2000s commercial space activity leveled off and governmental spending increased, but now commercial space activities are beginning to rebound. Two developments, namely the bursting of the dot-com bubble in 2000 and the attacks of 11 September 2001, interrupted the longer-term trend of the growth of private enterprise in space. This results from several factors including the technical maturity of space technologies, new market opportunities created by changed geopolitical circumstances, the overall impact of globalization, and supporting legislation to encourage private initiatives in space, particularly in the U.S.A.

During the late 1990s, the excessive investment in information technologies spilled over to commercial space, and many new, ambitious and technically risky projects were funded. These included projected mobile satellite systems like Iridium, ICO, Globalstar, and Orbcomm, and broadband systems such as Astrolink and Teledesic. These projects that anticipated the manufacture and launch of hundreds of satellites triggered an unprecedented surge in projected demand. However, the stock market valuations of these ventures collapsed, and the interest in new commercial space ventures declined. One reliable metric that covers this period is the number of commercial launches. Although this metric cannot capture the economic value of each launch, it can nevertheless act as a barometer measuring the health of commercial space. As can be seen in Figure 11.2, the number of commercial launches peaked in 1998 with 41 launches. The early 2000s were a period of consolidation and, after the initial shock of 2001 with only 16 commercial launches the numbers seem to be increasing again. The fact that the launchers now often carry heavier and heavier payloads (up to around 22,000 pounds, or 10,000 kilograms) and are lifting two or more satellites into orbit at the same time helps to put this graph in better context.

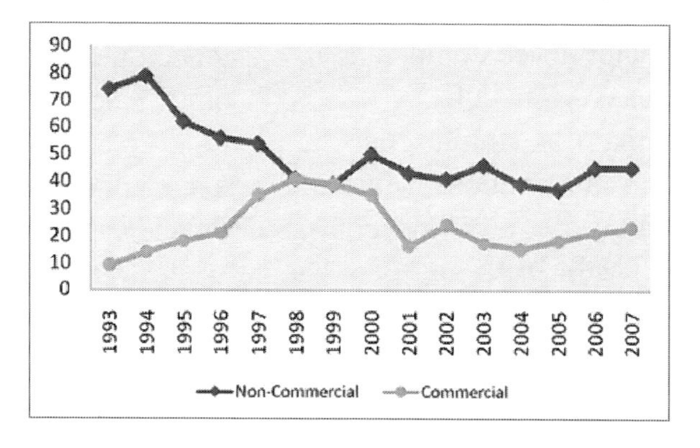

Figure 11.2. Commercial and non-commercial launch events between 1993 and 2007 (Data Source: FAA [7]).

While the commercial space market was going through its ups and downs, military space expenditures, especially in the United States, went into overdrive following the attacks of 11 September 2001 (see Figure 11.3). Although the space budgets of civilian space agencies have been largely stable, the overall government expenditures are on the rise, especially in the United States. It is conceivable that this increase can be perceived as a threat by other nations and potentially could trigger a new arms race, but it can also have a positive impact. This is because most space technologies can be used for both civilian and military purposes; this recent wave of investment in military space may eventually result in many civilian applications[8].

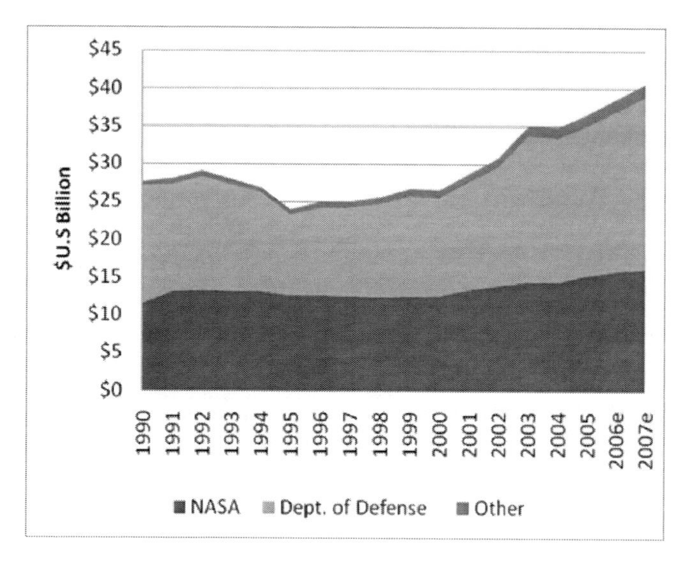

Figure 11.3. U.S. Government space expenditures between 1990 and 2007 (in US$ billions) (Note: 2006 and 2007 data are estimates; Data Source: NASA Aeronautics and Space Report of the President, Fiscal Year 2005).

Faced with the trend of increasing commercialization, the space industry embraced a more global and collaborative approach that triggered a series of consolidations, mergers and strategic alliances. In the U.S.A., out of the twenty major space companies in existence during the 1980s, only three 'prime' ones were left by 1997 (Boeing, Lockheed Martin and Northrop Grumman). A similar consolidation took place in Europe in the 1990s, leading to the creation of two major space conglomerates that are currently operating at the prime contractor level (EADS Astrium[9] and Thales Alenia Space). In Japan, there has not been so much consolidation due to the division of roles between various contractors in terms of subsystem design and manufacture. In China, virtually all space projects are carried out by CASC (the Chinese Aerospace Corporation) and, in Russia, industry has either consolidated or moved to partnership with aerospace corporations from other countries. In other countries with space industries, such as India, the Republic of Korea, Canada and Australia, various moves toward consolidation and global alliances have also occurred.

Although such strategic alliances decrease the number of players in the industry (and arguably decrease the level of competition as well), they also bring a number of benefits, including the creation of end-to-end capabilities, larger internal R&D divisions and broader access to technologies. Moreover, due to the increased financial strength of such conglomerates, they have better access to new capital and risk sharing mechanisms.

What Sets the Space Market Apart?

The space market is notably different from other sectors of economic activity. Some of these differences result from the challenging technical requirements demanded by this very harsh environment. Some are related to the origins of the space industry and the strong military rationale that remain an important part of global space activities. There are three main aspects that set the space market apart.

Meet the Number One Customer: The Government

Up until the 1990s, against the backdrop of the Cold War, public space markets were shaped by the Space Race as the U.S.A., the former U.S.S.R. and Europe managed their space programs as an extension of their political ambitions. Governments invested in launch vehicles in the 1950s, human spaceflight in the 1960s and space stations in the 1970s. These commitments were all perceived as national priorities closely related to "defense". The role of the private sector was limited to fulfilling the requirements set by public sector clients (with very high level military involvement on both sides of the Atlantic).

Thus, from the very early days of the space age, governments played a very active role, along with the military, in shaping the space industry. Faced with very significant technical challenges and fuelled by national pride, enormous amounts of taxpayer dollars were spent on space programs. At the height of the Apollo program, the U.S.A. spent nearly 0.8% of its Gross Domestic Product (GDP) on financing NASA's space activities[10]. However, compared to the Apollo era, today's government space budgets are more modest. Today, the U.S. government's total space budget, including both civilian and defense spending, corresponds to about 0.3% of the U.S. GDP, as reported by the OECD.

The motivations of private sector entities for investing in space activities are fundamentally different from the motivations of the public sector. For instance, if a company invests in a satellite system capable of covering a wide geographical area, it is natural for it to explore this whole region as a potential market for the products and services generated by this satellite, irrespective of national borders. Compared to the purely commercial motivation of the private sector, a government entity investing in a similar system may be motivated by building national industrial capability or enabling international cooperation through a similar satellite system.

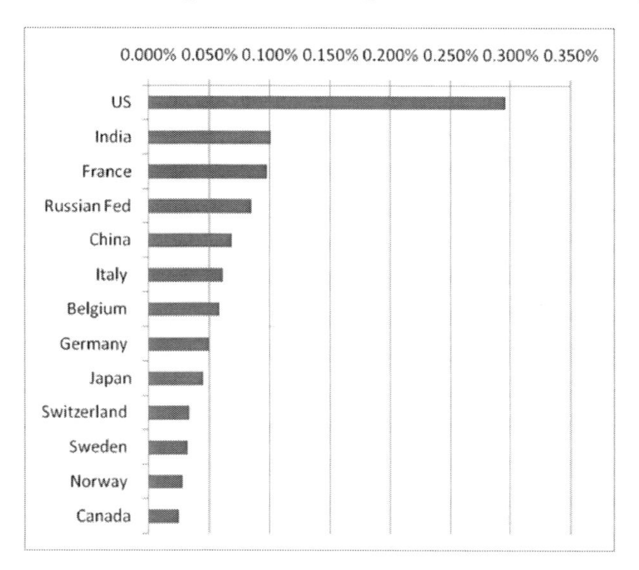

Figure 11.4. Public space budgets as a percentage of Gross Domestic Product for various countries (Data Source: OECD).

When the motivations of the private and the public sector intersect, there is a potential for cooperation in the form of a "public-private partnership", or PPP. The core concept of the PPP model is to involve the public sector in the early stages of a project as an investor and the private sector as the designer and manufacturer. During the operational phase, the project will then be exploited by the private companies that, in theory, will reimburse the initial public investments (e.g., via royalties, tax payments, or free access to satellite data for government entities). Although the PPP model can be very useful in sharing both the risk and benefits of space projects, as evidenced by the Galileo satellite navigation program or the Canadian Radarsat program, it can result in complications, such as a smaller return on investment for the government than was hoped for, as well as delays.

Government space budgets are still very substantial, as are the number of regulations that aim to control the diffusion of sensitive technologies. Given the scale and complexity of operations and the hefty price tag for reaching orbit, apart from commercial satellite telecommunications and the booming market for satellite navigation, most of the space business takes place between governments and businesses. In fact, it's not an exaggeration to say that the number one client of space products and services today is still the government. Although government budget allocations can translate into more stability for the space market and protect it against the whims of consumer spending patterns, they also limit the dynamism of space business. However, the new generation of space entrepreneurs is determined to bring more dynamism and to attract more private capital into the space industry.

The Double-edged Sword: The "Dual use" of Satellite Technology and Systems for Commercial and Military Purposes

The second special characteristic of space markets is the "dual use" of the technology for both military and civilian markets. Many building blocks of the space industry, such as launch vehicles and satellite technologies, have military origins and they continue to be used extensively for defense related purposes. The dual use of space technologies and applications is in the genes, or "DNA", of the space market. Just like any other technology, space technologies can be used for both civilian and military purposes. For companies involved in both commercial and defense space markets, this characteristic can provide a strong advantage, enabling some of the key technologies and capabilities gained through defense contracts to be commercialized into civilian markets[11]. However, dual use can also cause a lot of headaches for the industry and restrict many new commercial initiatives due to regulations such as the U.S. "International Traffic in Arms Regulations" (ITAR). The dreaded ITAR clearance process that is cumbersome and protracted in nature can be a showstopper for increased commercialization, especially for international contracts.

Investment Horizon

The very long lead-times associated with most space products, services and markets is the third key factor that sets space markets apart. The technical complexity of space projects generally results in an extended period between project kick-offs and the commencement of operations. Although some of this delay is inevitable, such as transit times to Mars, a large portion of it is driven by the complexity of space projects. The Iridium, Globalstar and ICO mobile satellite ventures, for instance, were victims of long lead times and changing market conditions. When these various systems were first conceived, the global market for wireless services looked and acted one way. By the time these space communications systems had been designed, built and deployed their terrestrial competition had evolved to provide a service that was technically different, more robust and available at lower cost. Managing the complexity associated with space markets requires the combination of several bodies of knowledge and maintaining a highly skilled labor force throughout a project's lifetime. As a result, achieving economic returns from space projects tends to take longer than in many other industries. Although that could be acceptable from a government's perspective, most private sector players are interested in shorter financing horizons.

The Space "Value Chain"

A value chain is a useful conceptual tool to understand the scale of operations in the space market as well as the link between major actors. In economics, "value-added" refers to the additional value created at each stage of production, as raw materials are transformed into finished goods and services through factors of production (e.g., labor and capital).

Consider the following example that highlights some of the characteristics of the space value chain. A direct-to-home TV broadcast subscriber is an end user of satellite telecommunications. The benefit that this end user derives from this service is entertainment and access to information, and this benefit is enabled, to a large extent, by the signals transmitted by a satellite in GEO. TV content providers lease satellite capacity from fixed satellite operators who ensure that the satellite is functioning in an optimal fashion, maintaining a high quality of service while maximizing operational lifetime. In turn, the satellite operators are dependent on the reliable services of launch service providers (who are in charge of placing the satellite in its planned orbit), ground equipment manufacturers (who provide the capability to operate the satellite), and satellite manufacturers who originally built the satellite. Thus, a vast space "value chain" works "behind the curtains" to ensure high-quality service when the subscriber tunes into the evening news with the flip of a switch.

In a similar fashion, all space-related goods and services follow a "value chain" on their way from blueprints to clean rooms to the launch pad. Therefore, when we estimate the size of the space industry, we have to include the economic value added of each chain as it contributes to the final value of goods and services produced.

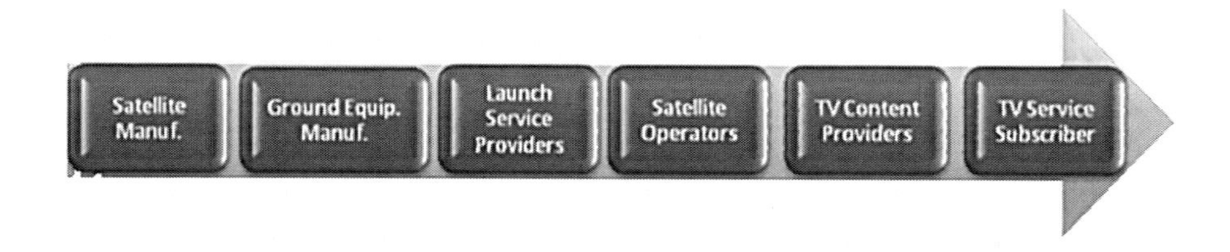

Figure 11.5. The value chain of direct-to-home broadcast services

(Courtesy of Turquoise Technology Solutions Inc.).

Measuring the Economic Impacts of Space Programs

Based on a purely economic perspective, the space industry is a relatively small sector of economic activity. Although the global size of the overall space industry, at US$178 to US$250 billion, is impressive, it corresponds to about 0.3-0.4% of the Global World Product of US$65 trillion as tallied for 2007.

However, given the strategic importance of space, the economists have for long devoted considerable attention to measuring the economic impacts of space programs. Their concern is attributable to a growing awareness of the economic significance of technological change, especially from complex and risky projects generally supported by public funding, as in the case of space projects. The economic effects of space programs can be broken down into three main components, a direct industrial effect, an indirect industrial effect and social effects.

Space products and services can be perceived as direct industrial effects. Since the 1950s, space programs have generated a whole range of space products and services including space hardware, software and a multitude of space-enabled services. These outputs would not have existed without the budgets allocated by space agencies and the wave of private sector investment that peaked at the turn of the millennium. These products and services also constitute our main statistical indicators and helped us to determine the magnitude of the space market, as discussed above.

In addition to these products and services that are directly attributable to space programs, an indirect industrial effect of these programs is the so-called *spin-off*. Most space projects require the integration of new bodies of knowledge as well as new methods of management. From these sources of novelty, contracting companies participating in space projects have generated new ideas, new products, opened up new markets, and learned new organizational methods as their production teams gained more expertise. The process then expands outwards, spreading first to other parts of the contracting companies themselves and subsequently throughout the society itself. As an example, composite materials developed for the European Ariane launcher have been successfully transferred to design the braking systems of high-speed trains. Additional examples are shown in Table 11.1.

Spin-offs represent all those indirect economic effects that were not anticipated and scheduled in the contracts between space agencies and the industry. Measuring spin-offs is difficult, but specific methods based on direct interviews with the industry tend to show that the benefits accrued from spin-offs is about three times the original government investments in the form of space industry contracts[12].

Space Program Technology and Commercial "Spin-offs"

Product	Space Origin
Tumor tomography	NASA scanner for testing
Battery powered surgical instruments	Apollo Moon program
Non-reflective coating on personal computer screens	Gemini spacecraft window coating
Emergency blankets (survival/anti-shock)	Satellite thermal insulation
Mammogram screening, plant photon-counting technology	Space telescope instruments
Skin cancer detection	ROSAT X-ray detection
Dental orthodontic spring	Space shape memory alloys
Early detection of cancerous cells	Microwave spectroscopy
Carbon composite car brakes	Solid rocket engine nozzles
Car assembly robots	Space robotics
Flameproof textiles, railway scheduling, fuel tank insulation	Various Ariane components, including software
Lightweight car frames, computer game controllers, fuel cell vehicles, coatings for clearer plastics, heart assist pump, non-skid road paint	Various Space Shuttle components
Fresh water systems	ISS technology
Corrosion free coating for statues	Launch pad protective coating
Flexible ski boots, light allergy protection, firefighter suits, golf shoes with inner liner	Various space suit designs
Healthy snacks	Space food

Table 11.1. Some examples of products arising from spin-offs from space programs.

The third significant effect of space programs is the social effect: this includes all the societal benefits that result from the existence of space products and services. For instance, space meteorological systems have significantly improved meteorological forecasting, which in turn has induced benefit in a considerable number of economic sectors (tourism, maritime operations, airlines, insurance, etc.). All the space programs dedicated to satellite applications have thus created a societal impact with associated economic benefits. For measuring these benefits, economists generally use cost-benefit analysis, a well-known method that identifies the net benefit for each well-identified market. A rough order of magnitude estimate indicates that the amount of social benefits from space projects for the decade 1995-2005 is far above US$500 billion.[13]

Instead of measuring each component individually, economists have also tried to capture "in one shot" all the *macroeconomic* effects of the space programs by using econometric models. For more background, see the more detailed explanation in the endnote.[14] These models, designed to identify and measure the portion of economic growth attributable to space activities, were particularly popular in the 1970s and 1980s. For instance, the Midwest Research Institute (MRI) study of the relationship between NASA R&D expenditures and technology-induced increases in Gross National Product (GNP) indicated that each dollar spent by NASA on R&D returns an average of slightly over *seven* dollars in GNP over an eighteen-year period following the expenditure. Assuming that NASA's R&D expenditures produce the same economic payoff as the average R&D expenditure, MRI concluded that the US$25 billion (in 1958 dollars) spent on civilian space R&D during the 1959-69 period returned US$52 billion through 1970 and continued to generate benefits through 1987, for a total benefit of US$181 billion. There is some danger in looking at such returns on an averaged basis. Some R&D activities have been projected to have returned twenty times their original public investment while other activities have shown much more modest returns. Further, some activities of a "public good" nature, such as preventing climate change or protecting the planet from cosmic destruction by a comet or meteorite, may not have a pay off in the commercial market but could prove to have an almost infinite value based on their societal returns.

Figure 11.6. The LZR Racer swimsuit of Speedo [15], a recent example of a spin-off from a NASA research program (Courtesy of NASA).

All of the above studies converge in proving that space activities have positively impacted the whole economy through the creation of new industrial activities, the generation of successful transfers of technology and spin-offs, and the formation of important social and environmental effects. However, through time, the nature and importance of the economic effects may change. For example, if spin-offs were an important argument at the beginning of the space era to the mid-1980s, it seems that, as time passes, there is less emphasis on spin-offs, and more on "spin-ins"; this is the adoption of technologies developed within various "high tech" terrestrial sectors by the space sector. This trend is a result of several factors including (i) the relatively limited circulation of technology in the space industry due to restrictions such as ITAR, (ii) the slower emergence of global players due to the limitations posed by national security priorities, and (iii) the longer lead times associated with designing, manufacturing and launching space assets that can severely limit the incorporation of recent innovations.

With the emergence of the knowledge-based economy, new economic benefits can be advocated for supporting space projects. Space is, for instance, a key vector of *globalization*. It is an ideal vector for deregulating and globalizing telecommunications and broadcasting. It has the capacity to be both local and worldwide, and offers large spectral coverage and virtually real time services. Also, space applications have an inherently *cohesive* impact on society. This is because space systems in Earth's orbit are largely independent of terrestrial infrastructures, provide very wide accessibility, and can be cost effective with high, medium or sparse population densities. Since satellite services can be accessible to large geographical areas at relatively low cost, their development can often be very beneficial to the peripheral and less developed zones of the globe. Finally, when considering the main issues facing humanity at the beginning of this new century[16], such as climate change and clean energy systems, space appears to be always involved as a key complement that adds value to critical problem-solving activities. However, such essential aspects of space are rarely measured, or even recognized. Thus, economists have avenues of research to accomplish in this domain.

Cost Engineering in the Space Sector

The hefty price tags of projects have become a defining characteristic of the space industry. Although stringent performance requirements are an important driver of cost, the one-of-a-kind nature of most space projects lies at the core of the problem. Estimating the cost of a project without precedence is certainly not an easy endeavor. In most other industrial sectors, some form of extrapolation or benchmarking is possible, even for projects that incorporate many novel processes and materials. The inability to use extrapolation from past projects as a reliable cost estimation method is still a major issue in space programs. Especially in the early days of the space age, during the first human spaceflight missions, cost estimation was a recurrent problem. Right from the start, cost overruns marred space programs, with both the Mercury and Apollo programs running over budget. A direct consequence of inaccurate cost estimates has been the increasingly strong criticism of cost overruns, as the final costs of the programs often exceeded the original estimates by a large factor. Although cost overruns are still observed in the industry, as evidenced by the Space Shuttle and International Space Station programs, significant progress has been made to develop reliable cost estimation methods and cost control practices.

By definition, a cost overrun implies deviation from an original estimate. Therefore developing accurate estimates at the beginning of a project's lifetime is one of the biggest cost engineering challenges. Moreover, changes imposed on the project definition in early project phases can have major consequences on the actual cost. There are four main types of cost estimation methods: a) Top Down Methods: i) cost by comparison, ii) cost by analogy, and iii) parametric costing, and b) Bottom Up Methods: i) grassroots costing.

Cost by comparison is the most risky estimation method from the perspective of the client. Without estimating the cost in advance, the client can issue a tender and then compare the cost of the received offers. Although some variations have been introduced, such as eliminating the lowest offer, the risk of cost overruns is still significant due to misinterpretation of the requirements by all bidders, or price fixing between competitors.

Cost by analogy may be applicable in sectors that have a more repetitive character with high similarity in performance characteristics. This methodology requires the involvement of costing experts with significant experience who can compare the cost of a new project to analogous ones in the past. In areas where a space industry has existed for decades, such as satellite telecommunications, this can be a very useful tool. In many aspects of the space industry, such an approach is often not viable due to differing technical and performance characteristics. For instance, we can use the Apollo program cost data to estimate the cost of a human exploration mission to Mars, but the difference in performance requirements, life support systems, and mission duration can skew the results significantly.

Grassroots costing, a commonly used method in the construction industry requires a very high degree of precision of the final design. For most space contracts, this is simply not possible in the early phases of development. In the space industry, the client and the contractor typically work hand in hand to complete the feasibility and design phases, and only after this stage is the design frozen. Also, contrary to other sectors such as the construction industry, unit costs have higher levels of uncertainty due to the uniqueness of each space project in terms of propulsion and fuels, unique materials, dissimilar thermal conditions, etc.

This leaves parametric costing as one of the most used tools for cost estimation in the space sector. Parametric costing is defined as a technique employing cost estimating relationships (CERs) to estimate costs associated with the development, manufacture, or modification of a project. A CER defines a quantifiable correlation between certain system costs and other system variables, either of a cost or technical nature (e.g., the mass of a subsystem versus its cost)[17].

Space Marketing and Public Outreach

Since the 1950s, marketing principles have been successfully employed for consumer products and services in many different sectors. However, due to some of the characteristics of the space industry examined above (such as the dominant role of the government and security related secrecy issues) during the Cold War, marketing principles were, by and large, absent from the space sector. Increased commercialization in the 1990s and competition for capital and talent with the other sectors (such as information technology (IT) and biotechnology) resulted in the introduction of marketing in a systematic way into the space sector.

The Concept of the 4P's of Marketing

In classical terms, a marketing strategy is composed of four main elements, known as the 4Ps. These stand for Product, Price, Promotion and Physical Distribution. The idea behind this simple classification is the ability to manage all four aspects in order to create successful marketing strategies. In other words, any project or service has to meet the quality and functionality expectations of the market (Product) at a fair, market-driven and competitive price (Price). Moreover, the customer must be made aware of the availability and features of this offering (Promotion) and the product must be brought, via appropriate channels, to the final user (Physical Distribution).

Depending on the nature of the project or service, more emphasis will be put on one of these elements. For example, for marketing automobiles, product characteristics such as power, acceleration, and fuel consumption are likely to be emphasized; on the other hand for marketing cosmetic products, promotion is more likely to be the main focus.

Therefore one of the most important objectives of a marketing strategy is to determine the ideal "Marketing Mix", thereby ensuring that all four elements are managed in a balanced manner, depending on the nature of the project or the service.

The Fifth Element in Marketing

When it comes to applying these principles to the way we manage space activities, we need to consider a fifth "P"; this is the philosophical element. The discipline of anthropology documents many historical and current perspectives that consider space as the "Final Frontier" to be conquered by humankind. Throughout history, when confronted by a frontier, our species - and rather uniquely so - seems to be driven by a desire to breach it. When we were confronted with the vast seas, we built more powerful ships. When we yearned to fly, we invented balloons and airplanes. In a very similar fashion, the challenges and mysteries posed by the space environment have once again fueled the human desire to reach out and explore.

This rationale has always been a driving force for space exploration and, if managed properly, it can unlock a high level of public support for space activities. In addition, the political climate at the time also plays a key role in shaping space activities. Just as international competition and the so-called "cold war" were the big motivating factors of the 1960s, today's space activities are driven by other priorities. For instance, recent concerns on climate change and the potential for space-driven solutions has increased the need for the space sector to attract more public attention, and to develop more structured marketing approaches.

Space professionals often assume that the general population usually views space exploration very positively. This assumption is not supported by structured surveys. Surveys made in Europe have shown that less than 50% of the Europeans find space important. Recent studies in the USA reveal a similar picture. Hence, there is an acute need for space professionals to "market" space activities better. They need to understand the components of an effective marketing mix and to communicate the benefits of space to the general public more effectively.

Strategies for the Improved Marketing of Space

Some of the past achievements of space exploration have certainly garnered high public interest, one of the highlights of the 20th century being the landing of Apollo 11 on the Moon. It is estimated that 538 million people watched the grainy image of Neil Armstrong's first steps on the lunar surface in July 1969 on TV, the first truly global satellite broadcast and the largest television audience ever to that point in history. This unprecedented diffusion was enabled by the Intelsat global satellite network with coverage of all three ocean regions of the world (Atlantic, Pacific and Indian Ocean areas) that was only completed just a little over a week before the lunar touchdown.

Given the significance of the event and the proliferation of telecommunications technologies in the meantime, it is likely that more than two billion people would tune in if something similar were to happen today. Similarly, in the early days of the Internet, NASA's Pathfinder mission sparked worldwide interest and various Pathfinder websites attracted more than 500 million web hits within a month. Today's space websites are even more popular. Besides these spectacular events, spin-off products and services that constitute a significant portion of space benefits for our society deserve to be better known by the general public.

Pricing Space Projects

Space projects suffer from an unfortunate reputation for being very expensive and prone to large cost overruns. There is indeed a basis for this reputation: space projects have to operate in very harsh and hostile environments, both on the ground as well as in space. Furthermore, barring very few exceptions such as the Iridium, Globalstar and larger constellation satellites, mass production practices are not applicable, and generally a high percentage of the costs are associated with developing a prototype and only a few production units. This is changing as commercial space satellite manufacturers are now developing basic platforms; this is similar to the way that automobile manufacturers streamline their production methods. Clearly satellite manufacturers that deal with production runs in the dozens are a long way away from automobile production runs that may run to millions.

Nevertheless, it would be unfair to declare space activities as being a waste of taxpayers' money. We should always consider the relative cost of space activities in the light of the associated benefits, discussed earlier. Consider the following cost comparisons. Annually, the ISS project costs the U.S. taxpayer the equivalent of one movie ticket, while the cost of European space programs is in the range of only three eurocents per day for each European taxpayer.

Furthermore, as discussed in the section on cost engineering, many methods have been developed to manage the cost of space projects and to reduce cost overruns. These methods have gained considerable successes recently and, if used properly, they can help manage costs much more effectively than was done in the past.

Physical Distribution of Space Related Results

Traditionally, a culture of secrecy was prominent in the space industry that limited the sharing of information and know-how, as well as the diffusion of best practices. Luckily this has changed and a more open approach has emerged. In particular, space agencies have realized that the taxpayer needs to be better informed about the benefits of space programs. This realization resulted in more interactive websites, targeted outreach efforts and the use of modern multimedia tools for establishing better communications with the general public. Figure 11.7 illustrates such an approach adopted by NASA for communicating the benefits of spin-off products and services. Other examples are the NASA TV channel that allows the general public to follow particular events, such as Space Shuttle launches and important ISS related operations. Specialized cable television channels such as Discovery, CNN, National Geographic Television and the History Channel also produce a number of shows with space themes, many of which are distributed internationally. European, as well as Japanese, Chinese, Canadian, and Australian, TV networks also have many productions with space-related shows. These have proven particularly effective when astronauts from various countries are in space, or in special instances such as when a Japanese network (NTT) sends a high definition camera into space.

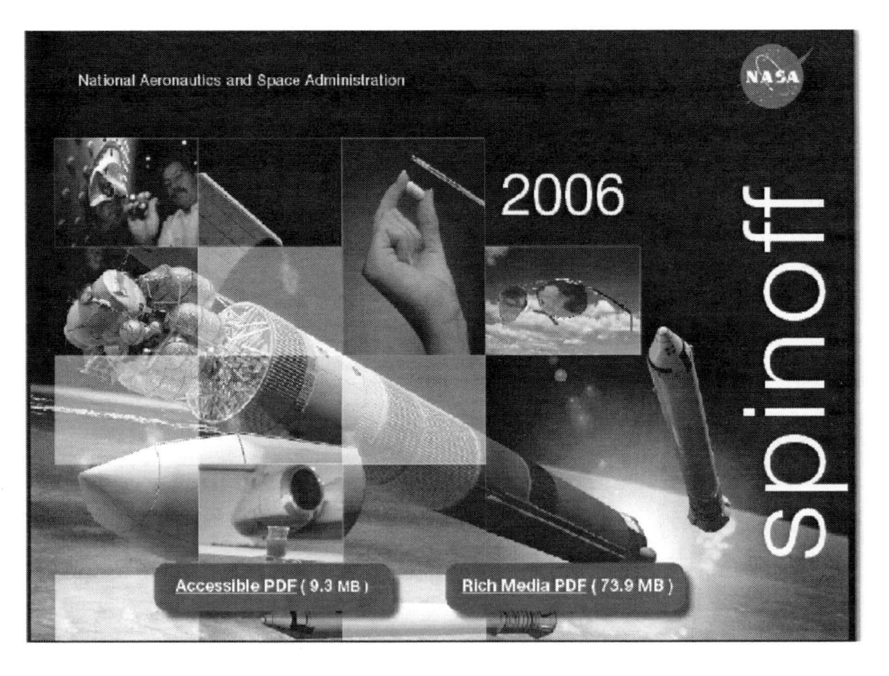

Fig.11.7. A NASA image showing spin-off results (Courtesy of NASA).

Space Promotion

These examples illustrate the growing awareness of space agencies about strategic communication. In particular the shift from printed material to multimedia and online content offers a range of options to capture the imagination of the general public and build support for space programs. Indeed the combination of appealing images (such as images from the surface of Mars, astronaut spacewalks or artist impressions of scientific missions) and a sustained interest in space exploration opens up many possibilities.

Space agencies also work with the media and film producers to reach a broader public. A good example of this was the Apollo 13 movie, where NASA gave advice to the filmmakers; an experienced astronaut was supporting the different scenes. In addition, space is often used as a theme in advertising, in some cases with the support of space agencies, as shown in Figure 11.8.

Figure 11.8. A space agency logo featured in commercial publicity (Image courtesy of Parfums Givenchy).

Following the end of the "space race", and confronted with new competitors for public funding and stagnant space budgets, the space sector finally realized the importance of a favorable image and learned not to take the public's support for granted. In the meantime, marketing has also become a socially accepted tool within the non-profit sector. The classical marketing mix methods are becoming an integral part of the outreach and communication efforts of the space agencies.

The philosophical foundations of the public support for and interest in space exploration are still strong, and our curiosity of the unknown and the desire to explore are natural allies of the space programs. However, the space agencies must always make a concerted effort to sustain this interest, help educate tomorrow's space workforce and convince taxpayers that space programs are wise investments.

Partnerships in Space:

Lessons Learned and Opportunities for the Future

The space industry can be characterized as a strategic industry based on both its economic importance and its key role in security. Still at the dawn of its commercialization, this industry both inspires the public and provides valuable services to businesses, households, and governments. Yet it is also one of the most secretive areas due to its national security significance.

The space industry participants in general and the space "integrators" (who synthesize various stages of manufacturing and services towards a final space product), in particular, are largely government-dependent entities. These industries are generally geographically confined. This is for a variety of reasons that include security, the need for highly specialized and costly testing and manufacturing facilities, and the recruiting demands associated with a highly skilled labor force. The first major multi-national aerospace integrator is the European EADS, but the trend towards global value chains that are based on distributed factors of production is here to stay.

Transnational alliances include:

- Alcatel (France), Loral (U.S.) and NPO-PM (Russia)
- Starsem: Aerospatiale and Arianespace (France) with RAKA and Progress (Russia)
- OHB (Germany) with Fiat-Avio (Italy) and Yuzhnoe (Ukraine).
- SES Global (ASTRA of Europe, New Skies, SES Americom, and AsiaSat)
- EurasSpace: a joint venture between EADS/Astrium and the China Aerospace Corporation (CASC).

Two distinct, but parallel, processes have taken place since the end of the Cold War. The first has to do with the consolidation of European and U.S. space capabilities through mega-mergers resulting in major space integrators. The second relates to marketing and, partnerships between U.S. and European space firms and their Russian and Ukrainian counterparts. The advantage of the first is that it is presumed to facilitate the commercialization of space, while safeguarding national interests. The drawback is that there could well be an incentives problem, whereby national champions may not have very strong incentives to be highly efficient locomotives of commercialization. Increasingly, commercial entities that are not government-controlled are making a mark in space; their success could well reshape the industrial landscape in space and rewrite certain regulations and policies across the world. When we look at how private and public sectors have transformed their roles with regards to space activities and how they are likely to proceed, the formation of further transatlantic partnerships driven by commercial pressures seems quite likely, and could result in major industrial restructuring within the next decade.

From a commercial perspective, the most mature segment of the space industry is satellite telecommunications, which is shaped by the traditional dynamics of consumer demand and multiple service providers. In contrast, less commercialized segments such as navigation and Earth observation are still heavily supported by the public sector. The cost of this support relates not only to the resources reallocated by the governments, but also to the closed nature of the markets and the formation of a space-industrial complex.

This has clear benefits in terms of safeguarding the technologies and enabling the scale necessary to develop new programs, but negatively affects new entrants and inhibits a competition-enhancing environment and, in the long run, it may have a negative impact on innovation. Examples are numerous, but a recent one is the case of the European Galileo program, where the public sector started with the intention of establishing competition for the choice of the contractor, and ended up supporting the merger of the two remaining competing consortia. In the U.S., similar problems have surfaced, as agencies find it increasingly difficult to follow competitive processes when their pool of contractors is declining.

Thus the remedy of increasing the number of national competitors by expanding the base of qualified contractors and allowing overseas competitors has its limitations. Not only are there political issues of moving jobs overseas, there are also security considerations and technology transfer issues associated with many of the government-led programs. The bottom line is that some form of government control over the resulting industrial partnerships and processes will almost undoubtedly be present.

In general, the public sector in most space-faring countries tends to favor the involvement of the private sector in space programs. However, the public policy on space commercialization cannot be taken for granted; it depends not only on the space programs, but also on "policy" factors, with the most prominent one being security concerns. An extreme case of this was observed after the collapse of the U.S.S.R., when ex-Soviet countries started to market their space infrastructure to international clients. This was only possible with the relaxation of security constraints imposed by the Cold War. Thus, there is a constantly changing dynamic balance between perceptions of security and commercial dynamics driven by economic realities.

Changing Patterns of Space Partnership — Partnerships in Space through Time: The Dawn of Commercialization?

There are two particularly interesting examples illustrating the scale and pace of change in the global space industry. One is the former International Telecommunications Satellite Organization (now known as Intelsat Ltd., a corporation licensed in Bermuda). The other is the former International Mobile Satellite Organization (now known as Inmarsat Ltd). Less than a decade ago, both of these global satellite operators were characterized as natural monopolies in terms of their efficient scale of operations. In addition, most of their members were government-owned entities. These national monopoly entities were seen as safeguarding national and security interests in the sensitive area of global telecommunications. Today, both former global monopolies are multinational companies, capable of buying and selling other corporations, assuming major corporate debt, and involved in aggressive corporate maneuvers. These organizations are owned by equity investment organizations and pay corporate taxes. Eutelsat, the European Telecommunications Satellite Organization, made a similar transition from a regional monopoly to a private commercial enterprise.

Much of this probably sounds like a case study from a history book to the younger generations, growing up in an era of explosive choice when it comes to telecommunication services. A new equilibrium between security considerations and commercialization has emerged, with private satellite operators and service providers capitalizing on new market segments and new commercial opportunities.

Another example comes from the Earth observation sector where once all satellite networks were owned and operated by governments, only a couple of decades back. Today the resolution of commercial images is sufficiently high to identify individual buses and trucks on highways, a capability that is light-years ahead of remote sensing systems of the past. There is even talk of (and some steps towards) a commercial satellite navigation system. New space markets are created, just as in any other sector, when product variations are introduced that can profitably segment the market, in particular in the presence of public goods (such as television broadcast and navigation services).

However, this is a necessary, but not sufficient, condition for commercialization, as the reactions to the development of the new Galileo satellite navigation system revealed both within and outside Europe. Who controls access to the services of these strategic assets in the event of a clash of interests between security and financial returns from operations?

Whoever it is had better take note of the fact that security is more likely to take precedence over profits.

The different objectives of stakeholders often result in compromises expressed in the form of partnerships. Historically, partnerships between governments have been used as a means of enhancing scientific and political common objectives of the different partners. Primary examples of such partnerships include the International Space Station and Spacelab. In fact, many scientific programs and even one of the most well-known space agencies, the European Space Agency (ESA), were created by "upgrading" the nature of collaboration from a programmatic to an institutional level.

Finally, private-public partnerships have often formed in the Earth observation sector, such as Landsat, with mixed results, but newly emerging prominent examples include the Terrasar-X Earth observation satellite program. Here the German public sector funds the program, with the private partner then having marketing rights to data. Radarsat 2 is a similar case involving the Canadian Space Agency and industry partners. In the latest version of the U.K. Skynet defense communications satellite system, Thales is manufacturing and deploying the system on the basis of providing military communications to the U.K. government on a long-term lease basis, but is able to sell excess capacity on a commercial basis. Variations on this theme can be seen in other European countries, as well as in the dual use (commercial and military) of Inmarsat, Intelsat and Eutelsat services.

Security, a Driver in Public-Private Space Partnerships

It has to be noted that many of these public-private partnerships are of a national or, at best, geopolitically confined nature. This is largely due to security considerations and the involvement of the military as a prominent end user at the public sector side. These "quasi-commercial" partnerships made by considering national security objectives in a unilateral manner face restricted markets which, given the scale and nature of the program, is not a critical drawback.

Such partnerships are less stable when they assume a multinational nature since balancing political and security considerations becomes more challenging; the political complexity of the partnership starts shadowing the technical challenges. Again, the European Galileo satellite navigation system is a case in point, where political and security considerations both from within and outside Europe played a key role in undermining this partnership which was meant to bring together public and private sector actors from multiple countries.

Is the Future Bright?

Space is by no means the only sector of economic activity with a prominent security dimension, where the private sector is expected to respect and work with national security concerns. Other sectors, such as the oil and gas industry, share similar characteristics in this regard. Their best practices and lessons learned can be very useful for the management of future partnerships in space.

The oil and gas industry has many privately held companies that control strategic interests. This structure is often viewed as a way to safeguard national interests, while limiting the direct involvement of governments in industrial affairs. In a similar fashion, space integrators across the Atlantic can establish Galileo-type partnerships of a multi-private-public partnership (MP^3) nature. Such a partnership structure can facilitate, rather than complicate, finding the delicate balance between security concerns and commercialization.

In the era of global markets and vast international supply chains, there are many other segments of the space industry that can benefit from new international partnerships bringing together businesses and governments. For instance, if the projections for the space tourism market are accurate, there will be great demand for reusable spacecraft that can reach suborbital trajectories. Operating these spacecraft requires not only expertise in design and manufacturing, but also a significant investment in the supporting infrastructure such as spaceports and air traffic management. The motivations of the public sector are not just limited to ensuring safe operations, but also to creating brand new industries and new jobs. The private sector can benefit from a decreased uncertainty in regulations, a streamlined business environment, and the provision of the necessary supporting infrastructure. Thus, if the involved parties are ready to sacrifice some autonomy, they can create a very favorable environment to stimulate the space tourism market by adopting a multi-private-public partnership model. This could result in many beneficial acts, including removing barriers to enhanced competition or controls on various security aspects, and achieving economies of scale.

Space Workforce in the Age of Collaboration

One of the enablers of international cooperation is the availability of human capital accustomed to working in diverse, globally distributed and interdisciplinary teams. In addition to the obvious requirements such as language skills, the smooth functioning of such teams also requires an intercultural spirit, even if socio-cultural differences continue to exist.

It is only natural that one of the by-products of increased international cooperation is increasingly complex legal and policy issues and processes. International endeavors such as the Sea Launch consortium, or compliance with ITAR (International Traffic in Arms Regulations) in the U.S.A. require space professionals with specialized skill sets built on top of a solid interdisciplinary foundation.

In addition to this "quality" dimension, there is also an imminent challenge with the quantity of qualified space professionals. During the 1960s and 1970s, aerospace was identified as a priority area in higher education and the subsequent hiring wave resulted in a large workforce mostly consisting of "baby boomers". Many of these pioneers are still part of today's workforce, but they are quickly approaching the retirement age. A further complication is the decline in the number of aerospace and space science graduates over the years and the smaller number of younger employees. Together, these two factors are posing a serious problem and require immediate attention, especially in the U.S.A. Figure 11.9 clearly demonstrates the extent of this problem.

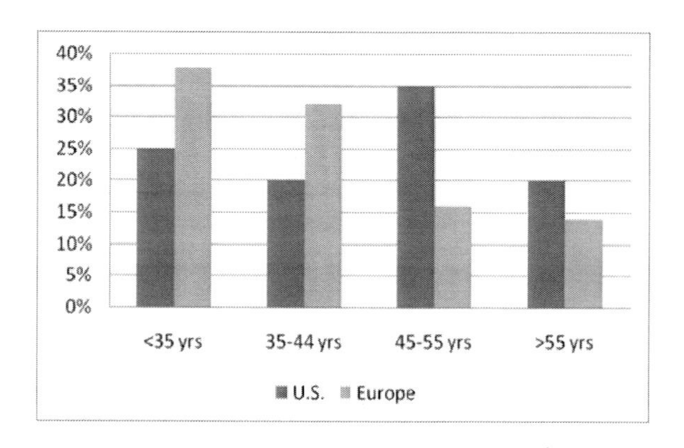

Figure 11.9. Age distribution of employees in the U.S. and European aerospace sectors.[18]

European policy makers were alarmed by this trend and studies were commissioned to evaluate the situation in Europe and to understand the demographics of the European space workforce. These European studies concluded that, although not ideal, the situation is less of a concern for Europe than for the U.S.A.

For most other sectors of economic activity, a common strategy to cope with this problem would be to attract skilled workforce from other sectors and train them for aerospace careers. The peculiarities of the space sector and the need for significant domain expertise render such a solution less powerful for space. The space business is not like any other business; space products and services have many unique characteristics. Therefore, it is not easy to attract specialists from other sectors. In fact, a certain degree of brain drain from the space sector to other industries is more common than the other way around.

As a result of increased commercialization, the space sector has considerably changed during the last few decades. The need for bigger and financially more powerful alliances has resulted in major conglomerates, prepared for worldwide competition. Initially, reduced public funding was the prime catalyst for this commercialization process, as companies started exploring new ways of doing business. Now that this trend is established, a key factor that will increasingly shape the future is the availability of a highly qualified workforce composed of individuals who can thrive in an international and interdisciplinary work environment. Several were taken to address this need. In the academic realm, higher education institutions such as the International Space University and other university programs were established to help to train tomorrow's space leaders. Companies have also started to invest in intercultural training to cope with the challenges of a global space industry. Hopefully, in a broader context, these efforts will also lead to an increase in mutual respect for different cultures, and the recognition of space as a global catalyst for better communications in a global society.

Space Entrepreneurship

Based on the characteristics of the space industry and various cases presented here, it is almost natural to think that space must be a very hostile environment for budding entrepreneurs. It is clear that the space market is still very much dominated by government contracts. Only satellite telecommunications and, to a limited extent, satellite navigation services resemble a truly commercial market.

Nevertheless, there is a change agent disrupting this relatively static picture: this is the small group of dedicated individuals who are determined to bring an entrepreneurial edge to the space industry. This group is slowly but surely starting to convince private investors to target specific areas of the space industry, such as space tourism. This new generation of space entrepreneurs, called "astropreneurs" by Peter Diamandis, has at least two traits in common: a fascination with space, and the know-how to raise capital in high-tech industries, most notably in the IT sector. Individual space investors such as Paul Allen, Jeff Bezos, Robert Bigelow, Sir Richard Branson, John Carmack, and Elon Musk, having achieved tremendous successes in other sectors have identified space as one of the next big challenges for their careers. Many of them also believe that the space market will expand significantly if individuals are given a chance to experience the space environment at first hand as opposed to being only spectators or distant end-users.

Compared to the rather conservative approach of space agencies, especially when it comes to human spaceflight, space entrepreneurs are prepared to take risks. They are also more interested in achieving efficiency at the company level, rather than fulfilling social objectives, such as creating employment. Therefore, their work teams tend to be much smaller than their space agency counterparts, reducing the labor cost of running these operations drastically.

Well known serial space entrepreneurs such as Peter Diamandis or Ed Tuck, as well as those that raise money to back innovative space ventures such as Space Vest and Burton Lee's new Space Angels, have a common message. They all argue that most breakthroughs come after many serial failures; however, the government does not accept failure easily or learn effectively from past mistakes. Moreover, they argue that when it comes to space projects a top priority for the governments is creating jobs, not achieving efficiency of operations and driving down costs.

Currently, space tourism (also called personal space travel) and entertainment attract the most attention of space entrepreneurs. Space tourism can be broken down into orbital, sub-orbital and terrestrial market segments, as discussed earlier. Space entertainment aims to expand the space experience to the general public through private missions to the Moon and other celestial bodies, various contests and races in space broadcast on TV/Internet, as well as other private space activities.

For many space visionaries, space tourism and entertainment are just the tip of the iceberg. They see space resources and space settlements as the ultimate market and believe that, in the vastness of space, almost all of the primary goods of economic importance can be found in abundance. The resources are real estate (such as International Telecommunications Union licenses for GEO orbital slots), minerals and metals (such as helium-3 on the lunar surface, and titanium and platinum rich asteroids) and solar energy (vast amounts of which exist in the inner solar system). To find innovative ways of using these resources will be the greatest economic challenge of the Space Age, and it will also create a very strong motivation for space exploration. Combined with the innate philosophical rationale for exploring space, this economic incentive should be a very powerful motivator for future generations.

It is widely believed that a healthy dose of competition can accelerate the development of space activities. The dynamism of the 1960s can be achieved in a more peaceful and efficient environment by focusing the energy of private capital on selected space activities. The founder of the Ansari X PRIZE, Peter Diamandis, believes that the most important accomplishment of this competition was to change the way the world thinks about spaceflight. The Ansari X PRIZE was designed to spawn, stimulate and support the commercial public spaceflight industry. It has created an emerging market. Further, it has demonstrated that prize-based incentives do work and can pave the way for personal space travel at an affordable price. As clearly shown by the success of the Ansari X PRIZE, and embodied in the words of Margaret Mead: "Never doubt that a few committed individuals can change the world. Indeed it is the only thing that ever has".

The Start of a New Era

There were certainly no accountants or management consultants among the twelve astronauts who set foot on the Moon (and it is probably a safe bet that this situation will not change when the crew of the first human spaceflight mission to Mars is chosen). However, without a thorough understanding of business and management aspects, we cannot build sustainable space exploration programs.

Furthermore, it is becoming increasingly clear that the space industry is on the cusp of change. The effects of globalization, increased commercialization and managerial innovations, such as the Ansari X PRIZE, are changing the landscape of the space industry. As the existing market segments such as satellite telecommunications and navigation mature and our technical and financial risk management skills improve, more private investors will be attracted to the space industry. Their presence is already transforming the satellite telecommunications business. We are also witnessing the first steps of disruptive change: as the suborbital space tourism market evolves, it will bring with it a new era of space products and services.

1 The discrepancy between the low-end and the high-end numbers is due to the different estimates of the U.S. public space budget (OECD estimates that the U.S. public space budget is US$36.6 billion for 2005, while The Space Report estimate is US$62.5 billion for 2007).

2 Compared to the US$36.6 billion U.S. public space budget for 2005, OECD reports a total of $11.3 billion for the rest of the world (this number includes the public space budgets of European countries, Russia, India, China and other space-faring nations).

3 OECD, "The Space Economy at a Glance 2007", Paris, 2007 and "The Space Report", http://www.thespacereport.org/content/overview/activity.php

4 OECD, The Space Economy at a Glance, International Futures Programme, Paris, 2007

5 Source: "Space Economics and Future Trends", lecture notes of Prof. Walter Peeters, International Space University, Strasbourg, France.

6 Union of Concerned Scientists maintains a database of operational satellites. As of April 2008, the database contains 873 satellites. For more information, please see http://www.ucsusa.org/global_security/space_weapons/satellite_database.html

7 Federal Aviation Administration, Year in Review Reports (1997-2007), available at http://www.faa.gov/about/office_org/headquarters_offices/ast/reports_studies/year_review/

8 Please see Chapter 9 for a detailed discussion of emerging security issues and space activities.

9 EADS stands for European Aeronautic Defence and Space Company.

10 Source: "Space Economics and Future Trends", lecture notes of Prof. Walter Peeters, International Space University, Strasbourg, France.

11 This transfer of know-how is more the norm in the industry rather than an exception: behind all the major elements of the industry, from launch vehicles to satellite subsystems and payloads, is a very prominent dual-use heritage.

12 L. Bach, P. Cohendet, G. Lambert, M.J. Ledoux, "Measuring and Managing Spinoffs: The Case of the Spinoffs Generated by ESA Programs", in Space Economics, edited by J.S. Greenberg and H. Hertzfeld, AIAA Publication, 1992

13 Euroconsult, World Prospects for Government Space Markets 2006/2007 edition, Paris, France, 2006

14 The field of economics deals with the allocation of scarce (limited) resources among competing uses, and studies how people make choices to cope with scarcity. Markets are the centerpieces of economic activity; they are the "places" where buyers and sellers of goods and services can interact. As the role of the Internet in commerce increases, markets are also being transformed, and virtual marketplaces are formed (e-commerce). Therefore, today's markets are no longer attached to a physical location. Traditional markets are complemented and, in some cases, replaced by e-commerce. This transformation has not changed the fundamental role of two economic concepts - supply and demand. For this reason, sound economic policy and successful business decisions are all based on a careful analysis of the trends that affect the supply and demand for space related goods and services. Microeconomics is the branch of economics concerned with the decisions made by individuals, households, and firms, and how these decisions interact to determine the prices of goods and services and the factors of production. As the name suggests, the level of analysis is at the "micro" level: we study the interaction of individual units. Typical areas of study include consumer behavior and choices, demand and supply interaction and game theory. By way of contrast, macroeconomics is the study of the entire economic system in terms of the total (aggregate) amount of goods and services produced, total income earned, the level of employment of productive resources, and the general behavior of prices. The analysis is performed at the "macro" level. Here we strive to understand and manage the behavior of the whole system, and not just of the individual parts. Typical areas of study include economic growth, inflation, unemployment, trade balances and fiscal (taxation) policy.Business administration/management is the process of leading and directing all or part of an organization, often a business, through the allocation of resources. These resources include all the traditional factors of production (labor, capital, etc.) as well as human capital, intellectual/ intangible resources and technology. In management, we deal with the operational, tactical and strategic aspects of leading an organization. Typical areas of study include strategic management, marketing, finance and human resources management.

15 NASA helped Speedo to improve the performance of the LZR Racer swimsuit by performing surface drag testing and applying expertise in the area of fluid dynamics. This spin-off provided "golden" results: 94 percent of gold medals in swimming at the 2008 Olympics were won in the new Speedo suit.

16 Such as access to water, environmental issues, crop monitoring, treaty verification / surveillance of conflicts, increasing need of mobility, protection against natural disasters, energy issues, equal access to knowledge, etc.

17 NASA, Parametric Cost Estimating Handbook, available at http://cost.jsc.nasa.gov/PCEHHTML/pceh.htm

18 Squeo, A., "Attracting Top Engineering Talent Requires Some Creative Thinking", The Wall Street Journal Europe, (25 April 2000) p.30. European data is from the European Science Foundation, Demography of European Space Science, Strasbourg, April 2003.

Chapter Twelve
The Meaning of It All

"... And when [Humanity] ... has conquered all the depths of space, and all the mysteries of time, still [we] ... will be beginning."

— H.G. Wells, "Things to Come" (1936)

Why should there be space programs? Going into space is complicated, and consumes a great deal of people's time and energy. We all know that it is still hugely expensive. But we do not go into space solely to satisfy our idle curiosity. Increasingly, we are finding that new space technologies and capabilities represent core capabilities that we need to survive as a species as our global population expands. We must ensure that we have space solutions to the global problems that surely will continue to grow in scope and seriousness.

The truth is that the reasons for going into space are just as numerous as the diverse disciplines that an active space program requires. Here we have learned why and how space is international, intercultural and interdisciplinary. Space is about exploration. It is about science. And it is about an ever-expanding range of applications that helps us to combat global warming, keep our oceans clean, link our world together, provide us instant news, and ensure that many things are truly global—whether it be a Rule of Law, the chance to learn, or the Internet. We now know that space is about much, much more than building and launching rockets. It is truly a mistake to think that space is about a single thing, or a single discipline.

Space is about the future of humankind and about defining the role of *Homo sapiens* in the cosmos. In previous chapters of *The Farthest Shore* we have reviewed, at least to some degree, the following space disciplines and areas of interest:

- The arts, including literature, poetry, the visual arts, dance, plays and movies.
- Astrophysics, astronomy and cosmology.
- Space sciences, the Sun-Earth connection and space weather.

- Space applications, including telecommunications, radio and TV broadcasting, navigation, Earth observation, remote sensing, geomatics, meteorology, search and rescue, and disaster detection, warning and recovery.

- Space transportation, rocketry, satellite design and engineering, and orbital mechanics.

- Space systems engineering, tracking, telemetry, and command systems.

- Space exploration and space habitats.

- Space life sciences and biology.

- Space business and commerce, marketing, entrepreneurship, risk management, capital formation, innovation, and intellectual property acquisition and protection.

- Space law, regulation, policy, and security.

- New space industries and opportunities, including such diverse areas as solar power satellites, new space transportation and space access systems, space tourism, space colonies, space materials, space robotics, mining meteorites, and even terra-forming extraterrestrial bodies.

When people say that space programs are a waste of taxpayers' money, especially when there are so many vital unmet needs here on Earth, it is important to correct the record. Each year our "eyes in the skies" save tens of thousands of lives, or even more, by warning people of the need to flee from the paths of oncoming hurricanes, typhoons, tropical storms and other natural disasters. Each year satellite search and rescue systems allow hundreds of stranded pilots, fishermen and explorers to be rescued. Without Earth observation satellites we would not know enough about the ozone layer in the stratosphere that protects all animal and plant life on our planet from extinction by radiation from the Sun and the cosmos. Of even more practical importance are our satellites that monitor pollution, husband water resources, locate key resources, spot crop disease and forest fires, carry out "smart farming", and combat the most troublesome aspects of global warming.

Scientific satellites and probes such as the Hubble Space Telescope and the Chandra X-Ray Telescope have enabled us to peer into the very depths of the Universe, some 12 billion light years into the past. In future years, the James Webb space telescope and constellation telescope systems will tell us more about the structure of galaxies, black holes, quasars, nebulae, and star systems with planets, giving us insight into the Universe and how it works. Explorer satellites have revealed amazing detail in the residual microwave radio noise from the Big Bang. Surveyor satellites have plumbed the surface of Mars and the Moon. Life scientists have used space stations such as Skylab, Mir and the International Space Station (ISS) to understand the mysteries of life in a micro-gravity (near zero gravity) environment.

Until recently, governments and national space agencies have driven progress in space exploration, space sciences and space applications. One of the key reasons that governments invested so significantly in space was for military and strategic purposes, starting with the development of ballistic missiles. The delicate balance of space-related investment for civil and military purposes has continued now for fifty years. It is ironic that international cooperation in space has been one of the most wonderful opportunities for nations to cooperate together, yet military space programs, by way of contrast, generate international tensions.

Some of humankind's most elaborate and high profile cooperative programs have been carried in the high frontier of space. Not only has there been the International Space Station, or the Apollo-Soyuz and Shuttle-Mir joint missions, but also scores of joint or multi-lateral cooperative satellite programs have been stimulated around the world. The most successful commercial space program was the Intelsat Global Satellite System that involved over 100 countries. It is significant to note that, with the end of the so-called Cold War, many missile systems have been adapted to cooperative purposes to perform civil and commercial satellite launches.

Today fourteen space agencies are committed to the "Global Exploration Strategy" that was adopted in May 2007 (see Figure 12.1). The International Space Station, despite its difficulties and cost overruns, remains a testimonial to international cooperation. It shows that the international dimension of space more often than not pulls the nations of the world together, rather than apart.

Countless commercial space applications have evolved from telecommunications satellite services. In the 1960s, the success of the global Intelsat system led to the deployment of hundreds of commercial satellites, generating US$ hundreds of billions in revenues. Today there are many "spin-off" technologies and applications from space research and development, R&D. These are in areas such as advanced construction materials, medical applications, consumer appliances, transportation and communications systems, household goods, robotics and even sporting equipment.

Although no commercial space venture has to date been as successful as communications satellites, remote sensing, navigation and geomatics-related applications continue to grow in importance; they now generate significant commercial revenues. There are currently high hopes that entirely new commercial space ventures, such as space tourism, space habitats and solar power satellites, may one day grow to be multi-billion dollar ventures. Opportunities for many people to "fly into the dark sky", to experience weightlessness, to see the curvature of Earth, and to witness our planet from space are one of the most exciting opportunities of our times.

Figure 12.1. Many of the world's space agencies cooperate on international space programs. A testimonial to such cooperation is the International Space Station program.

The vision for space going forward is a large and ambitious one. In future decades space systems may allow us to preserve human, animal and plant life on Earth against the stresses and strains of a global population that exceeds 10 billion people. Future space systems may be needed for reasons that we, today, neither fully understand nor appreciate. We may need important new space capabilities to help us preserve the ozone layer, to reduce the effects of increasing amounts of greenhouse gases, to monitor atmospheric pollution, or to restore the oceans to their earlier pristine state. Only space systems can allow us to have and to operate a unified planetary monitoring system. Future space systems may help us to obtain clean and green energy, to obtain new valuable resources or even to terraform other celestial bodies in space. That would make possible human colonies beyond the Earth's "gravity well" that holds us close to our home planet—a hold that retards our ultimate destiny.

Our future, both on and off planet Earth, is about people working together. It is about people from North and South America, from Africa and Asia, from the Middle East and Europe, and from Oceania working together to create new knowledge with a new level and spirit of cooperation. Likewise it is about life scientists and doctors working with engineers and cosmologists. It is about journalists and futurists, and people of arts and letters, working with people in satellite applications and with space lawyers. It is about business people and entrepreneurs working with sociologists, politicians and educators. It is about the synergy that comes from the whole being greater than the sum of the parts.

It is clear is that the exploration and understanding of space remains our long-term destiny. R. Buckminster Fuller elegantly reminded us that we humans are all space travelers, and that the Earth is our space ship. Let us remember those who chided, or even scolded, Thomas Jefferson when he paid US$ millions, for the Louisiana Purchase, to double the size of the United States. When, in 1867, U.S. Secretary of State William Seward spent even more to acquire Alaska, it was seen as a totally crazy acquisition of a worthless icebox. There is, however, no doubt that these investments paid off.

Today, our investment in space is just that, a bet on the future worth of space and of realizing the future potential of humanity. All the peoples of the Earth need to join together to make the investment, which is essential for creating new technologies and systems. These will be the key technologies required one day to save our species from climate change and extinction, or to meet Earth's energy needs, or to prevent a meteorite from catastrophically destroying our civilization. When one invests in the future it is always a gamble. We do not know what the long term rewards will be. If we look only to tomorrow we will not succeed. We must look to the day after tomorrow to sustain ourselves, and to realize our future dreams. We encourage our readers to join us in our mission. And we must not fail.

Appendix

Biographical Sketches of the Editorial Board and Contributing Author Affiliations

Editorial Board

Joseph N. Pelton (Executive Editor)

Dr. Joseph N. Pelton is Director Emeritus of the Space and Advanced Communications Research Institute (SACRI) at George Washington University. He is the founder and Vice Chairman of the Arthur C. Clarke Foundation that works in partnership with space and telecommunications research institutes and foundations in Europe, North America and Asia. He is the author of over 25 books in the fields of telecommunications, space policy, and satellite communications, including the Pulitzer Prize nominated book, *Global Talk*. He has served as Chairman of the Board (1992-95) and Vice President of Academic Programs and Dean (1996-97) of the International Space University of Strasbourg, France. Dr. Pelton has also been Director of the Interdisciplinary Telecommunications Program at the University of Colorado at Boulder, and Director of Strategic Policy at Intelsat. He was the founding president of the Society of Satellite Professionals International in 1983. He is also the former Executive Editor of the Journal of International Space Communications. He has been elected an Associate Fellow of the AIAA, elected to be a member of the International Academy of Astronautics, chosen for the Arthur C. Clarke Lifetime Achievement Award and the Satellite Hall of Fame. He also serves as associate editor of the Academy's Journal *Ad Astra*. Dr. Pelton was the 1997-1998 chairman of a NASA and the National Science Foundation Panel of Experts on Satellite Communications and from 1998 to 2001 on the External Evaluation Committee for the Japanese Space Agency (NASDA). Dr. Pelton holds BS, MA and Ph.D. degrees from the University of Tulsa, New York University and Georgetown University respectively.

Angelia P. (Angie) Bukley (Executive Editor)

Dr. Angie Bukley is Associate Dean, Professor and Space Studies Program Director for the International Space University in Strasbourg, France. Prior academic appointments include Associate Vice President and Chief

Administrator for the University of Tennessee Space Institute in Tullahoma, Tennessee and Associate Dean for Research and Graduate Studies in the Russ College of Engineering at Ohio University in Athens, Ohio. Prior to entering academe, Dr. Bukley gained nearly two decades of experience working with a number of space systems and defense contractors on a wide variety of flight programs including five years with The Aerospace Corporation in Albuquerque, New Mexico. She spent seven years working at the NASA Marshall Space Flight Center in Huntsville, Alabama where she directed the Large Space Structures Laboratory and developed a number of remote sensing applications aimed at disaster management. Dr. Bukley has also been a member of and chaired the ISU Academic Council and has held department and team project chair positions in the ISU Space Studies Program since 1994. In addition to her administrative duties, she also teaches Space Engineering courses in the ISU Master's Program. Dr. Bukley has over 70 technical publications including the 2007 book *Artificial Gravity*, which she co-wrote and co-edited with Dr. Gilles Clément. The book received the International Academy of Astronautics 2008 Life Sciences Book of the Year Award. Dr. Bukley holds a Ph.D. in Electrical Engineering, with a specialty in Control Theory, from the University of Alabama in Huntsville. She is an Associate Fellow in the American Institute for Aeronautics and Astronautics and an active member in the American Astronautical Society, International Federation for Automatic Control, National Space Society, and American Society for Engineering Education. In 2003, Dr. Bukley received the University of Alabama in Huntsville Distinguished Engineering Alumni Award.

Michael Rycroft (Manuscript Editor)

Dr. Rycroft holds a B.Sc. in Physics from Imperial College, London, a Ph.D. from Cambridge University for studies in meteorological physics, and a D.Sc. (honoris causa) from De Montfort University, in Leicester, U.K. After a NAS/NRC Associateship at the NASA Ames Research Center, California, he became a Lecturer at Southampton University, then Head of the Atmospheric Sciences Division at the British Antarctic Survey in Cambridge at the time of the discovery of the "ozone hole", then Professor of Aerospace and Head of the College of Aeronautics at Cranfield University, and finally Professor, Head of the School of Sciences and Applications and Director of Research at the International Space University in Strasbourg, France. With two colleagues there he prepared the curriculum for the new Master of Space Studies Program. Currently he is a Visiting Professor at Cranfield and a part-time faculty member at ISU, a Senior Research Fellow at the University of Bath, Editor-in-Chief of the overview journal Surveys in Geophysics, and Proprietor of CAESAR Consultancy (Cambridge Atmospheric, Environmental and Space Activities and Research Consultancy). He has published over 200 papers in refereed journals on topics in atmospheric and solar-terrestrial physics; he has written or edited over 40 books or special issues of journals, including the Cambridge Encyclopedia of Space published in 1990. From 1989 to 1999 he was Editor-in-Chief of the Journal of Atmospheric and Solar-Terrestrial Physics. He has been elected a member of the International Academy of Astronautics and of Academia Europaea. He is a Member of the Institute of Physics, a Fellow of the Institute of Mathematics and its Applications, a fellow of the Royal Astronomical Society (and Council Member from 2003 to 2007), a Member of the American Geophysical Union and of the European Geosciences Union (of which he was General Secretary from 1996 to 2003). He is a member of a European team proposing to study sprites using instruments aboard the International Space Station.

Gilles Clément (Editorial Board)

Dr. Gilles Clément received Doctoral Degrees in Neurobiology from the University of Lyon in 1981 and in Natural Science from the University of Paris in 1986. He is Director of Research from the French National Center for Scientific Research (CNRS) Toulouse, France. Research in space life sciences has been his primary focus with experiments on Salyut-7 (1982), MIR (1988), and on more than 25 Space Shuttle flights (1985-present). His research topics include influence of microgravity on posture, eye movements, spatial orientation, and cognition in humans. To date, he gathered data on more than 70 astronauts during and after space missions. Dr. Clément is currently the principal investigator of life sciences experiments being conducted before and after Space Shuttle flights, and on board the International Space Station. Dr. Clément is a faculty member of the International Space University since 1989. He recently published three books entitled 'Fundamentals of Space Medicine' (Springer: Dordrecht, 2005), 'Fundamentals of Space Biology' (Springer: New York, 2006), and 'Artificial Gravity' (Springer: New York, 2007).

Jim Dator (Editorial Board)

Dr. Jim Dator is Professor and Director of the Hawaii Research Center for Futures Studies, Department of Political Science, and Adjunct Professor in the Program in Public Administration, the College of Architecture, and the Center for Japanese Studies, of the University of Hawaii at Manoa; Co-Chair, Space and Society Division, International Space University, Strasbourg, France; former President, World Futures Studies Federation; Fellow and member of the Executive Council, World Academy of Art and Science. He also taught at Rikkyo University (Tokyo, for six years), the University of Maryland, Virginia Tech, the University of Toronto, and the InterUniversity Consortium for Postgraduate Studies in Dubrovnik, Yugoslavia. He received a BA in Ancient and Medieval History and Philosophy from Stetson University, an MA in Political Science from the University of Pennsylvania, and a PhD in Political Science from The American University. He did post graduate work at Virginia Theological Seminary (Ethics and Church History), Yale University (Japanese Language), The University of Michigan (Linguistics and Quantitative Methods), and Southern Methodist University (Mathematical Applications in Political Science). He is a Danforth Fellow, Woodrow Wilson Fellow, and Fulbright Fellow. He consults widely on the futures of law, governance, education, tourism, and space.

Ozgur Gurtuna (Editorial Board)

Dr. Ozgur Gurtuna is active in both professional and academic domains, and he has a keen interest in developing innovative solutions by merging multiple technology areas. He is the founder and president of Turquoise Technology Solutions Inc., a Canadian company providing services in the fields of aerospace, energy an environment. Ozgur launched a second company, Teknosfer Ltd., in order to start operations in Turkey. Teknosfer aims to develop integrated solutions for both private and public sector clients in Turkey with a specific focus on space and energy. He is a faculty member at the International Space University, lecturing in topics related to the business and management aspects of space activities. He obtained his Ph.D. in Operations Research from the joint Ph.D. program in Montreal (this program is administered by four universities: Concordia, HEC, McGill and UQAM). He has a B.A. degree in Business and Management from Bogazici University (1997), Istanbul, Turkey and a Master of Space Studies degree from International Space University, Strasbourg, France (1999). His areas of expertise include the economic analysis of space applications, emerging technology markets and quantitative analysis in decision making (covering areas such as optimization, simulation and mathematical modeling).

Isabelle Scholl (Editorial Board)

Dr. Isabelle Scholl received first her Master and Postgraduate degrees in Computer Sciences in 1993 from the Conservatoire National des Arts et

Métiers, and in 1994 at the Université Louis Pasteur, Strasbourg, France, respectively, and then her PhD in Astrophysics and Space Techniques in 2003 at the Université Paris VI. She recently joined the Institute for Astronomy of the University of Hawaii as Assistant Astronomer. She previously held a position of Assistant Professor at the International Space University for five years where she taught Information Technologies and Astrophysics. She was formerly a Research Engineer with the French National Center for Scientific Research (CNRS) at the Institut d'Astrophysique Spatiale (IAS) near Paris for nine years after initial activities in industry and lecturing at the Strasbourg Institute of Technology. At IAS, she was Technical Manager for the European Center MEDOC (Multi-Experiments, Data and Operations Centre for the SOHO mission). She has been a CO-I for the EGSO (European Grid for Solar Observations) project for which she led the software development. She also served as expert member during CNES/CNRS reviews for archive centers. Her research interests are in Solar Physics (solar features study and evolution, automatic recognition and classification) and in Data Management of Scientific Data (models, archives, dissemination, presentation and Virtual Observatories).

Lucy Stojak (Editorial Board)

Dr. Lucy Stojak holds an undergraduate law degree (LL.L) from the Université de Montréal, and a Masters and Doctorate degree in law (LL.M and D.C.L.) from the Institute of Air and Space Law, McGill University. Her fields of interest include international space law, space commercialization, space and security, Earth observation data policy issues, and team building/leadership skill development. She was a Senior Researcher at the Centre for Research in Air and Space Law, McGill. Dr. Stojak served as the first Director of the International Space University's (ISU) Summer Session Program (SSP) from 1994-2000, and served as co-chair of the Policy and Law Department (most recently at the SSP'05 in Vancouver, SSP'07 in Beijing, and SSP'08). She is a Faculty Member of the ISU in Strasbourg, France, and a regular invited lecturer at the Institut du droit de l'espace et des télécommunications (IDEST) - Université de Paris Sud -IX, and the HEC Montréal-ESCP Executive MBA. Dr. Stojak is currently involved in the Space Security Index (SSI), a research partnership between several governmental, non-governmental and academic institutions, including McGill's Institute of Air and Space Law. Since 2001, her consulting firm, which is based in Montreal, carries out in-depth studies for the Canadian Space Agency (CSA) and Foreign Affairs Canada on Earth observation data policies and commercialization policies, legal issues of space commercialization, the non-weaponization of space, and legal and institutional developments related to space and security. She served as the elected Chair of the ISU Academic Council from 2004-2006, and continues to be a member of the Council. She is a member of the Board of Trustees of the Canadian Foundation for ISU (CFISU), and the Board of Advisors of the Space Security Index. She is a member of the European Centre for Space Law (ECSL), the International Institute of Space Law (IISL), the International Astronomical Academy (IAA), and the Board of Editors of *Space Policy*.

Contributing Authors

Eric Choi Manager of Business Development, Missions Development Group, COM DEV, Cambridge, Ontario, Canada

Charles Cockell Professor of Geomicrobiology, Open University, UK

Patrick Cohendet Professor of Economics, University Louis Pasteur, Strasbourg, France

John Connolly Director of Vehicle Design & Engineering, Altair Lunar Lander Project Office, NASA Johnson Space Center, Houston, Texas, USA

Eric Dahlstrom President, International Space Consultants, Washington, DC, USA

Peter Diamandis Founder, International Space University and Founder and CEO, XPRIZE

Giovanni Fazio Senior Physicist, Harvard-Smithsonian Center for Astrophysics; Lecturer, Astronomy Department, Harvard University, Cambridge, Massachusetts, USA

Lauren Fletcher Engineer-Scientist, Space Science Division, NASA Ames Research Center, California, USA

James Green Director of the Planetary Science Division, NASA Headquarters, Washington, DC, USA

Arthur Guest Graduate Student, Massachusetts Institute of Technology Department of Aeronautics and Astronautics, Cambridge, Massachusetts, USA

Douglas Hamilton Flight Surgeon supporting the Shuttle and Space Station Operations, Wyle Life Sciences, NASA Johnson Space Center, Houston, Texas, USA

He Linshu Professor, School of Astronautics and Lead Professor of Education in the Asian Pacific Space Education Center, BeiHang University (the Beijing University of Aeronautics and Astronautics or BUAA), Beijing, China

Jeffrey Hoffman Former Astronaut and MIT Professor, Cambridge,

Massachusetts, USA

Ram Jakhu Associate Professor at the Institute of Air and Space Law, Faculty of Law, McGill University, Montreal, Canada

Tarik Kaya Professor, Mechanical and Aerospace Engineering, Carleton University, Toronto, Canada

John Logsdon Former Director, Space Policy Institute, George Washington University and Lindbergh Chair in Aerospace History, National Air and Space Museum, Bethesda, Maryland, USA and former Director, Space Policy Institute, GW University

Scott Madry Professor, University of North Carolina and Renaissance Super Computer Center, Chapel Hill, North Carolina, USA

Mikhail Marov Department Head, Planetary Physics, M.V. Keldysh Institute of Applied Mathematics, Moscow, Russia

Christopher P. McKay Scientist in the Space Science Division, NASA Ames Research Center, California, USA

Emeline Paat-Dahlstrom Partner and Consultant, International Space Consultants, Washington, DC, USA

Walter Peeters Dean of the International Space University, Strasbourg, France

Robert Richards Founder, International Space University and CEO, Odyssey Moon, Toronto, Ontario, Canada

Rusty Schweickart Apollo 9 Astronaut & Chairman of the Board of the B612 Foundation, Sonoma, California, USA

Kazuya Yoshida Professor, Department of Aerospace Engineering, Tohoku University, Japan

Vasilis Zervos Resident Faculty at the International Space University, Strasbourg, France

INDEX

B

C

D

E

F

G

H

I

J

K

L

N

O

P

Q

R

S

T

U

V

W

X

Y

Z